T0310447

Ultra-Low Energy Wireless Sensor Networks in Practice

Ultra-Low Energy Wireless Sensor Networks in Practice

Theory, Realization and Deployment

Mauri Kuorilehto, Mikko Kohvakka,
Jukka Suhonen, Panu Hämäläinen,
Marko Hännikäinen, and Timo D. Hämäläinen

Tampere University of Technology, Finland

John Wiley & Sons, Ltd

Published by John Wiley & Sons Ltd, The Atrium, Southern Gate, Chichester, West Sussex PO19 8SQ, England

Telephone (+44) 1243 779777

Email (for orders and customer service enquiries): cs-books@wiley.co.uk
Visit our Home Page on www.wileyeurope.com or www.wiley.com

Other Wiley Editorial Offices

John Wiley & Sons Inc., 111 River Street, Hoboken, NJ 07030, USA

Jossey-Bass, 989 Market Street, San Francisco, CA 94103-1741, USA

Wiley-VCH Verlag GmbH, Boschstr. 12, D-69469 Weinheim, Germany

John Wiley & Sons Australia Ltd, 42 McDougall Street, Milton, Queensland 4064, Australia

John Wiley & Sons (Asia) Pte Ltd, 2 Clementi Loop #02-01, Jin Xing Distripark, Singapore 129809

John Wiley & Sons Canada Ltd, 6045 Freemont Blvd, Mississauga, Ontario, L5R 4J3, Canada

Wiley also publishes its books in a variety of electronic formats. Some content that appears in print may not be available in electronic books.

Library of Congress Cataloging-in-Publication Data

Ultra-low energy wireless sensor networks in practice / Mauri Kuorilehto ... [et al.].
p. cm.
Includes bibliographical references and index.
ISBN 978-0-470-05786-5 (cloth)
1. Sensor networks. 2. Wireless LANs. I. Kuorilehto, Mauri.
TK7872.D48U48 2007
681′.2 – dc22

2007033349

British Library Cataloguing in Publication Data

A catalogue record for this book is available from the British Library

ISBN 978-0-470-05786-5 (HB)

Typeset in 10/12pt Times by Laserwords Private Limited, Chennai, India
Printed and bound in Great Britain by Antony Rowe Ltd, Chippenham, Wiltshire
This book is printed on acid-free paper responsibly manufactured from sustainable forestry in which at least two trees are planted for each one used for paper production.

Contents

Preface xiii

List of Abbreviations xv

PART I INTRODUCTION 1

1 Introduction 3
1.1 Overview of Wireless Technologies 3
1.2 TUTWSN 5
1.3 Contents of the Book 6

PART II DESIGN SPACE OF WSNS 7

2 WSN Properties 9
2.1 Characteristics of WSNs 9
2.2 WSN Applications 11
2.2.1 Commercial WSNs 12
2.2.2 Research WSNs 14
2.3 Requirements for WSNs 16

3 Standards and Proposals 19
3.1 Standards 19
3.1.1 IEEE 1451 Standard 19
3.1.2 IEEE 802.15 Standard 21
3.2 Variations of Standards 28
3.2.1 Wibree 28
3.2.2 Z-Wave 28
3.2.3 MiWi 28

4 Sensor Node Platforms 29
4.1 Platform Components 29
4.1.1 Communication Subsystem 30
4.1.2 Computing Subsystem 33

4.1.3 Sensing Subsystem . . . 33
4.1.4 Power Subsystem . . . 34
4.2 Existing Platforms . . . 36
4.3 TUTWSN Platforms . . . 39
4.3.1 Temperature-sensing Platform . . . 39
4.3.2 SoC Node Prototype . . . 43
4.3.3 Ethernet Gateway Prototype . . . 44
4.4 Antenna Design . . . 46
4.4.1 Antenna Design Flow . . . 46
4.4.2 Planar Antenna Types . . . 48
4.4.3 Trade-Offs in Antenna Design . . . 49

5 Design of WSNs **51**
5.1 Design Dimensions . . . 51
5.2 WSN Design Flow . . . 54
5.3 Related Research on WSN Design . . . 56
5.3.1 WSN Design Methodologies . . . 56
5.4 WSN Evaluation Methods . . . 60
5.5 WSN Evaluation Tools . . . 61
5.5.1 Networking Oriented Simulators for WSN . . . 61
5.5.2 Sensor Node Simulators . . . 62
5.5.3 Analysis of Evaluation Tools . . . 63

PART III WSN PROTOCOL STACK **67**

6 Protocol Stack Overview **69**
6.1 Outline of WSN Stack . . . 69
6.1.1 Physical Layer . . . 70
6.1.2 Data Link Layer . . . 71
6.1.3 Network Layer . . . 71
6.1.4 Transport Layer . . . 71
6.1.5 Application Layer . . . 72

7 MAC Protocols **73**
7.1 Requirements . . . 73
7.2 General MAC Approaches . . . 75
7.2.1 Contention Protocols . . . 75
7.2.2 Contention-free Protocols . . . 77
7.2.3 Multichannel Protocols . . . 78
7.3 WSN MAC Protocols . . . 80
7.3.1 Synchronized Low Duty-cycle Protocols . . . 80
7.3.2 Unsynchronized Low Duty-cycle Protocols . . . 85
7.3.3 Wake-up Radio Protocols . . . 87
7.3.4 Summary . . . 88

8 Routing Protocols **91**
8.1 Requirements 91
8.2 Classifications 92
8.3 Operation Principles 93
8.3.1 Nodecentric Routing 93
8.3.2 Data-centric Routing 94
8.3.3 Location-based Routing 95
8.3.4 Multipath Routing 97
8.3.5 Negotiation-based Routing 97
8.3.6 Query-based Routing 98
8.3.7 Cost Field-based Routing 99
8.4 Summary 101

9 Middleware and Application Layer **103**
9.1 Motivation and Requirements 103
9.2 WSN Middleware Approaches 105
9.3 WSN Middleware Proposals 106
9.3.1 Interfaces 106
9.3.2 Virtual Machines 107
9.3.3 Database Middlewares 107
9.3.4 Mobile Agent Middlewares 108
9.3.5 Application-driven Middlewares 108
9.3.6 Programming Abstractions 109
9.3.7 WSN Middleware Analysis 110

10 Operating Systems **115**
10.1 Motivation and Requirements 115
10.1.1 OS Services and Requirements 116
10.1.2 Implementation Approaches 117
10.2 Existing OSs 119
10.2.1 Event-handler OSs 120
10.2.2 Preemptive Multithreading OSs 121
10.2.3 Analysis 121

11 QoS Issues in WSN **125**
11.1 Traditional QoS 125
11.2 Unique Requirements in WSNs 125
11.3 Parameters Defining WSN QoS 126
11.4 QoS Support in Protocol Layers 128
11.4.1 Application Layer 128
11.4.2 Transport Layer 128
11.4.3 Network Layer 129
11.4.4 Data Link Layer 130
11.4.5 Physical Layer 131
11.5 Summary 131

12 Security in WSNs **133**
12.1 WSN Security Threats and Countermeasures 133
12.1.1 Passive Attacks 134
12.1.2 Active Attacks 134
12.2 Security Architectures for WSNs 135
12.2.1 TinySec 135
12.2.2 SPINS 136
12.2.3 IEEE 802.15.4 Security 136
12.2.4 ZigBee Security 137
12.2.5 Bluetooth Security 139
12.3 Key Distribution in WSNs 140
12.3.1 Public-key Cryptography 140
12.3.2 Pre-distributed Keys 140
12.3.3 Centralized Key Distribution 141
12.4 Summary of WSN Security Considerations 142

PART IV TUTWSN **143**

13 TUTWSN MAC Protocol **145**
13.1 Network Topology 145
13.2 Channel Access 147
13.3 Frequency Division 149
13.4 Advanced Mobility Support 152
13.4.1 Proactive Distribution of Neighbor Information 153
13.4.2 Neighbor-discovery Algorithm 154
13.4.3 Measured Performance of ENDP Protocol 158
13.5 Advanced Support for Bursty Traffic 159
13.5.1 Slot Reservations within a Superframe 160
13.5.2 On-demand Slot Reservation 161
13.5.3 Traffic-adaptive Slot Reservation 161
13.5.4 Performance Analysis 162
13.6 TUTWSN MAC Optimization 165
13.6.1 Reducing Radio Requirements 165
13.6.2 Network Beacon Rate Optimization 170
13.7 TUTWSN MAC Implementation 179
13.8 Measured Performance of TUTWSN MAC 180

14 TUTWSN Routing Protocol **183**
14.1 Design and Implementation 183
14.2 Related Work 183
14.3 Cost-Aware Routing 184
14.3.1 Sink-initiated Route Establishment 185
14.3.2 Node-initiated Route Discovery 185
14.3.3 Traffic Classification 186

14.4 Implementation . . . 187
14.4.1 Protocol Architecture . . . 187
14.4.2 Implementation on TUTWSN MAC . . . 188
14.5 Measurement Results . . . 188
14.5.1 Network Parameter Configuration . . . 189
14.5.2 Network Build-up Time . . . 189
14.5.3 Distribution of Traffic . . . 190
14.5.4 End-to-end Delays . . . 192

15 TUTWSN API **193**
15.1 Design of TUTWSN API . . . 194
15.1.1 Gateway API . . . 194
15.1.2 Node API . . . 196
15.2 TUTWSN API Implementation . . . 197
15.2.1 Gateway API . . . 198
15.2.2 Node API . . . 198
15.3 TUTWSN API Evaluation . . . 200
15.3.1 Ease of Use . . . 200
15.3.2 Resource Consumption . . . 200
15.3.3 Operational Performance . . . 201

16 TUTWSN SensorOS **203**
16.1 SensorOS Design . . . 203
16.1.1 SensorOS Architecture . . . 204
16.1.2 OS Components . . . 204
16.2 SensorOS Implementation . . . 206
16.2.1 HAL Implementation . . . 206
16.2.2 Component Implementation . . . 207
16.3 SensorOS Performance Evaluation . . . 210
16.3.1 Resource Usage . . . 210
16.3.2 Context Switch Performance . . . 210
16.4 Lightweight Kernel Configuration . . . 211
16.4.1 Lightweight OS Architecture and Implementation . . . 211
16.4.2 Performance Evaluation . . . 212
16.5 SensorOS Bootloader Service . . . 213
16.5.1 SensorOS Bootloader Design Principles . . . 213
16.5.2 Bootloader Implementation . . . 213

17 Cross-layer Issues in TUTWSN **217**
17.1 Cross-layer Node Configuration . . . 217
17.1.1 Application Layer . . . 219
17.1.2 Routing Layer . . . 219
17.1.3 MAC Layer . . . 219
17.1.4 Physical Layer . . . 220
17.1.5 Configuration Examples . . . 220

17.2 Piggybacking Data 223
17.3 Self-configuration with Cross-layer Information 224
17.3.1 Frequency and TDMA Selection 224
17.3.2 Connectivity Maintenance 224
17.3.3 Role Selection 225

18 Protocol Analysis Models **227**
18.1 PHY Power Analysis 227
18.2 Radio Energy Models 229
18.2.1 TUTWSN Radio Energy Models 230
18.2.2 ZigBee Radio Energy Models 232
18.3 Contention Models 234
18.3.1 TUTWSN Contention Models 234
18.3.2 ZigBee Contention Models 235
18.4 Node Operation Models 238
18.4.1 TUTWSN Throughput Models 238
18.4.2 ZigBee Throughput Models 239
18.4.3 TUTWSN Power Consumption Models 240
18.4.4 ZigBee Power Consumption Models 243
18.5 Summary 245

19 WISENES Design and Evaluation Environment **247**
19.1 Features 247
19.2 WSN Design with WISENES 248
19.3 WISENES Framework 249
19.3.1 Short Introduction to SDL 251
19.3.2 WISENES Instantiation 252
19.3.3 Central Simulation Control 253
19.3.4 Transmission Medium 253
19.3.5 Sensing Channel 254
19.3.6 Sensor Node 254
19.4 Existing WISENES Designs 256
19.4.1 TUTWSN Stack 258
19.4.2 ZigBee Stack 260
19.5 WISENES Simulation Results 263
19.5.1 Simulated Node Platforms 264
19.5.2 Accuracy of Simulation Results 266
19.5.3 Protocol Comparison Simulations 268

PART V DEPLOYMENT **277**

20 TUTWSN Deployments **279**
20.1 TUTWSN Deployment Architecture 280
20.1.1 WSN Server 281

20.1.2 WSN and Gateway 282
20.1.3 Database 282
20.1.4 User Interfaces 282
20.2 Network Self-diagnostics 283
20.2.1 Problem Statement 283
20.2.2 Implementation 284
20.3 Security Experiments 290
20.3.1 Experimental KDC-based Key Distribution and Authentication Scheme 291
20.3.2 Implementation Experiments 291

21 Sensing Applications **293**
21.1 Linear-position Metering 293
21.1.1 Problem Statement 293
21.1.2 Implementation 294
21.1.3 Results 296
21.2 Indoor-temperature Sensing 297
21.2.1 WSN Node Design 298
21.2.2 Results 298
21.3 Environmental Monitoring 300
21.3.1 Problem Statement 300
21.3.2 Implementation 300
21.3.3 Results 306

22 Transfer Applications **313**
22.1 TCP/IP for TUTWSN 313
22.1.1 Problem Statement 313
22.1.2 Implementation 314
22.1.3 Results 316
22.2 Realtime High-performance WSN 318
22.2.1 Problem Statement 318
22.2.2 Implementation 318
22.2.3 Results 324

23 Tracking Applications **327**
23.1 Surveillance System 327
23.1.1 Problem Statement 328
23.1.2 Surveillance WSN Design 328
23.1.3 WSN Prototype Implementation 331
23.1.4 Surveillance WSN Implementation on TUTWSN Prototypes . . . 332
23.2 Indoor Positioning 334
23.2.1 Problem Statement 335
23.2.2 Implementation 335
23.3 Team Game Management 342
23.3.1 Problem Statement 343

23.3.2 Implementation 343
23.3.3 Example Application Scenario 345

PART VI CONCLUSIONS 349

24 Conclusions 351

References 353

Index 369

Preface

Wireless short-range networking bloomed in late 1990s when the first WLAN and Bluetooth standards were completed and the technology migrated to early consumer products. The first standards targeted simple wireless applications like the file transfer between a limited number of devices. Dreams and visions about ubiquitous networking had already started at that time with the concept of thousands of communicating gadgets in our everyday life. It was natural to try the first experiments of ubiquitous networking with the current existing standards, but soon it turned out that the commercial devices were not feasible for such applications. Later on several proposals were presented for wireless sensor networks, but there are still many application domains where a single, or even a couple of standards, can not completely fulfill all the requirements. For this reason many proprietary wireless sensor network (WSN) technologies have emerged.

Soon after the advent of the first WLAN standards much of the research focused on improving and enhancing known deficiencies especially for Quality of Service (QoS) and security. Another branch of research attempted to adapt the standard to fit completely new set of applications not previously intended for the purpose of the original standard. In both cases such gradual developments can improve something but not necessarily make a major scientific breakthrough.

Our approach has been different. We started from scratch, focusing on what we wanted to do with WSNs and then began to search the technology base for what we needed. No WSN standards were available when we started our short-range wireless activities in 1997, only the first visions and ideas about ubiquitous networking. Over the years, we persistently developed our own WSN technology with the help of a large group of talented PhD and MSc students.

Right from the beginning we realized that WSNs introduced a far greater challenge, well exceeding that of mobile phone networks, which were the hot topic of the 1990s. We started with a broad frontier of knowledge from theoretical analysis to full-scale prototype implementations and real-life deployment experiments. Our strength has been the ability to realize the inventions in practical terms, taking into account the real-world non-ideas for the purposes of getting the design to really work. We have also developed new design tools to support the research since none of the existing simulator frameworks were complete enough to meet our wide design scope.

We have experienced how long the road is from taking a new algorithm on a scratchpad to developing a working mesh WSN and would like more realism to be reflected in scientific WSN publications in general. One of the most severe problems is the lack of holistic view. We have learned the hard way in that any real WSN cannot be simplified for a couple of algorithms and considered in a vacuum either. One slight detail can have a drastic effect on the whole network, multiplying the energy consumption, or driving the whole network in an unstable state. Surprisingly, it is not self-evident what information should be probed from a deployed WSN pertaining to its operation. For that purpose we have developed mechanisms for performing WSN self-diagnostics and automated it with our tool support.

At this point it should be noted that we have focused on WSNs consisting of embedded, resource-limited nodes with small-to-moderate physical size. Such nodes can be used on their own or attached to many kinds of devices or activities. Target lifetime is years of operation, during which large amounts of data is collected. It is also important to note that we have targeted completely autonomous WSNs that do not need any external control. For example, each node computes mesh routing independently, but in collaboration with other nodes and without any central router or network coordinator. This is important to make network deployment very fast without preliminary planning and manual configuration. We think that the best WSN is one that is invisible WSN to users.

We are proud to present this book that details our findings, inventions, and experiments in low-power mesh WSNs. We are confident this volume will provide a fresh outlook to the key design issues and show how they can be approached. This book will also serve as teaching material, although it is not written in the form of a textbook with homework problems.

The research work has been funded by several research projects in collaboration with a number of companies, Tekes (Finnish funding agency for technology and innovations) and Academy of Finland.

Abbreviations

ACK	Acknowledgment
ACL	Access Control List
ACQUIRE	Active Query forwarding In Sensor Networks
ADC	Analog-to-Digital Converter
AES	Advanced Encryption Standard
AJAX	Asynchronous JavaScript and XML
ANSI	American National Standards Institute
API	Application Programming Interface
APS	Application Support
ASIC	Application Specific Integrated Circuit
ATEMU	Atmel Emulator
ATM	Asynchronous Transfer Mode
BER	Bit Error Rate
BI	Beacon Interval
B-MAC	Berkeley Media Access Control
BO	Beacon Order
CAN	Controller Area Network
CAP	Contention Access Period
CBC	Cipher Block Chaining
CBR	Constant Bit Rate
CCA	Clear Channel Assessment
CCM	CTR with CBC-MIC

CDMA Code Division Multiple Access

CFP Contention-Free Period

CMAC Cipher-based Message Authentication Code

CORBA Common Object Request Broker Architecture

COTS Commercial Off-The-Shelf

CPU Central Processing Unit

CRC Cyclic Redundancy Check

CSMA Carrier Sense Multiple Access

CSMA-CA Carrier Sense Multiple Access with Collision Avoidance

CTR Counter

CTS Clear-To-Send

DAC Digital-to-Analog Converter

DCA Dynamic Channel Assignment

DCF Distributed Coordination Function

DD Directed Diffusion

DECT Digital Enhanced Cordless Telecommunications

DLL Data Link Layer

DoS Denial-of-Service

DSAP Data Service Access Point

DSL Digital Subscriber Line

DSR Dynamic Source Routing

DSSS Direct Sequence Spread Spectrum

DVM Distributed Virtual Machine

DVS Dynamic Voltage Scaling

EAR Eavesdrop-And-Register

ECB Electronic Codebook

ECC Elliptic Curve Cryptography

ED Energy Detection

EEPROM	Electrically Erasable Programmable Read-Only Memory
EFSM	Extended Finite State Machine
ENDP	Energy-efficient Neighbor Discovery Protocol
ESB	Embedded Sensor Board
FAMA	Floor Acquisition Multiple Access
FAR	Face Aware Routing
FDMA	Frequency Division Multiple Access
FFD	Full Function Device
FHSS	Frequency Hopping Spread Spectrum
FIFO	First-In-First-Out
FPGA	Field Programmable Gate-Array
FSM	Finite State Machine
FTP	File Transfer Protocol
GDI	Great Duck Island
GFSK	Gaussian Frequency Shift Keying
GPRS	General Packet Radio Service
GPS	Global Positioning System
GPSR	Greedy Perimeter Stateless Routing
GRAB	Gradient Broadcast
GSM	Global System for Mobile Communications
GSN	Global Sensor Network
GTS	Guaranteed Time Slot
GUI	Graphical User Interface
GW	Gateway
HAL	Hardware Abstraction Layer
HCI	Host Controller Interface
HIPERLAN/2	High-Performance Radio Local Area Network type 2
HIPERMAN	High-Performance Radio Metropolitan Area Network

HomeRF	Home Radio Frequency
HRMA	Hop Reservation Multiple Access
HTML	Hypertext Markup Language
HTTP	Hypertext Transfer Protocol
HVAC	Heating, Ventilation & Air Conditioning
HW	Hardware
IC	Integrated Circuit
I2C	Inter-Integrated Circuit
ICMP	Internet Control Message Protocol
ID	Identifier
IEEE	Institute of Electrical and Electronics Engineers
I/O	Input/Output
IP	Internet Protocol
IPC	Inter-Process Communication
IR	Infrared
IREQ	Interest Request
ISM	Industrial, Scientific, Medicine
JDBC	Java Database Connectivity
JMS	Java Message Service
JSR	Java Specification Request
JVM	Java Virtual Machine
KDC	Key Distribution Center
L2CAP	Logical Link Control and Adaptation Protocol
LAN	Local Area Network
LEACH	Low-Energy Adaptive Clustering Hierarchy
LED	Light Emitting Diode
LFSR	Linear Feedback Shift Register
LIFS	Long Inter-Frame Spacing

LLC	Logical Link Control
LM	Link Manager
LMP	Link Manager Protocol
LOS	Line-of-Sight
LQI	Link Quality Indication
LR-WPAN	Low-Rate Wireless Personal Area Network
LWA	Linux Wireless sensor network Adaptation
MAC	Medium Access Control
MACA	Multiple Access with Collision Avoidance
MACAW	Media Access protocol for Wireless LANs
MARE	Mobile Agent Runtime Environment
MCU	Micro-Controller Unit
MIC	Message Integrity Code
MiLAN	Middleware Linking Applications and Networks
MIPS	Million Instructions Per Second
MMAC	Multichannel MAC
MMI	Mixed-Mode Interface
MoC	Model of Computation
MOM	Message Oriented Middleware
MOS	Mantis Operating System
MPDU	MAC Protocol Data Unit
MSAP	Management Service Access Point
MSDU	MAC Service Data Unit
MTS	More-to-Send
NAMA	Node Activation Multiple Access
NCAP	Network Capable Application Processor
NoC	Network-on-Chip
NP	Neighbor Protocol

NWK	Network
OMG	Object Management Group
ORB	Object Request Broker
OS	Operating System
OSI	Open Systems Interconnection
PACT	Power Aware Clustered TDMA
PAMAS	Power Aware Multi-Access protocol with Signaling
PAN	Personal Area Network
PC	Personal Computer
PCB	Printed Circuit Board
PDA	Personal Digital Assistant
PDSAP	Physical Data Service Access Point
PDU	Protocol Data Unit
PHY	Physical
PHP	Hypertext Pre-Processor
PID	Process Identifier
PIN	Personal Identification Number
PIO	Parallel Inout/Output
PIR	Passive Infrared
PLL	Phase Locked Loop
PMSAP	Physical Management Service Access Point
POSIX	Portable Operating System Interface
PRNET	Packet Radio Network
PSoC	Programmable System-on-Chip
PWM	Pulse-Width Modulation
QoS	Quality of Service
RADV	Route Advertisement
RF	Radio Frequency

RFD	Reduced Function Device
RFID	Radio Frequency Identification
RPC	Remote Procedure Call
RREQ	Route Request
RSSI	Received Signal Strength Indicator
RTOS	Realtime Operating System
RTS	Request-To-Send
RTT	Round Trip Time
RX	Receive
SAP	Service Access Point
SAR	Sequential Assignment Routing
SD	Superframe Duration
SDL	Specification and Description Language
SDU	Synchronization Data Unit
SEE	Sensor Execution Environment
SEP	Schedule Exchange Protocol
SF	Superframe
SIFS	Short Inter-Frame Spacing
SIG	Special Interest Group
SINA	Sensor Information and Networking Architecture
SKKE	Symmetric-Key Key Exchange
S-MAC	Sensor-MAC
SMACS	Self-Organizing Medium Access Control for Sensor Networks
SMD	Surface Mount Device
SMS	Short Message Service
SNAP	Sensor Network Asynchronous Processor
SNEP	Secure Network Encryption Protocol
SO	Superframe Order

SoC	System-on-Chip
SPI	Serial Peripheral Interface Bus
SPIN	Sensor Protocols for Information via Negotiation
SpeckMAC	Speck Medium Access Control
SpeckMAC-B	Speck Medium Access Control Backoff
SpeckMAC-D	Speck Medium Access Control Data
SQL	Structured Query Language
SQTL	Sensor Querying and Tasking Language
SRAM	Static Random Access Memory
SRSA	Self-Organizing Slot Allocation
SSF	Scalable Simulation Framework
SSL	Secure Sockets Layer
SSP	Security Service Provider
STEM	Sparse Topology and Energy Management
SW	Software
SWAN	Simulator for Wireless Ad-hoc Networks
SYNC	Synchronization
TBF	Trajectory-Based Forwarding
TC	Trust Center
TCB	Thread Control Block
TCL	Tool Command Language
TCP	Transmission Control Protocol
TDMA	Time Division Multiple Access
TDOA	Time Difference of Arrival
TEDD	Trajectory- and Energy-Based Data Dissemination
TEDS	Transducer Electronic Data Sheet
TII	Transducer Independent Interface
TIM	Transducer Interface Module

T-MAC	Timeout-MAC
TML	Token Machine Language
TOSSF	TinyOS Scalable Simulation Framework
TOSSIM	TinyOS Simulator
TRAMA	Traffic-Adaptive Medium Access
TTDD	Two-Tier Data Dissemination
TUTWSN	Tampere University of Technology Wireless Sensor Network
TUTWSNR	TUTWSN Routing Protocol
TX	Transmit
UART	Universal Asynchronous Receiver/Transmitter
UI	User Interface
UMTS	Universal Mobile Telecommunications System
USB	Universal Serial Bus
VM	Virtual Machine
WEP	Wired Equivalent Privacy
WG	Work Group
WiseMAC	Wireless Sensor MAC
WISENES	Wireless Sensor Network Simulator
WLAN	Wireless Local Area Network
WMAN	Wireless Metropolitan Area Network
WPAN	Wireless Personal Area Network
WSN	Wireless Sensor Network
WWAN	Wireless Wide Area Network
XML	Extensible Markup Language
ZDO	ZigBee Device Object
Z-MAC	Zebra MAC

Part I

INTRODUCTION

1

Introduction

During recent years, wireless network technologies have achieved a key role as the media for telecommunications. Whereas wired networks provide only fixed network topologies, wireless networks support low-cost and effortless installations, ad hoc networking, portability of network devices, and mobility of network users. Together with the growth of network and processing capacities, the application area of wireless networks has extended from limited speech and broadcast TV services into high-speed data transfer and multimedia. At the other end of the wireless technology spectrum, where no real-time multimedia is present, the need for low-cost, low-rate, and very low-power technologies has emerged. Devices supporting multiple wireless technologies and objects with embedded networking capabilities are appearing and envisioned to provide ubiquitous services.

1.1 Overview of Wireless Technologies

From the technology spectrum point of view, wireless communication can be categorized according to their typical applications, data rates, and coverage. Table 1.1 illustrates the generally known classification that originates from the Institute of Electrical and Electronics Engineers (IEEE). The values presented in the table are not definitive; rather they are provided for perceiving the relationships of the different classes. The wireless transceiver is assumed to be a radio although other wireless physical layers, such as infrared, can be used as well.

Wireless Wide Area Networks (WWANs) and Wireless Metropolitan Area Networks (WMANs) provide the widest geographical coverage. The highly utilized WWANs mainly consist of traditional digital cellular telephone networks and their extensions for data services and higher speeds, such as Global System for Mobile Communications (GSM) and Universal Mobile Telecommunications System (UMTS). Communication satellites belong to this class as well. WMANs are emerging technologies developed for broadband network access as an alternative to cable networks and Digital Subscriber Lines (DSLs) in homes and enterprises. Examples of WMANs are IEEE 802.16, its mobile extensions and the High-Performance Radio Metropolitan Area Network (HIPERMAN).

Ultra-Low Energy Wireless Sensor Networks in Practice: Theory, Realization and Deployment

Table 1.1 A classification of wireless communication technologies.

Class	Data rate	Radio coverage	Typical applications	Exemplar technologies
WWAN	<10 Mbps	>10 km	Telephony, mobile Internet	GSM, UMTS, satellite
WMAN	<100 Mbps	<10 km	Broadband Internet	IEEE 802.16, HIPERMAN
WLAN	<100 Mbps	<100 m	Wired LAN replacement	IEEE 802.11, HIPERLAN/2
WPAN	<10 Mbps	<10 m	Personal data transfer	Bluetooth, IEEE 802.15.3
WSN	<1 Mbps	<1 km	Monitoring, control	proprietary, IEEE 802.15.4, RFID

Wireless Local Area Networks (WLANs) have rapidly gained popularity on the wireless markets. WLAN was originally developed for extending or replacing wired computer Local Area Networks (LANs), in cases where fixed cabling was costly or impossible due to mobility, short network lifetime, or historic value of the buildings. At the moment, WLAN is widely employed for providing network access in public buildings and enterprises, and for municipal network implementations. WLAN has also extended to home networking, including consumer electronics and household appliances.

The de facto WLAN technology is IEEE 802.11, and its numerous extensions for higher communication speeds, Quality of Services support, security, and mesh networking. Different standards and industry specifications, such as High-Performance Radio Local Area Network type 2 (HIPERLAN/2), Home Radio Frequency (HomeRF) and Digital Enhanced Cordless Telecommunications (DECT) have mainly remained at the level of standardization with small product volumes.

The class in close relation to WLANs is comprised of Wireless Personal Area Networks (WPANs), such as Bluetooth and IEEE 802.15.3. WPANs are generally targeted at data communications between personal devices, including Personal Digital Assistants (PDAs), mobile phones, headsets, and laptops. WPANs are also used for low-rate and low-power communications, e.g. in automation and alarm systems. Furthermore, WPAN technologies can be used for providing wireless access to an infrastructure LAN (Bluetooth) and for enabling high-speed multimedia content delivery (IEEE 802.15.3). Hence, WPANs are not clearly distinct from WLANs, sharing the same operational environments and application domains. The differences are in the non-functional requirements, such as cost, power, and networking range.

Wireless Sensor Networks (WSNs) (Kuorilehto et al. 2005b; Stankovic et al. 2003) is an emerging class of wireless technologies that has recently aroused much interest in industry and academic research, and which is envisioned to create massive markets in next few years. WSNs consist of independent, collaborating, highly resource-constrained nodes and actuators that sense, process, store, and deliver data.

In contrast to WLANs and WPANs, WSNs are seen as larger scale, self-organizing, and strictly application-oriented rather than measured by the coverage of a single radio cell or the nominal capacity of a link. Thus, the network coverage can vary from centimeters to hundreds of meters and kilometers and the network can grow to thousands of nodes, while the data rates are in the order of bits/s (Kuorilehto et al. 2005b). The network performance is measured by its capability to serve the implemented applications.

Compared to traditional communication networks, there is no pre-existing physical infrastructure that restricts the topology. Recourses are constrained in the means of, for example, size, cost, memory, and especially energy, which is seen as the most limiting factor for mass applications and is the area that we have chosen as a special topic in this book.

WSNs have been implemented as proprietary solutions (Kuorilehto et al. 2005b). The recent standardized technology supporting WSN implementations is IEEE 802.15.4 (IEE 2003b), which is utilized by networks implemented according to the ZigBee specification (Zig 2004). The Radio Frequency Identification (RFID) technology can be seen to belong to the class of WSNs as well with limited scaling and networking performance.

This book concentrates on WSNs, specifically on the design, implementation, and deployment of ultra-low energy WSNs for different application domains.

WSNs are typically ad hoc networks (Stallings 2004) but there are major conceptual differences. First, WSNs are data-centric with an objective to deliver time-sensitive data to different destinations. Second, a deployed WSN is application-oriented and performs a specific task. Third, messages should not be sent to individual nodes but to geographical locations or regions defined by data content. In WSNs, quantitative requirements in terms of latency and accuracy are strict due to the tight relation to the environment. In general, the capabilities of an individual sensor node are limited, but the feasibility of WSN lies on the joint effort of the nodes (Stankovic et al. 2003).

1.2 TUTWSN

The DACI (Design, Applications, Communication, and Implementation) research group at Tampere University of Technology has been actively developing novel short-range wireless network technologies and applications since 1997. The Tampere University of Technology Wireless Sensor Network (TUTWSN) research contains both theoretical and experimental methods, in the extent that we implement full-scale prototypes.

Also, we have been concentrating on the whole problem area, from applications to low-level power optimization algorithms. We use the term TUTWSN framework to address a variety of general problems, open questions, and real-life deployments in the WSN field. The use of TUTWSN should be considered as best practice according to our analysis. Non-inclusion of the term "TUTWSN" suggests that the same fundamental research problems and best solutions apply as presented in this book.

The framework comprehends the methods, algorithms, and implementation that we have found to be suitable or optimal according to certain requirements. Thus, TUTWSN is not a single WSN realization, but contains numerous design choices and practices that are applicable or necessary when moving from theory to real-life deployments. We are not concentrating on theory for the sake of theory, but to present a path from theory to practical implementations. We believe that an indepth coverage of the commonly known features of WSN is not needed, and thus kept in minimum. Due to book's emphasis on practice, we need to have real-life implementations for making real-life comparisons between technologies and design choices.

Also, as we will conclude, WSNs need cross-layer design for achieving performance or operability in the first instance. The cross-layer means that we cannot design separate protocol layers and set their parameters independently of each other, since a simple function on the upper layer protocol can result in massive transactions with the lower layer, consuming the available energy and blocking the existing network operation. Having said that, this does not mean that a protocol stack has to be designed from scratch every time we need to support a new application, just that we need to efficiently reuse the available functional components. These components are standard protocols and their parts; operating systems, application programming interfaces, databases, radio transceivers, etc. The management and optimization of functionality creates a large design space, which requires methods and tools to meet the functional and non-functional needs of the applications.

We need to address entire protocol stacks, applications, and operation environments. Most weight is still attributed to the layers that significantly affect the end result, i.e. layers 2 and 3 (MAC and routing) and their implementations.

1.3 Contents of the Book

This book is divided into five parts. Part I; just discussed, provides an introduction to the various wireless technologies and WSNs. Part II presents the design space of WSNs by defining the requirements, a WSN design flow, and WSN standards. This part also covers sensor platform design approaches and lists various platforms in example. Part III describes the WSN protocol stack and presents related WSN protocol proposals for each layer. Software architecture is considered by examining middleware and application layers, as well as light-weight operating systems suited for WSNs. Quality of Service (QoS) and security issues are discussed in respect of WSNs. Part IV presents TUTWSN as an extensive example of a protocol stack and software architecture that is completely designed for WSNs. The part also presents Wireless Sensor Network Simulator (WISENES) design and evaluation environment targeted at developing WSN protocols. A separate chapter is devoted to protocol analysis models comparing TUTWSN and ZigBee protocols. Part V presents the TUTWSN deployment architecture ranging from sensor network to back-end tools for refining the gathered data. Several sensor application use cases are discussed, while presenting extensive experiences on real-life deployments related to each use case. Part VI concludes the book.

Part II

DESIGN SPACE OF WSNS

2

WSN Properties

The term *Wireless Sensor Network* can be comprehended differently depending on the context and the correspondent. Literally, the term defines that there is a *network*, meaning multiple end devices, communicating through *wireless* medium, and at least some of the devices *sense*, i.e. measure some physical quantity, in their environment.

When considering the broadest sense of the WSN term, the devices can be anything from laptops, PDAs, or mobile phones to very tiny and simple sensing devices. The communication method adopted may vary from WLAN with 54 Mbps (megabits per second) bandwidth to a proprietary technique with only few (kilobits per second) throughput.

Especially in the research community, WSNs are nowadays considered for networks consisting of an extensive number of tiny sensor nodes operating in batteries. Such WSNs are deployed to perform a specific task for a long period of time, even years. In general, these kinds of WSNs are only a small subset of the whole concept. Of course, if there are more powerful or mains-powered devices in the vicinity, it is beneficial to utilize their computation and communication resources for complex algorithms and as gateways to other networks.

The integration of different kinds of devices to a same seamless network is part of the ubiquitous computing paradigm (Weiser 1993, 1999). The integration of different kinds of devices to WSN is only a matter of incorporating similar communication interfaces. The implementation of communication protocols, applications, and management features in a Personal Computer (PC) environment is not a challenging task. Instead, the implementation of similar functionality in the tiny, resource-limited sensor nodes poses several characteristic challenges. Therefore, the discussion is limited to WSNs with such limited resource nodes.

2.1 Characteristics of WSNs

Consider the example WSN scenario depicted in Figure 2.1, in which a large number of nodes are randomly deployed in the vicinity or inside an inspected phenomenon (Akyildiz et al. 2002b). The nodes self-organize and collaboratively coordinate the sensing process depending on the phenomenon (Römer et al. 2002). Instead of sending raw data, each node

Ultra-Low Energy Wireless Sensor Networks in Practice: Theory, Realization and Deployment

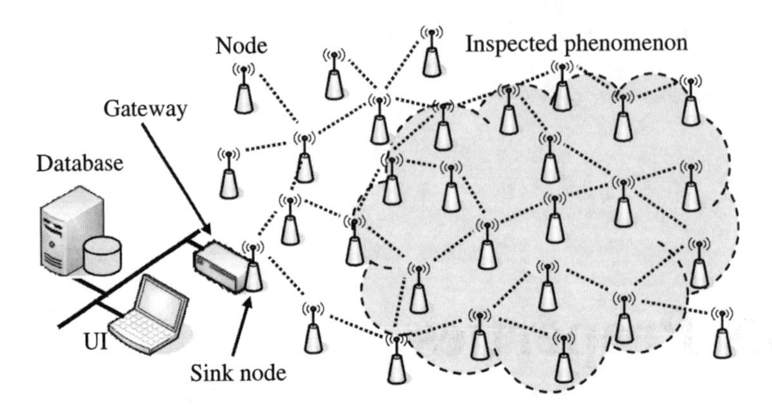

Figure 2.1 An example WSN scenario.

preprocesses its measurement results. The results are further aggregated or fused to obtain high-level application results while forwarding them on to a data gathering point, called a sink node (Elson and Estrin 2004; Stankovic et al. 2003). The sink node typically acts as a gateway to other networks and to human users (Karl and Willig 2005). The backbone infrastructure may contain components for data storing, visualization, and network control (Suhonen et al. 2006a).

The above scenario highlights some the unique characteristics that set WSNs apart from other communication networks. Even though the generalization of the characteristics may not be valid in all cases, the following lists the typical properties of envisioned WSNs (Karl and Willig 2005; Stankovic et al. 2003).

- **Communication paradigm**: Compared to traditional communication networks, individual node Identifiers (IDs) are not important. Instead, WSNs are *data-centric* meaning that the communication should be targeted to nodes in a given location or with a defined data content. As an example, data can be queried from an area defined by Global Positioning System (GPS) coordinates or from nodes that have a temperature exceeding the defined limit value. The identities of nodes that respond to the query is meaningless unlike their location.

- **Application-specific**: A single end device in a computer network may have multiple applications and the network must serve each application according to the defined quality requirements. Instead, WSN is deployed to perform a specific task. This allows application-dependent node platforms, communication protocols, data aggregation, and in-network processing and decision making.

- **Dynamic nature**: In a typical WSNs, node platforms are error-prone due to harsh operating conditions. Communication links between nodes are not stable due to node errors, unreliable and simple modulations, mobility of nodes, and environmental interferences. These aspects together with in-built dynamic decisions and data structures in protocols make the operating environment of even static WSN deployments unpredictable.

- **Scale and density**: Compared to other wireless networks, the number of nodes comprising WSNs may be huge. WSNs consisting of tens of thousands of nodes are envisioned. Further, the density of nodes can be high. These aspects depend on the requirements of the application for sensing coverage and robustness (redundancy).

- **Resource constraints**: A typical WSN node is small in physical size and battery powered. Thus the components, such as Micro-Controller Unit (MCU), data and code memories, transceiver, and sensing unit, are selected to minimize their area and power consumption. This implies that computation, communication, and memory resources in nodes are very limited. Further, batteries are typically small-sized; thus energy resources are also scarce.

- **Deployment**: In large-scale WSNs, the deployment of nodes is random and their maintenance and replacement is impractical. Still, the requirements and applications of the deployed WSN may alter, which implicate that runtime reconfiguration and reprogramming are needed.

2.2 WSN Applications

Applications for WSNs may either be traditional ones, in which the wired cabling is replaced by WSN, or completely new. Small-size nodes and maintenance-free operation of WSNs allow their rapid and easy deployment even to rough and dangerous environments. Further, the integration and local processing of data measuring several different quantities enable a completely new types of applications.

The physical quantities that can be measured are diverse. The most common measured quantities are temperature, humidity, pressure, acceleration (vibration), sound, light (luminance, image), infrared, magnetic fields (compass), radiation, location (GPS), chemical compositions, and mechanical stress.

As a consequence of extensive possibilities, the envisioned applications for WSNs are diverse. The miniaturization of sensing devices allows the embedding of small WSN nodes as an unnoticeable part of the everyday life. The most typical application domains for WSNs are (Akyildiz et al. 2002b; Chong and Kumar 2003; Karl and Willig 2005):

- **Home automation**: WSNs are key building blocks for smart homes. Heating, Ventilation & Air Conditioning (HVAC) control, and local and remote management of home appliances are examples of usage possibilities.

- **Environmental monitoring**: Random deployments on large-scale areas make the monitoring of agriculture and wildlife easier. Further, WSNs may be used in catastrophe (e.g. wildfire, earthquake, tsunami) warning systems and in disaster relief.

- **Industrial monitoring and control**: WSNs have a potential for replacing traditional cabling in monitoring and control systems in factories.

- **Military**: In military applications, the rapid deployment of WSNs allows instant use of data. Usage scenarios for WSN data are wide-ranging in intelligence, surveillance, reconnaissance, and targeting. Further, WSNs can be a part of the communication infrastructure.

- **Security**: Compared to existing wired alarm systems, WSNs allow easier deployment, adaptivity, and error robustness. The increasing demand for security and alarm systems is evident in home, office, and other public environments such as airports and factories.
- **Traffic control**: WSNs allows the easy monitoring and control of traffic conditions, especially at peak times. Temporary situations such as roadworks and accidents can be monitored in situ. Further, the integration of monitoring and management operations, such as signpost control, is facilitated by a common WSN infrastructure.
- **Health care**: Body sensor networks can be used to monitor physiological data of patients. These can be integrated to WSNs that monitor and control the locations of doctors, patients, and specialist equipment.

Independent of the application domain, the main functionalities of WSN applications are data gathering and processing. The type of gathered data and the nature of processing depend on the application. In spite of the diversity of applications and domains, four main tasks that are independent of application domain can be identified (Akyildiz et al. 2002b; Karl and Willig 2005; NIST Advanced Network Technologies Division 2006):

1. **Monitoring**: Determine the value of a parameter in a given location or the coverage area of the network. Typically, the task is completed using periodic measurements.
2. **Event detection**: Detect the occurrence of events of interest and their parameters.
3. **Object classification**: Identify an object or event. In general, this task requires the combination of data from several sources and a collaborative processing to conclude the result.
4. **Object tracking**: Trace the movements and position of a mobile object within the coverage area of the network.

In addition to the data gathering, WSNs may also implement control or actuating functionality. The control can be either integrated into the network or controlled by a central entity. The integrated control makes the control decisions based on the information available locally whereas the central control entity can exploit the data available from the entire network. On the other hand, the integrated control operations are faster because of the lack of data routing delay. The control operations are performed either by integrated actuators, through external Input/Output (I/O), or by separate actor nodes (Akyildiz and Kasimoglu 2004).

2.2.1 Commercial WSNs

In recent years, WSNs have attracted attention from industry. Alternatives for the traditional cabling infrastructure have been looked at from the WSN domain. As a consequence, hundreds of companies that offer products from plain WSN nodes to complete wireless sensing systems have emerged.

Table 2.1 gives a list of companies with their key WSN products with a short description. The table is not comprehensive and merely visualizes the composition of the commercial

Table 2.1 Some commercial companies offering WSN products.

Company	Products	Description
Accsense	Nodes, GWs, starter kits	*PODS* nodes and access points to node data (either Web or direct)
Ambient Systems	Nodes, GWs, UIs	Ambient product line
Arch Rock	Web-enabled WSN, application WSNs	Web-enabled IP-based WSN, tailored WSNs for monitoring and automation
BTnodes	Nodes	*BTnode* platforms with Bluetooth stack
Chip45	Nodes	*iDwaRF* node families
Cincinnati Technologies	Nodes	ZigBee compliant nodes
Crossbow Technology	Nodes, GWs, UIs, application WSNs	Rich set of *Mote* node platforms and accessories and sensing systems
Ember	Nodes	ZigBee compliant nodes
Emerson	Application WSNs	Wireless measurement modules
Freescale	Node ICs	Node ICs with ZigBee compliant interface
Jennoc	Nodes	ZigBee modules
MAXFOR Technology	Nodes, GWs	Different node families
Millennial	Monitoring and control application	*MeshScape* system
Moteiv	Nodes, GWs	*Tmote* node platforms
Riga Development	HVAC control application	*WiSuite* wireless environmental control
Sensicast Systems	Monitoring application WSN	*SensiNet* monitoring WSN
SunSpotWorld	Nodes	*SunSPOT* nodes with integrated Java VM
Ubisense	Nodes, GWs, UIs	*SmartSpace* system
Ubiwave	Nodes, GWs, UIs	*UbiSentry*, *UbiGuard*, *UbiCare* products

WSN development. The companies that offer pure Software (SW) solutions or tailored SW implementations are not included. The number of possible commercial WSNs has increased considerably in the past couple of years. Furthermore, the nature of products has shifted from nodes towards networks targeted to specific application domains.

Most of the companies still offer plain nodes with varying networking capabilities. The networking is either implemented with standardized networking technologies (e.g. ZigBee) allowing interoperability with other platforms with similar applications, or customized networking stacks (e.g. BTnodes). Complete WSN solutions are available mainly for monitoring but home and industrial control applications are also emerging.

A large part of the companies listed originate from university research groups. On the other hand, several large companies such as Freescale, Sun Microsystems, Nokia, and Intel operate actively in the WSN field. Freescale and Sun Microsystems already offer products, whereas Intel and Nokia are currently actively supporting several research projects.

2.2.2 Research WSNs

WSNs have been actively researched since the late 1990s. Over that time research groups from a variety of universities have experimented with different kinds of deployment scenarios. The first deployments were test networks with few nodes and a limited network lifetime, but state-of-the-art research networks consist of hundreds of nodes and target lifetimes in years. A listing and short analysis of 50 environmental monitoring WSNs is given in Hart and Martinez (2006). The list incorporates commercial WSNs but also gives a good overview of environmental sensor networks.

The past advances in manufacturing technologies and availability of commercial WSN nodes have made it easier for research units to explore WSN deployments. Still, most of the research is based on the analytical or simulation studies. When considering requirements and constraints of real-world WSNs, such methods may not always be sufficient (Andel and Yasinsac 2006).

In order to illustrate existing WSN deployments, the research efforts for WSN deployments are listed in Table 2.2. The list is not exhaustive but gives an overview of current deployments and their scale. The scale defines the number of nodes and lifetime length of the deployment. The data interval illustrates the activity of the network by giving the frequency of the data communication. In addition, the main purpose of each deployment is given.

Most of the presented WSNs are for environmental monitoring. The environmental aspects and the occupancy of seabird burrows is monitored in Great Duck Island (GDI) (Szewczyk et al. 2004). PODS in Hawaii monitor the weather conditions and rare plants. CORIE (Steere et al. 2000) measures the temperature and pressure at stations located in the Columbia River and other environmental aspects from offshore stations. NIMS (Batalin et al. 2004) monitors the solar radiation in forest environments. Tolle et al. (2005) present "Macroscope" WSN for monitoring temperature and humidity in a redwood. The agricultural WSN deployment to a vineyard setting is discussed in Beckwith et al. (2004). A temperature-sensing WSN consisting of 25 nodes is deployed in the PicoRadio project (Reason and Rabaey 2004). An environmental monitoring network deployed to heathlands in Northern Germany is presented in Turau et al. (2006). GlacsWeb (Martinez et al. 2005) measures several different aspects in glaciers.

Table 2.2 Examples of WSNs deployments published in the research community.

Deployment	Scale	Lifespan	Data interval	Purpose	Reference
Environmental monitoring WSNs					
CORIE	18	–	1–15 min	Environmental monitoring	`http://www.ccalmr.ogi.edu/CORIE/`
GDI	150	4 months	20 min	Environmental monitoring	Szewczyk et al. (2004)
GlacsWeb	8	Few months	1 day	Glacier monitoring	Martinez et al. (2005)
Heathland	24	16 days	1 hour	Environmental monitoring	Turau et al. (2006)
Macroscope	33	44 days	30 seconds	Environmental monitoring	Tolle et al. (2005)
NIMS	7	–	3–20 per minute	Environmental monitoring	Batalin et al. (2004)
PicoRadio	25	1–2 months	5 seconds	Environmental monitoring	Reason and Rabaey (2004)
PODS	30–50	–	Infrequent	Environmental monitoring	`http://www.pods.hawaii.edu/`
Vineyard	65	6 months	5 minutes	Environmental monitoring	Beckwith et al. (2004)
ZebraNet	7	12 months	8 minutes	Tracking wildlife	Zhang et al. (2004)
Shellfish	4	<1 day	5 minutes	Catch temperature monitoring	Crowley et al. (2005)
Object tracking WSNs					
Multi-target	144	<1 day	Few seconds	Tracking multiple objects	Oh et al. (2006)
PEG	100	<1 day	0.5 seconds	Tracking moving vehicle	Sharp et al. (2005)
PinPtr	56	<1 day	<1 second	Locating a shooter	Simon et al. (2004)
Tracking	6	–	<1 second	Tracking moving object	Römer (2004)
VigilNet	70	<1 day	<1 second	Tracking vehicles	He et al. (2006)
Building monitoring WSNs					
SensorScope	20	2 months	5 minutes	Building monitoring	Schmid et al. (2005)
Wisden	10	<1 day	<1 second	Structural monitoring	Xu et al. (2004)

ZebraNet (Zhang et al. 2004) aims for tracking wildlife movements using a WSNs. Crowley et al. (2005) present a WSN for short-period monitoring of shellfish catches.

Another widely adopted application domain is object tracking. Yet, due to its relative complexity with resource constrained nodes, the number of proposals is quite limited and the lifetime of the deployments is not long. However, in these cases, the given lifetime is not restricted by the battery capacity but the fact that the deployment is merely an experiment performed only for a short period of time.

PEG (Sharp et al. 2005) tracks a moving vehicle within the sensor network coverage using ultrasound. Römer (2004) present a system for tracking the movements of an object emitting Infrared (IR) light. PinPtr (Simon et al. 2004) instead locates a shooter based on the Time Difference of Arrival (TDOA). VigilNet (He et al. 2006) uses multiple different type of sensors, including magnetometers, to track vehicles. Multiple targets are tracked simultaneously in Oh et al. (2006) using Passive Infrared (PIR) sensors.

Wisden (Xu et al. 2004) experiments WSN for the monitoring of vibrations in building structures. SensorScope (Schmid et al. 2005) measures light, sound, and temperature in indoor environment.

Basically, the size of the largest deployments is still in the scale of hundreds of nodes. The lifetime of the continuous deployments rarely reaches a year, but battery replacements or other types of maintenance are required sooner. These apply mainly to static environmental monitoring deployments, whereas mode dynamic tracking and surveillance WSNs are experimented in considerably smaller scale and in short-term tests. Further, the activity in monitoring WSNs is quite infrequent whereas the tracking networks are typically very data-intensive. Nonetheless, the rapid development in this area results in the constant widening and lengthening of deployments.

2.3 Requirements for WSNs

The requirements for a deployed WSN depend on the application and the operating environment. Hence, the generalization of the requirements in detail is not practical. Nonetheless, WSN applications possess several common characteristics, based on which general requirements for the node platforms, protocols, and applications can be defined (Akyildiz et al. 2002a,b; Karl and Willig 2005; Stankovic et al. 2003):

- **Fault tolerance**: WSNs must be robust against failures of individual nodes or even all nodes in a certain location. The network functionality must be maintained even though the built-in dynamic nature and failures of nodes due to harsh environment, depletion of batteries, or external interference make networks prone to errors.

- **Lifetime**: The network lifetime is a crucial issue in WSNs. The nodes are battery-powered or the energy is scavenged from the environment and their maintenance is difficult. Nevertheless, networks are expected to be fully functional for long periods of time. Thus, energy saving and load balancing must be taken into account in the design and implementation of WSN platforms, protocols, and applications.

- **Scalability**: The number of nodes in WSN is typically high. While the scale partly depends on the covered area, the replication of nodes, limited sensing coverage, and application requirements also call for dense deployments in small areas. Thus, the WSN protocols must deal with high densities and numbers of nodes.

- **Realtime**: WSNs are tightly related to the real world. Therefore, strict timing constraints for sensing, processing, and communication are present in WSNs. For example, on a realtime identification of an event, first the stimulations from sensors must be captured, the obtained data processed, and the event identified. Finally, the result should be passed instantly through the network.

- **Security**: The need for security in WSNs is evident, especially in health care, security, and military applications. Most of the applications relay data that contain private or confidential information. In general, the security requirements for WSNs are difficult to fulfill due to complex and time-consuming algorithms and the limited resources of nodes.

- **Production cost**: The number of nodes in WSNs is high, and once nodes run out of batteries they are replaced by new ones. Further, WSNs are envisioned to be everywhere. Therefore, to make the deployments possible, the nodes should be extremely low cost.

The application requirements define the goals for a deployed WSN that must be met in the target environment and with the resource constraints set by WSN nodes. In order to fulfill the objectives of a network, cross-layer design and optimization are required. A traditional, strictly layered protocol stack is not suitable for WSNs but the protocols must consider the effects of their operations on the other layers, and the essential information must be shared seamlessly.

3

Standards and Proposals

3.1 Standards

Standards are produced by groups of companies for obtaining interoperability between products from different manufacturers. For example, in the sensors industry there exists a vast number of different sensors to measure physical parameters, such as temperature, pressure, humidity, illumination, gas, flow rate, strain, and acidity. The sensors may be accompanied by manufacturer's proprietary Application Programming Interface (API), and connected to a computing device using a variety of different proprietary communication technologies. To enable effective integration, access, fusion and use of sensor-derived data, open and standardized sensor interfaces and sensor data formats are needed. The objective is for as many different sensors from different manufacturers to work together without human intervention and customization. Thus, also small sensor manufacturers can participate in control system implementation.

IEEE defines two standard families that can be used for WSNs, namely the IEEE 1451 standard family for a smart transducer interface for sensors and actuators, and the IEEE 802.15 standard family for wireless personal area networks. The standard families complement each other. While the IEEE 802.15 defines the interface for wireless communication, the IEEE 1451 defines interface for sensors and actuators.

3.1.1 IEEE 1451 Standard

For enabling the interoperability of sensors and actuators from multiple vendors, the National Institute of Standards and Technology proposed a plug-and-play smart sensor protocol in 1994. This protocol has since been formally adopted as IEEE 1451 standard. IEEE 1451 is a suite of smart transducer interface standards, which describes a set of open, network-independent communication interfaces for connecting sensors and actuators (transducers) to microprocessors, instrumentation systems, and networks.

The architecture of IEEE 1451 is presented in Figure 3.1. IEEE 1451 defines two terms: Transducer Interface Module (TIM) and Network Capable Application Processor

Ultra-Low Energy Wireless Sensor Networks in Practice: Theory, Realization and Deployment

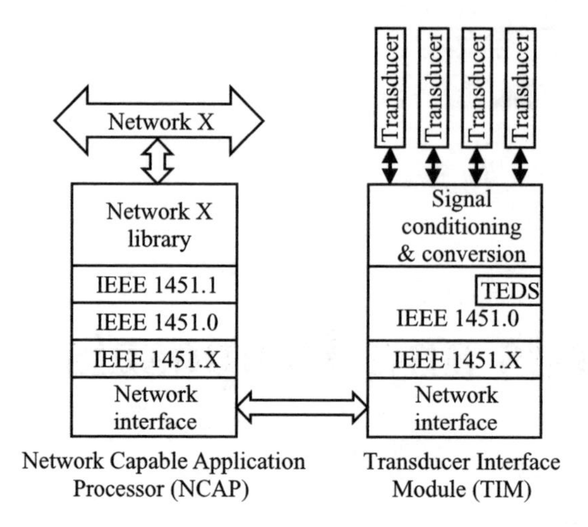

Figure 3.1 The NCAP and TIM defined by IEEE 1451.

(NCAP). TIM is a device that contains a set of transducers, signal conditioning and data conversion circuitry, and software modules. They consist of IEEE 1451 standard modules and standardized wireless or wired network technology. NCAP is any kind of network-connected computing device, such as a PC, laptop, PDA, or an embedded processor, which receives the sensed data from a set of TIMs, depending on the utilized communication technology. NCAP contains similar modules to TIM. The NCAP information model for smart transducers is defined by IEEE 1451.1 standard.

One of the most important features that IEEE 1451 provides is the definition of Transducer Electronic Data Sheet (TEDS) specified by IEEE 1451.0 standard. TEDS is used to describe the entire TIM including transducer, signal conditioner, and data converter. TEDS is practically a memory device attached to the transducer, which stores transducer identification, calibration, correction data, measurement range, and manufacturer-related information. Hence, TEDS virtually eliminates error prone, manual entering of data and system configuration and allows transducers to be installed, upgraded, replaced, or moved by the plug-and-play principle. In addition, IEEE 1451.0 defines the functional specification of TIM, the discovery and management of TIMs, and a set of sensor API calls with message exchange protocols and commands required for interfacing with transducers.

The supported ways to establish wired and wireless connections between NCAP and TIMs are presented in Figure 3.2. IEEE 1451 leaves Physical (PHY) and Medium Access Control (MAC) layers unspecified, but provides interfaces with numerous standardized technologies by IEEE 1451.2 through IEEE 1451.6.

IEEE 1451.2 defines wired point-to-point communication between NCAP and TIM through Universal Asynchronous Receiver/Transmitter (UART) or a Transducer Independent Interface (TII). IEEE 1451.3 defines distributed multidrop system, where a large number of TIMs may be connected along a wired multidrop bus. IEEE 1451.4 specifies mixed-mode communication protocols, which carry analog sensor values with digital TEDS

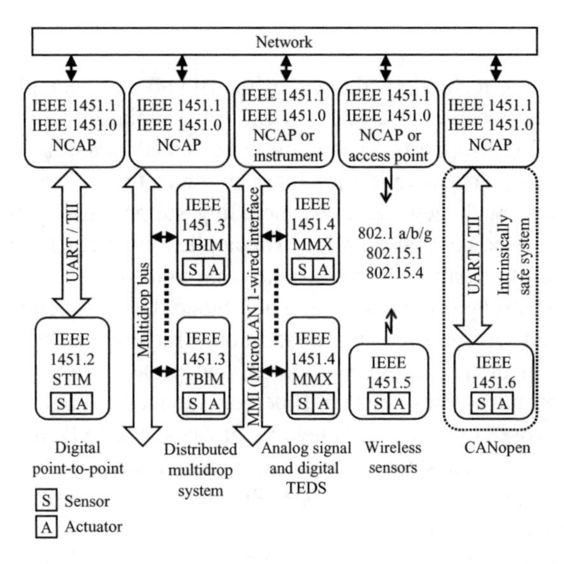

Figure 3.2 IEEE 1451 network architecture.

data. Multiple 1451.4 compliant TIMs are connected along a Mixed-Mode Interface (MMI) bus utilizing MicroLAN 1-wire interface. IEEE 1451.6 defines a high-speed Controller Area Network (CAN) bus between NCAP and TIMs.

IEEE 1451.5 standard (IEEE 2006) defines wireless sensors and thus, it is most closely related with WSNs. Supported communication technologies (including PHY and MAC) are IEEE 802.11a/b/g, IEEE 801.15.1, and IEEE 802.15.4. For each of these communication technologies, IEEE 1451.5 provides a higher layer API for seamless interoperability with IEEE 1451.0. Yet, the handling of security and key exchange are left for the communication technologies.

3.1.2 IEEE 802.15 Standard

The WPAN Working Group initially focused on creating the IEEE 802.15.1 standard for PHY and MAC layers based on Bluetooth technology (IEE 2002). The Working Group soon formed two other subgroups, firstly IEEE 802.15.3 focusing on high-speed WPAN (IEE 2003a) for multimedia applications. In December 2000, IEEE 802.15.4 Low-Rate Wireless Personal Area Network (LR-WPAN) (IEE 2003b) was initiated for providing low-complexity, low-cost, and low-power wireless connectivity among inexpensive devices (Gutiérrez et al. 2004). The IEEE 802.15.4 standard was ratified in the summer of 2003. ZigBee Alliance endorsed the standard shortly afterwards. While IEEE 802.15.4 sub-standard defines the MAC and PHY layers, the ZigBee specification defines the upper protocol layers and develops application profiles that can be shared among different manufacturers (Kohvakka et al. 2006b).

LR-WPAN together with ZigBee are one of the most potential technologies enabling WSNs. Hence, they are discussed in detail.

Bluetooth

Bluetooth (Blu 2004) is a technology defined by the Bluetooth Special Interest Group (SIG). Originally, it was only intended as a simple serial cable replacement for electronic devices. Presently, the technology supports various more advanced functionalities, such as ad hoc networking and access point operation for Internet connections. The ongoing development is extending Bluetooth with new features, including support for QoS, higher data rates, multicasting, and lower power consumption. The application areas keeps expanding as new products with Bluetooth capability are constantly introduced.

The Bluetooth technology consists of several protocol layers ranging from the PHY layer to object exchange and service discovery. In addition, Bluetooth defines a number of profiles that define a selection of messages, procedures, and protocols required for supporting a specific service. Communications between Bluetooth devices can be point-to-point or point-to-multipoint. A network composed of Bluetooth devices is called a *piconet*, which consists of a master and up to seven slave devices. More units can remain synchronized to the network while being in a power-save mode. Piconets can be linked together to form a larger network, known as *scatternet*.

Bluetooth operates in the 2.4 GHz (gigahertz) unlicensed ISM band and utilizes the Frequency Hopping Spread Spectrum (FHSS) technique in the radio interface. Gaussian Frequency Shift Keying (GFSK) modulation is used, resulting in a gross link speed of 1 Mbps and a maximum of 721 Kbps for the user data rate in asymmetric link. The voice channel bandwidth is 64 Kbps. The nominal link range is 10 cm to 10 m, but can be extended to more than 100 m by increasing the transmit power (Blu 2004).

IEEE has adopted the lowest Bluetooth layers, i.e. the radio, the baseband, Link Manager Protocol (LMP), Link Manager (LM), Host Controller Interface (HCI), and Logical Link Control and Adaptation Protocol (L2CAP), and standardized them in its IEEE 802.15.1 standard (IEE 2002). The PHY layer is constructed of the radio and part of the baseband. MAC layer corresponds to the rest of the baseband and the higher layers up to L2CAP.

IEEE 802.15.4 standard*

The IEEE 802.15.4 standard has been developed to enable low-cost and low-power applications in the fields of industrial, agricultural, vehicular, residential, and medical sensors and actuators (Gutiérrez et al. 2004). The standard defines the channel access mechanism, acknowledged frame delivery, network association and disassociation. The standard supports two Direct Sequence Spread Spectrum (DSSS) PHY layers operating in Industrial, Scientific, Medicine (ISM) frequency bands, presented in Table 3.1 (Kohvakka et al. 2006b). A low-band PHY operates in the 868 MHz (megahertz) or 915 MHz frequency band and has a raw data rate of 20 kbps or 40 kbps, respectively. A high-band PHY operating in the 2.4 GHz band specifies a data rate of 250 kbps and has nearly worldwide availability. The 2.4 GHz frequency band has the most potential for large-scale WSN applications, since the high radio data rate reduces frame transmission time and usually also the energy per transmitted and received bit of data.

Table 3.1 IEEE 802.15.4 frequency bands and data rates. (Kohvakka et al. 2006b, © 2006 ACM. Adapted by permission.)

Band	868 MHz	915 MHz	2.4 GHz
Region	EU, Japan	US	Worldwide
channels	1	10	16
data rate	20 kbps	40 kbps	250 kbps

The IEEE 802.15.4 network supports three types of network devices: a Personal Area Network (PAN) coordinator, coordinators, and devices. The PAN coordinator is the primary controller of PAN, which initiates the network and often operates as a gateway to other networks. Each PAN must have exactly one PAN coordinator. Coordinators collaborate with each other for executing data routing and network self-organization operations. Devices do not have data routing capability and can communicate only with coordinators.

Due to the low performance requirements of devices, they may be implemented with very simple and low-cost platforms. The standard designates these low-complexity platforms as Reduced Function Devices (RFDs). Platforms with the complete set of MAC services are called as Full Function Device (FFD).

The standard supports 64-bit long addresses and 16-bit short addresses. The long addresses are typically used for identifying devices accurately. Once the devices associate with a network, they get a short address for operating in the network. Thus, the maximum number of nodes in a IEEE 802.15.4 network is around 65 000.

MAC protocol transfers data as frames with a maximum size of 128 (bytes), which enables a maximum MAC payload of 104 (bytes). This is adequate for the low data rate systems for which the standard is designed.

The MAC protocol can operate on both beacon-enabled and non-beacon modes. In the non-beacon mode, a protocol is a simple Carrier Sense Multiple Access with Collision Avoidance (CSMA-CA). Coordinators are required to constantly receive for possible incoming data. Yet, devices may transmit data on demand and sleep the rest of time achieving high energy-efficiency.

The power saving features that are critical in WSN applications are provided for the entire network by the beacon-enabled mode with inactive time. All communications are performed in a superframe structure presented in Figure 3.3 (Kohvakka et al. 2006b). A

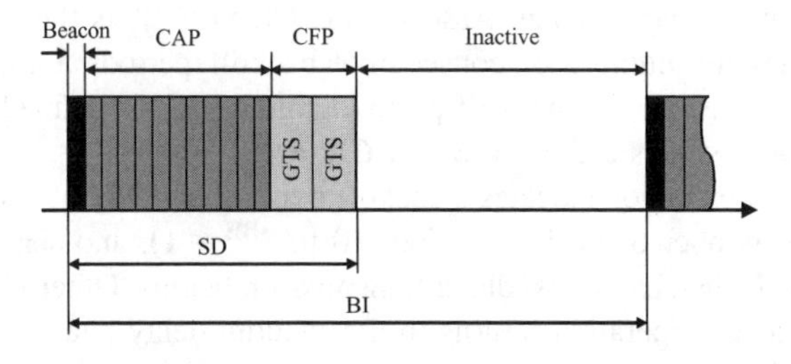

Figure 3.3 Superframe structure in beacon-enabled mode. (Kohvakka et al. 2006b. © 2006 ACM. Adapted by permission.)

superframe is bounded by periodically transmitted beacon frames, which allows the nodes to synchronize to the network. An active part of a superframe is divided into 16 contiguous time slots that form three parts: the beacon, Contention Access Period (CAP), and optional Contention-Free Period (CFP). Coordinators are required to listen to the channel for the whole CAP to detect and receive any data from their child nodes. At the end of the superframe is an optional inactive period, when nodes may enter into a power-saving mode. End devices may only receive beacons and transmit data on-demand, which increases their energy efficiency.

The Beacon Interval (BI) and the active Superframe Duration (SD) are adjustable by IEEE 802.15.4 parameters Beacon Order (BO) and Superframe Order (SO) as

$$BI = t_{\text{basesuperframe}} \times 2^{BO},$$

$$SD = t_{\text{basesuperframe}} \times 2^{SO},$$

where $0 \leq SO \leq BO \leq 14$ and $t_{\text{basesuperframe}}$ equals to 960 radio symbols (i.e. 15.36 ms (milliseconds) in the 2.4 GHz band). Hence, BI and SD may be between 15.36 ms (milliseconds) and 251.7 s (seconds). In cluster-tree networks, all coordinators transmit beacons and maintain their own superframe structures. In cluster-tree networks, all coordinators may transmit beacons to maintain synchronization with their children.

Downlink data from a coordinator to its child node are sent indirectly requiring a total of four transmissions. The availability of pending data is signaled in beacons. First, a child requests the pending data by transmitting a data request message. A coordinator responds to the request with Acknowledgment (ACK) frame, and then transmits the requested data frame. Finally, a child transmits ACK if the data frame is successfully received.

IEEE 802.15.4 utilizes a modified slotted CSMA-CA scheme during the CAP, except for ACK frames that are transmitted without carrier sensing. The scheme is modified from the IEEE 802.11 Distributed Coordination Function (DCF) protocol (IEE 1997). The major differences to legacy CSMA-CA are that a channel is not sensed during a backoff time, and that a new random backoff is selected if a channel is busy during a carrier sensing. In dense networks this may lead to inefficient channel utilization and long channel access delay. The protocol is prone to hidden node collisions, when two nodes outside the range of each other are transmitting data to a common coordinator (Hwang et al. 2005; Kohvakka et al. 2006b).

For accessing a channel, each node maintains three variables: *NB*, *BE*, and *CW*. *NB* is the number of CSMA-CA backoff attempts for the current transmission, initialized to 0. *BE* is the backoff exponent, which defines the number of backoff periods a node should wait before attempting Clear Channel Assessment (CCA). *CW* is the contention window length, which defines the number of consecutive backoff periods a channel needs to be silent prior to a transmission. The backoff period length (t_{BOP}) is defined as 0.32 ms in the 2.4 GHz band. Default values are $BE = 3$ and $CW = 2$.

Before a transmission, a node locates a backoff period boundary by the received beacon, waits for a random number of backoff periods (0 to $2^{\text{BE}} - 1$), and senses the channel by CCA for *CW* times. If the channel is idle, a transmission begins. Otherwise *NB* and *BE* are increased by one and the operation returns to the random delay phase. When *NB* exceeds a limit value (default: is 4), transmission terminates with a channel access failure.

By default, a node may try to retransmit the frame a maximum three times before a MAC issues a frame transmission failure. The standard defines that each frame is followed by an interframe spacing. The spacing depends on the length of a MAC Protocol Data Unit (MPDU). A Long Inter-Frame Spacing (LIFS) defined as 640 μs (microseconds) is used for frames containing longer than 18 B MPDU. Shorter frames are followed by a Short Inter-Frame Spacing (SIFS) defined as 192 μs.

CFP is an optional feature of IEEE 802.15.4 MAC, in which a channel access is performed in dedicated time slots. A node may reserve bandwidth for delay critical applications by requesting Guaranteed Time Slot (GTS) from a PAN Coordinator. The GTS allocations are signaled in beacon frames. In star networks, a device may obtain better QoS by the use of GTS, since contention and collisions are avoided. Yet, the applicability of GTS in peer-to-peer or cluster-tree networks is poor, since GTS may be used only between a PAN Coordinator and its one-hop neighbors. Moreover, intercoordinator collisions degrade QoS in GTS, since no collision avoidance mechanism is used in CFP.

IEEE 802.15.4 supports a *BatteryLifeExtension* option, which effectively reduces the coordinator energy consumption. When the option is selected, BE is initialized to 2, and any transaction should begin during *macBattLifeExtPeriods* backoff periods (default is 6 equaling 1.92 ms at 2.4 GHz band) after the beacon. This allows the coordinator to sleep the rest of CAP and conserve energy. In practice, a collision probability is high due to shorter backoff time. In addition, throughput is limited because of to a very short CAP. Thus, this option is suitable only for small and very low data rate networks.

ZigBee

ZigBee (Zig 2004) is an open specification for low-power wireless networking. ZigBee Alliance is an independent, open and nonprofit corporation founded in 2002. The first version of the ZigBee specification was announced in December 2004. A refinement of the specification was been launched in December 2006. In addition, ZigBee Alliance develops certification and compliance programs, branding, market development, and user education. At the end of year 2006, over 200 member-companies have joined the alliance including major names in the semiconductor, software developer, end product manufacturer, and service provider industries (Heile 2006).

ZigBee targets control and monitoring applications, where low-power consumption is a key requirement. The candidate applications are wireless sensors, lighting controls, and surveillance. Target market areas are residential home control, commercial building control, and industrial plant management.

The ZigBee protocol stack is presented in Figure 3.4 (Kuorilehto et al. 2006a). An application layer at the top of the stack contains application profiles, which determines device relationships and supervises network initiation and association functions. The profiles define which messages are sent over the air, application environment, and the types of utilized devices and clusters.

Each ZigBee device should have one or more application profiles, which may consists of ZigBee-specified public profiles and manufacturer-specific private profile. Only the devices equipped with the same application profiles interoperate end-to-end. Yet, manufacturers may add features on public profiles and implement additional profiles at the application layer, which live on different endpoints within the device. This allows the creation of manufacturer-specific extensions on the ZigBee.

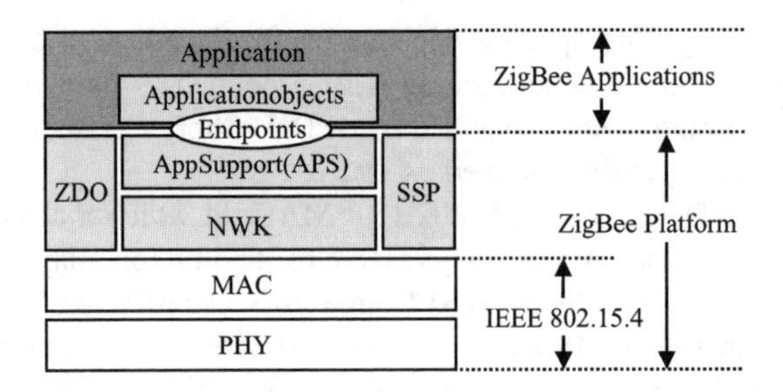

Figure 3.4 ZigBee protocol stack.

Network devices are implemented in ZigBee as application objects, which are connected to the ZigBee stack by endpoints (Garcia 2006). The endpoint is an addressable component within a device. Communication between devices is performed from endpoint to endpoint. ZigBee defines two special endpoints: 0 and 255. Endpoint 0 is used for the configuration and management of an entire ZigBee device, while endpoint 255 is used for broadcasting to all endpoints. Through the endpoint 0 attached to the ZigBee Device Object (ZDO), application can communicate with other layers of the ZigBee stack to configure their parameters. ZDO is a management module, which performs service discovery, device discovery, and device management functions.

Security Service Provider (SSP) performs security functions. Security is a major stepping stone for industrial applications (Sikora 2006). Security includes encryption of data, key generation and distribution, and authentication. ZigBee provides key generation and distribution services, which utilize 128-bit Advanced Encryption Standard (AES) encryption provided by IEEE 802.15.4. In addition, ZigBee supports access control lists and packet freshness timers. Security modes are supported for residential, commercial, and industrial applications.

An Application Support (APS) layer connects together the endpoints, Network (NWK) layer, and SSP. APS also helps the endpoints with data exchanges and security, and performs binding to match different but compatible devices together; for example, a remote controller and a controlled device (Garcia 2006).

The network layer is right above MAC specified by IEEE 802.15.4, and it is responsible for network self-organization, route discovery, and message-relaying functions.

ZigBee denotes the network device types differently from IEEE 802.15.4. A PAN coordinator is called a ZigBee coordinator, while coordinators and devices are called ZigBee routers and ZigBee end-devices, respectively. The NWK layer supports three network topologies: star, peer-to-peer (mesh), and cluster-tree, which are presented in Figure 3.5 (Kuorilehto et al. 2006a). In the star topology, communication is controlled by a ZigBee coordinator that operates as a network master, while ZigBee end-devices operate as slaves and communicate only with the ZigBee coordinator. This network is most suitable for delay critical applications, where a large network coverage area is not required.

A peer-to-peer topology allows mesh-type of networks, where any ZigBee router may communicate with any other router within its range. By routing data through nodes using multiple hops, network coverage can be extended far longer than a radio range. This enables the formation of complex self-organizing network topologies. In addition, peer-to-peer topologies have high robustness against node failures and interferences, since routes can be

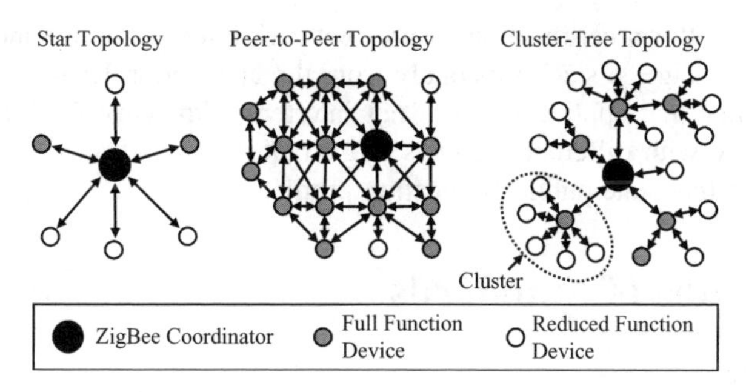

Figure 3.5 Star, peer-to-peer, and cluster-tree topologies.

freely changed. Peer-to-peer topologies are suitable for industrial and commercial applications, where efficient self-configurability and large coverage are important. A disadvantage is the increased network latency due to message relaying (Kohvakka et al. 2006b).

A cluster-tree or hybrid network is a combination of star and mesh topologies. The network consists of clusters, each having a ZigBee router as a cluster head and multiple ZigBee end devices as leaf nodes. A ZigBee coordinator initiates the network and serves as the root. The network is formed by parent–child relationships, where new nodes associate as children with the existing routers. A ZigBee coordinator may instruct a new child to become the head of a new cluster. Otherwise, the child operates as an end device.

This well-defined structure of a cluster-tree simplifies multi-hop routing and allows effective energy saving; each node maintains synchronization of data exchanges only with its parent router. The rest of the time, nodes may save energy in sleep mode. This is not possible in the mesh peer-to-peer networks, where routers need to receive continuously to be able to receive data from any node in the range. A disadvantage is that a failure in one ZigBee router may cause a large amount of orphaned child and grandchild nodes causing energy wasting during network re-associations (Kohvakka et al. 2006b).

ZigBee products should go through a product certification program to guarantee interoperability between products from different manufacturers and for assuring that products perform as promised. ZigBee implements two certification programs: Compliant Platforms and Certified Products (Sikora 2006), as presented in Figure 3.6. Due to the large number of choices the standard allows in each level of protocol stack, the certification is done

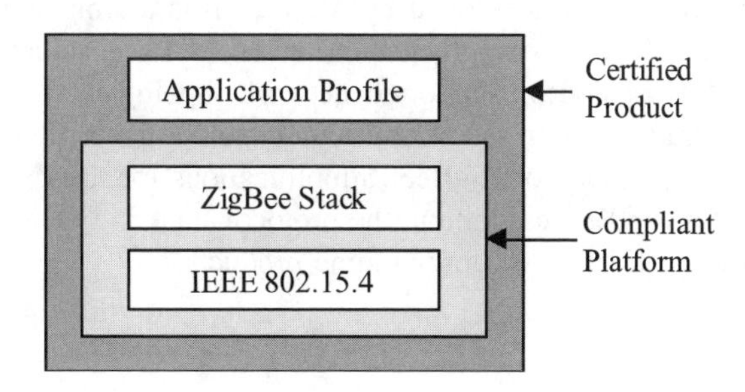

Figure 3.6 ZigBee compliant platform and certified product.

in two stages. A platform certification assures network interoperability and it includes the IEEE 802.15.4 and ZigBee stack separately from the application layer. A ZigBee product certification ensures that application does not interfere with other ZigBee networks. Yet, the interoperability with different products is guaranteed only if all the utilized products are equipped with the same public application profile.

3.2 Variations of Standards

3.2.1 Wibree

Wibree (Wibree 2007) can be considered as a light-weight Bluetooth. It was announced by the Nokia Corporation in October 2006. Contributing companies in the Wibree development have been, at the very least, Broadcom, CSR, Nordic Semiconductor, Epson, Suunto, and Taiyo Yuden.

Wibree is operating at 2.4 GHz frequency band and it supports a star network topology with one master and seven slave nodes. For the purposes of reducing the power consumption and expenses of Bluetooth, Wibree utilizes lower transmission power and lower symbol rate. It is expected that Wibree does not utilize frequency hopping, either. According to Nokia, Wibree can reduce the power consumption of Bluetooth to one tenth.

Wibree may have a common RF part with Bluetooth making its integration into cellular phones and laptop computers cheaper. Yet, small devices, such as watches and sport sensors, may utilize just a Wibree transceiver. Hence, Wibree can connect together two market segments: devices having Bluetooth and simple devices for which Bluetooth is too powerful and energy consuming.

3.2.2 Z-Wave

Z-Wave (Sachs 2006) can be considered as a lighter version of ZigBee operating at 868 MHz and 915 MHz frequency bands. Z-Wave is targeted for the control of building automation and entertainment electronics. The maximum number of nodes in a network is 232. Supported network topologies are star and mesh. Z-Wave has been developed by at least 120 companies including Zensys, Intel, and Cisco.

3.2.3 MiWi

MiWi (Flowers and Yang 2007) developed by Microchip is a simpler version of ZigBee operating above a IEEE 802.15.4 compliant transceiver. MiWi is suitable for smaller networks having at most 1024 nodes. Supported network topologies are star and mesh. At the moment, MiWi operates only in the non-beacon enabled mode. Hence, MiWi does not support the low duty-cycle mode of ZigBee. Simplifications for the ZigBee stack reduces the cost of MCU by 40%–60%. In addition, the protocol stack is free and does not require a certification being cost effective for low-volume products.

4

Sensor Node Platforms

In conventional networks, hardware has very significant affects on the achieved performance and the energy consumption of an entire network. In WSNs, the energy efficiency is implemented mostly by MAC and network layers of a protocol stack. An energy-efficient protocol stack may reduce the activity of a hardware below 1%. Yet, the activity occurs typically at very short periods of time. Clearly, the low power consumption of active operation modes is important. In WSNs, even more essential is the minimization of power consumption in idle and sleep modes. The sleep-mode energy consumption is dominating and limiting to a network's lifetime in very low data-rate applications. Next, we will discuss in detail the components of a sensor node hardware and their energy behavior.

4.1 Platform Components*

Due to the large number of nodes deeply entrenched in our living environments, the hardware realization need to be small and cheap, while the operating time should stretch to years powered by small batteries. The given requirements for hardware are best fulfilled by highly integrated Application Specific Integrated Circuits (ASICs). Since the hardware is application-specific and may contain application-specific hardware accelerators, computation is performed powerfully and energy efficiently (Rabaey et al. 2000). Since a single chip may integrate all essential digital circuits, the physical size and manufacturing cost of a node may be significantly reduced. Due to the high design and initial costs incurred, ASICs are cost-effective only for very high production volumes. Since WSN deployments are typically under development or in the pilot phase having small to medium production quantities, Commercial Off-The-Shelf (COTS) components are often the only feasible option. We will focus on COTS-based hardware, which will most probably be the dominating

*This section has been reproduced from Kohvakka et al. 2003, © 2003 IEEE. Reprinted with permission, from *Proceedings of 2003 IEEE International Conference on Industrial Electronics, Control and Instrumentation*.

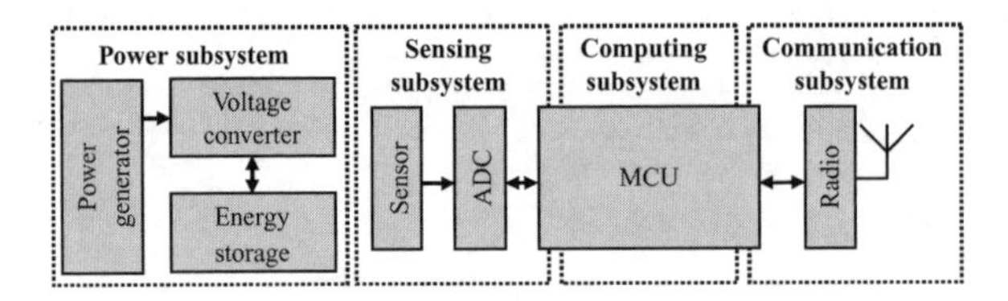

Figure 4.1 General hardware architecture of a WSN node. Kohvakka et al. 2003, © 2003 IEEE. Reprinted with permission, from *Proceedings of 2003 IEEE International Conference on Industrial Electronics, Control and Instrumentation.*

hardware in up-and-coming years. A general hardware architecture for a WSN node is presented in Figure 4.1 (Kohvakka et al. 2003). The architecture can be divided into four subsystems (Raghunathan et al. 2002):

- Communication subsystem enabling wireless communication
- Computing subsystem allowing data processing and the management of node functionality
- Sensing subsystem, which link's the wireless sensor node to the outside world
- Power subsystem, which provides the system supply voltage

The central component of the platform is a Micro-Controller Unit, which (MCU) forms the computing subsystem. In addition, part of the communication and sensing subsystems are executed on the MCU including device drivers and network protocols.

4.1.1 Communication Subsystem

The communication subsystem consists of a radio transceiver and an antenna, which enable wireless communication with neighboring nodes.

A wireless transceiver is based either on Radio Frequency (RF) or optical waves. Optical transceivers (Doherty et al. 2001; Wolf and Kress 2003) have higher energy efficiency and no need for external antenna. However, radiation is highly directional and needs a Line-of-Sight (LOS) condition. Hence, the alignment of a transmitter to a receiver is difficult or even impossible in large-scale WSN applications. RF antennae are typically slightly directional or omnidirectional and they do not require LOS, which makes them more suitable for WSN. Disadvantages are larger physical size, due to an antenna, and higher energy consumption compared to optical technology.

The frequency band significantly affects the radio properties. The operating frequency determines RF wave propagation characteristics and interferences from the surrounding machinery. For industrial environments the interference emitted from machinery is significant below a few hundred megahertz and drops rapidly above 1 GHz frequency (Rappaport 1989). On the other hand, higher frequency is beneficial for WSNs since a shorter wavelength enables small and efficient antennae. Most commercial short-range radio transceivers

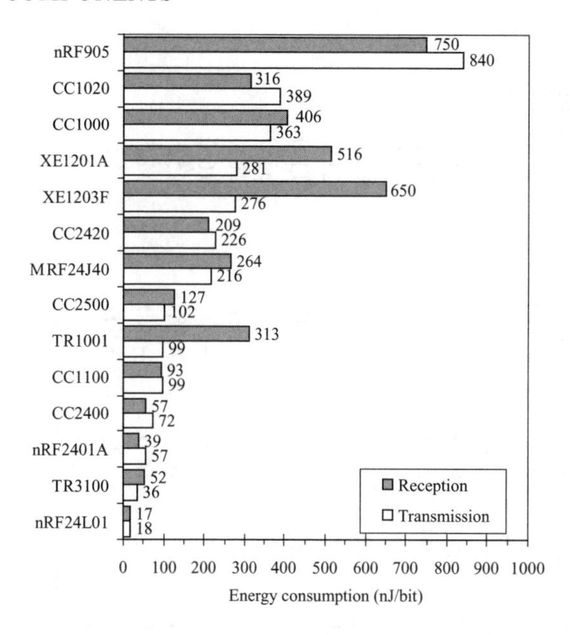

Figure 4.2 Transceiver energy consumption comparison. © 2003 IEEE. Reprinted with permission, from *Proceedings of 2003 IEEE International Conference on Industrial Electronics, Control and Instrumentation.*

in European region operate in the 433, 868, and 2400 MHz license-free ISM frequency bands. Depending on the frequency band, the operating range for a typical 1 m W Transmit (TX) power is from a few meters to over one kilometer (Kohvakka et al. 2003).

In general, an RF transceiver has four operation modes: transmit, receive, idle, and sleep. A radio is active in transmit and receive modes when power consumption is also at its highest. In idle mode most of circuitry is shut down but the main oscillators remain active. Thus, transition to the active mode is rapid. The lowest power consumption is achieved in sleep mode when all circuitry is switched off. However, the wake-up time is also at its longest (Kohvakka et al. 2003).

The characteristics of some of the most potential low-power RF transceivers currently available are summarized in Table 4.1 (Kohvakka et al. 2003).

The energy consumption of radios can be compared by dividing their power consumptions by maximum data rates, which yields energy per bit ratio (E/bit). This can be determined for both transmit or receive modes.

The E/bit is computed to the radios listed in Table 4.1 and the results are presented in Figure 4.2 (Kohvakka et al. 2003). The supply voltage is fixed to 3.0 V (volts) for comparability. As seen in the figure, radios operating at the 2.4 GHz frequency band are the most energy efficient due to high data rates. Yet, these high data rates require also a high-performance MCU with a high-speed serial bus. This problem is solved in some of transceivers by the inclusion of an on-chip data buffer (see Table 4.1), which adapts a low data rate MCU interface with a high data rate radio.

Table 4.1 Transceiver features.

Manufacturer	Model	Data rate (Kbps)	Band (MHz)	Data buffer (B)	Sleep (μA)	Idle (mA)	RX[a] (mA)	Sensitivity[b] (dBm)	TX @ 0 dBm[a] (mA)
Microchip	MRF24J40	250	2400	128	2	–	18	–91	22
Nordic[c]	nRF2401A	1000	2400	32	0.9	0.01	19.0	–85	13.0
Nordic[c]	nRF24L01	2000	2400	32	0.9	0.03	12.3	–82	11.3
TI	CC2400	1000	2400	32	1.5	1.2	24.0	–87	19.0
TI	CC2420	250	2400	128	0.02	0.4	18.8	–95	17.4
TI	CC2500	500	2400	64	0.4	1.5	17.0	–82	21.2
Semtech	XE1201A	64	433	–	0.2	0.06	6.0	–102	11.0
Semtech	XE1203F	152.3	433/868/915	–	0.2	0.85	14.0	–101	33.0
TI	CC1000	76.8	433/868/915	–	0.2	0.1	9.3	–101	10.4
TI	CC1020	153.6	433/868/915	–	0.2	0.08	19.9	–81	16.2
TI	CC1100	500	433/868/915	64	0.4	1.6	16.5	–88	15.5
Nordic[c]	nRF905	50	433/868/915	32	2.5	0.03	14.0	–100	12.5
RFM	TR1001	115.2	868	–	0.7	–	3.8	–91	12
RFM	TR3100	576	433	–	0.7	–	7.0	–85	10

[a] At the lowest frequency band.
[b] At the highest data rate.
[c] Nordic Semiconductor.

4.1.2 Computing Subsystem

The computing subsystem consists of an MCU, which integrates a processor core with program and data memories, timers, configurable I/O ports, Analog-to-Digital Converter (ADC), and other peripherals. The choice of MCU significantly affects the power consumption of a wireless sensor. The characteristics of potential MCU from different manufacturers are compared in Table 4.2.

MCU energy efficiency can be analyzed by dividing processor performance, using Million Instructions Per Second (MIPS), by the power consumption resulting in MIPS/W ratio (Piguet et al. 1997). The energy efficiency of the selected MCUs is compared in Figure 4.3. The supply voltage is fixed to 3.0 V for fairness. We should note that the instruction set affects the performance to some extent; thus, only orders of magnitude are important. As seen in the figure, Semtech XE8802 and TI MSP430F1611 MCUs are the most energy efficient. On the other hand, 8051-based AT89C51RE2 performs worst having 24 times lower energy efficiency (Kohvakka et al. 2003).

4.1.3 Sensing Subsystem

The sensing subsystem consists of a set of sensors and actuators. Sensors are equipped with an analog or digital output for reading sensor values. Analog output requires the use of an ADC also included in the sensing subsystem. MCUs typically have an integrated ADC, which can be used.

A disadvantage of analog output is the noise coupled to the wiring, limited resolution or sample rate of the integrated ADC, and the inaccuracy of low-budget voltage references required for the circuitry. When requiring a higher accuracy or sample rate, an external ADC located near the sensors can be utilized.

A more simple and lower cost solution is to use sensors with digital output. Digital sensors are often small-sized, factory-calibrated and interfaced with a standard serial bus found in most MCUs.

There exists a large variety of low-power sensors suitable for WSNs. For example, temperature, humidity, acceleration, compass, light, pressure, motion detection, acoustic and image sensors are available. Important requirements for sensors are adequate accuracy

Table 4.2 MCU features.

Manufacturer	Model	FLASH (kB)	SRAM (kB)	EEPROM (B)	Sleep (μA)	1 MIPS[a] (mA)
Atmel	AT89C51RE2 (8051)	128	8	0	75	7.4
Atmel	ATmega1281 (AVR)	128	8	4096	5	0.5
Atmel	AT91SAM7X (ARM)	128	32	0	26	1.1
Freescale	M68HC08	61	2	0	22	3.75
Microchip	PIC18LF8722	128	3.9	1024	2.32	1.0
Microchip	PIC24FJ128	128	8	0	21	1.6
Semtech	XE8802 (CoolRisc)	22	1	0	1.9	0.3
TI	MSP430F1611	48	10	0	1.3	0.33

[a] At 3.0 V supply voltage.

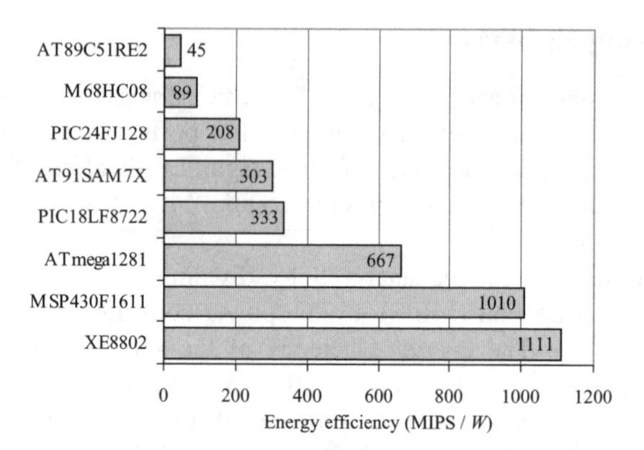

Figure 4.3 MCU energy-efficiency comparison. © 2003 IEEE. Reprinted with permission, from *Proceedings of 2003 IEEE International Conference on Industrial Electronics, Control and Instrumentation.*

over the required temperature range, low-power consumption in sleep mode, and low-energy consumption during a start-up time and an actual measurement.

Due to the scarce energy budget, the MCU cannot continuously gather sensors values. In practice, sensors are activated at intervals of several seconds, or even minutes. This is adequate for magnitudes altering slowly, for example temperature and humidity. Yet, the detection of temporary and rapidly occurring events is very difficult, such as acceleration, motion, and acoustic impulse. The problem is solved in advanced digital sensors by a low-power monitoring state where the sensor can wake up the MCU when a predetermined event occurs.

The utilization of actuators with WSNs has been rare. One of the most important reasons is that the utilization of an actuator actually means the physical connection between a WSN node and an existing system, such as lighting and HVAC, which requires activity from component manufacturers. Typically, the actuator is implemented by a relay or transistor circuit, which switches power to external circuitry, e.g. a lamp, a radiator, or an electric lock.

4.1.4 Power Subsystem

The power subsystem provides and regulates system supply voltage. The scarce energy budget of WSN applications sets strict energy-efficiency requirements also for the power subsystem. The power subsystem must minimize quiescent currents while being able to supply short-term, relatively high (tens of milliamperes) peak current levels, typically drawn by radio during frame receptions and transmissions. This is illustrated in Figure 4.4, which presents a typical current consumption waveform measured from two TUTWSN nodes. A WSN node typically spends around 99% of the time in sleep mode, which means that the sleep-mode power dominates the power consumption of the entire WSN node.

The power subsystem consists of an energy storage, voltage regulation, and optionally an energy scavenging unit. The energy storage is usually a primary battery, for example an alkaline battery. If an energy scavenging unit is used, rechargeable (secondary) batteries

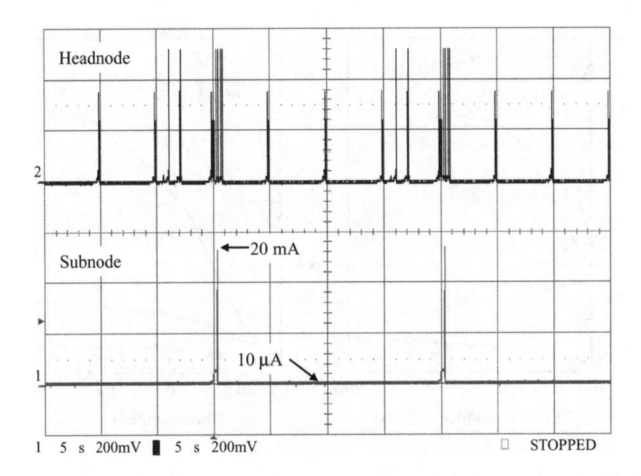

Figure 4.4 Measured current waveform from TUTWSN subnode and headnode.

are more suitable. An important requirement of the energy storage is a current sourcing capacity, which should exceed the peak current requirement of the WSN node. For example, small-sized lithium coin cells can usually source only 100 μA (microamperes) maximum current, thus alone being unsuitable for the energy storage of a WSN node. However, they have very high energy density. A cubic centimeter-sized lithium battery provides about 2880 J (joules) energy. If a one year node lifetime is required, this allows an average power consumption of 90 μW (microwatts) (Roundy et al. 2004).

For ambient energy scavenging, most potential sources are solar power and vibrations. Solar cells are commonly available, mature technology. As an example, the measured performance of a Panasonic BP-376634 solar panel is presented in Figure 4.5 (Kohvakka et al. 2003). The panel consists of small and serial connected silicon-based solar cells gaining totally 5.5 V output voltage. The output power of the solar panel is measured with various loading impedances at the same time as the light intensity is varied. As shown in the figure, optimal loading impedance is 400 ohms and the measured maximum output power is 36 mW (milliwatts) at 80 klx (kilolux) light intensity, which corresponds approximately to bright sunlight. Generated energy corresponds to the power of 1.5 mW/cm^2. Indoors with normal office lightning (500 lux), the obtained power decreases to around 10 μW/cm^2. Thus, solar panels are very applicable for outdoor use, where sunlight enables high power generation.

Indoors, vibration may be a more suitable source of power. A promising method with this source is a piezoelectric conversion. Commonly occurring vibrations can provide hundreds of microwatts of power per cubic centimeter (Roundy et al. 2004).

A voltage regulator stabilizes the system supply voltage to an appropriate level determined by the voltage requirements of the system components. For energy efficiency, it is reasonable to minimize system supply voltage, which typically also minimizes static current consumption.

Voltage regulation can be implemented either by a switched-mode regulator or a linear regulator. Switched-mode regulators have high 85%–95% conversion energy efficiency at currents above 1 mA (milliamperes) almost independently of input and output voltage

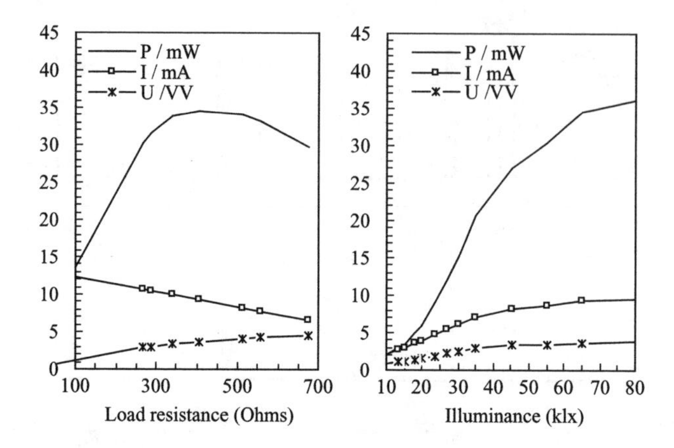

Figure 4.5 Measured solar panel power as a function of illumination and load resistance. © 2003 IEEE. Reprinted with permission, from *Proceedings of 2003 IEEE International Conference on Industrial Electronics, Control and Instrumentation.*

levels. Disadvantages are high, at least tens of microamps quiescent current, electromagnetic interferences due to high-frequency current switching, and more complex architecture due to required coil and capacitors.

Although, linear regulators suffer from low energy efficiency, when a voltage reduction is significant, they usually perform better than switch-mode regulators in WSN applications. One of the most important reasons is a very low, even below 1 μA quiescent current, which efficiently minimizes the sleep-mode power consumption. In addition, linear regulators are very small, simple, cheap and have insignificant electromagnetic interferences.

4.2 Existing Platforms

WSN nodes have been designed by several research groups over the last decade. Due to the advances in low-power processing and communication technology, platforms have been improved significantly with time. Due to the strict energy constraints of WSN nodes combined with visions of complex networking and data fusion in nodes, the requirements for platforms are very challenging. Since it is very difficult to fulfill all the requirements of WSNs with current technology, the platform research has been divided into two branches: high-performance platforms, and low-power platforms.

The high-performance platforms are typically developed for researching complex data processing and fusion in sensor nodes. Typically, the design target has been the reduction of transmitted data by efficient data processing, for example by recognizing and classifying events and tracking objects in a distributed manner without external computing. These platforms typically utilize high-performance MCUs having tens of MIPSs processing performance and hundreds of kilobytes program and data memories. Unfortunately, the energy consumption for long-lived battery operation is not adequate and usually a mains power adapter is required. However, these high-power platforms can be used as a part of a WSN for providing data processing and a backbone network. Examples of the high-performance WSN platforms are Piconode (Reason and Rabaey 2004), μAMPS (Min et al. 2002), and Stargate (Crossbow Technology 2004b).

Low-power platforms aim to maximize the lifetime and minimize the physical size of nodes. These are obtained by minimizing hardware complexity and energy consumption. These platforms are capable of performing low data-rate communication and data processing required for networking and simple applications.

The most well-known work among low-power WSN platforms has been conducted in the University of California, Berkeley. They have built a number of platforms called motes by a SmartDust project. The SmartDust (Warneke et al. 2001) project was aimed at fabricate a cubic millimeter-sized WSN node containing sensing, computing, and communication systems. An important goal of the project was to explore the limitations of microfabrication technology. Before that, Mote platforms were built to approximate the capabilities of an envisioned SmartDust node with COTS components.

In 2001, the first mote called Mica (Hill and Culler 2002) was released. Mica is a general-purpose platform for WSN research having an Atmel ATmega103L MCU and RFM TR1000 transceiver operating at 915 MHz frequency band. The MCU contains 128 kB program memory and 4 kB data memory and has a 4 MIPS processing speed. In addition, the platform contains a 4-Mbit external non-volatile memory and Atmel AT90LS2343 coprocessor for handling wireless reprogramming of the main processor. The most critical shortcoming of the platform is a step-up type switched-mode voltage converter, which provides a stable voltage, but uses 200–300 μA quiescent current significantly reducing the maximum node lifetime (Polastre et al. 2005).

An improved version of Mica called Mica2 (Crossbow Technology 2007a) was released in 2002. Sleep mode current consumption was reduced to about 17 μA by discarding step-up converter and changing MCU to ATmega128L. Radio was changed to Chipcon CC1000 enabling tunable frequency channel from 300 MHz to 900 MHz. A smaller version of Mica2 called Mica2Dot (Crossbow Technology 2007b) was also released in the same year. These two platforms were very popular in WSN research for several years.

MicaZ (Crossbow Technology 2004a) was the next model after Mica2Dot in the evolution of motes. MicaZ replaces the CC1000 transceiver with an IEEE 802.15.4 compatible Chipcon CC2420 transceiver. This modification increased the radio data rate from 76.8 kbps to 250 kbps. In addition, a wideband DSSS modulation scheme provided better tolerance against noise and interferences.

Telos (Polastre et al. 2005) is a completely new type of mote released in 2004. Telos consists of a very low-power MSP430 MCU from TI and the IEEE 802.15.4 compatible CC2420 transceiver. A special feature of the platform is an Universal Serial Bus (USB) connector, which is used for programming the platform. USB can also be used for powering the platform. This hardware architecture provides very low 5.1 μA sleep-mode current consumption, which together with a very low duty-cycle communication protocol enables up to multiple years of node lifespan. An improved version of Telos is called Tmote Sky.

Additionally, other research projects have been developing WSN platforms intensively. Medusa MK-2 (Savvides and Srivastava 2002) is a versatile WSN platform released in 2002 by the University of California, Los Angeles. The platform combines ATmega128L MCU with a TR1000 transceiver. Hence, it is quite similar to Mica and Mica2 motes. A special feature of Medusa MK-2 platform is a higher performance 16/32-bit Atmel AT91FR4081 ARM7TDMI co-processor, which is responsible for the processing of sensor data and localization algorithms. The platform is equipped with a set of on-board sensors including light, temperature, and acceleration. The sleep mode current consumption is 27 μA, which is adequate for long-lived WSN applications.

ETH Zurich has developed through their Smart-Its project (Smart-its 2007) a WSN platforms called BTnodes. A BTnode combines Bluetooth radio with a low-power Atmer ATmega128L MCU. A BTnode rev3, released in 2004, is a dual-radio platform, which can employ simultaneously a Zeevo ZV4002 Bluetooth radio and a Chipcon CC1000 low-power radio transceiver. Hence, the BTnode rev3 (Beutel et al. 2004a) platform can form a backbone network with Bluetooth technology, and exchange data with low-power Mica2 and Mica2Dot platforms by the CC1000 radio. This enables the creation of tiered architectures with high-bandwidth nodes bridging low-power devices to Bluetooth-enabled gateway appliances (Hill et al. 2004). A disadvantage is that the sleep-mode power consumption of ZV4002 radio is 140 μA, which limits the achievable node lifetime very significantly.

In recent years, most new platforms has been developed for using TI's new very low-power MSP430 MCU family. These MCUs combine efficient 16-bit processor core with low-power sleep modes and an adequate amount of memory for most WSN applications. Depending on a MCU model, around 60 kB of memory is divided into program and data memories.

A ScatterWeb Embedded Sensor Board (ESB) (Schiller et al. 2005) is a simple and low-power WSN platform developed by Freie Universität Berlin, Germany in 2003. The platform consists of TI MSP430 MCU, RFM TR1001 radio and sensors for measuring light, acoustic noise, vibration, and movement by a PIR sensor. In addition, a microphone, a speaker, and an IR transceiver are included. Long-lived WSN applications are enabled since sleep mode current consumption is only 8 μA. Communication with external networks is enabled by various gateway nodes having interfaces, for example, to Ethernet, GSM/GPRS, USB, and RS-485. A very similar hardware architecture is also employed in nodes (Reijers et al. 2004) developed by a European EYES research project.

A TinyNode (Dubois-Ferriere et al. 2006) developed by Shockfish SA, Switzerland, is a very small and low-power platform developed for research and industrial applications. The platform contains only the core components, while additional functionality and sensors are provided by extension boards. In addition, the platform contains a 512 kB external non-volatile memory for storing several firmware images and for logging data. Included main components in the TinyNode are MSP430 MCU, Semtech XE1205 transceiver, and a temperature sensor. The transceiver can operate in 433 MHz, 868 MHz, and 915 MHz frequency bands having up to 152 kbps data rate. A very high −116 dBm (decibels per minute) sensitivity at 4.8 kbps data rate combined with +15 dBm transmission power yields up to 1800 m radio range. Yet, at higher data rates the sensitivity decreases. For example at 76.8 kbps the sensitivity is −106 dBm resulting in 600 m range. Also, this platform can obtain a long lifetime, since a sleep-mode current consumption is only 5.1 μA.

A ProSpeckz (Arvind and Wong 2004) is a very small and simple WSN platform, which is equipped with a CY8C29666 Programmable System-on-Chip (PSoC) from Cypress Microsystems instead of MCU. PSoC combines a simple 8-bit MCU core having 16 kB of program memory and 256 B of data memory with versatile software reconfigurable analogue circuities to external interfaces and components. Included analog circuities are, for example, programmable gain amplifiers, low-pass or high-pass filters, switched capacitor blocks, and ADCs. Employed radio is IEEE 802.15.4 compliant Chipcon CC2420. Although, the PSoC platform provides versatile processing capabilities for analog sensor values, high 330 μA current consumption in sleep mode significantly reduces the achievable lifetime of WSN applications.

Besides the COTS platforms previously presented, a lot of research work has been conducted for developing System-on-Chip (SoC) platforms targeting even smaller size and higher energy efficiency. For example, a WiseNET SoC sensor node (Enz et al. 2004) developed in the Swiss Center for Electronics and Microtechnology integrates an ultra-low power dual-band (433/868 MHz) radio transceiver with Cool-RISC MCU core, low-leakage memories, two ADCs and power management blocks. Measured consumption in reception mode is only 2 mA, which is nearly one order of magnitude less than typical low-power transceivers. Yet, a data rate is only 25 kbps. A transmission mode current consumption at 10 dBm output power is 24 mA, which is lower than typical. A sleep-mode current consumption of the radio block is 3.5 μA. The properties of the low-power platforms discussed in this section are summarized in Table 4.3.

4.3 TUTWSN Platforms

In this section we present some examples of TUTWSN prototype platforms, which are used in various applications. Application-specific TUTWSN platforms will be discussed later.

4.3.1 Temperature-sensing Platform

A temperature-sensing platform is implemented for verifying and demonstrating the operation of TUTWSN protocols and algorithms. In addition, the platform is used for general temperature-sensing applications for buildings and the environment. The platform combines rather small size with energy-efficient hardware and adequate memory resources for most of applications. The platform is designed to fit in pipe type waterproof plastic enclosures. The platform is equipped with an accurate temperature sensor and efficient antenna for maximizing hop length with low-transmission power levels improving the network energy efficiency. For maximizing applicability in various operational environments, all selected components are specified to operate in an extended temperature range starting from −40 °C.

The architecture of the platform is presented in Figure 4.6

The platform contains a Nordic Semiconductor nRF2401A RF transceiver that operates at 2.4 GHz frequency and supports transmit powers of −20 dBm to 0 dBm. The transceiver includes a 16-bit CRC module and configurable preamble that is used as a network address. A high-frequency radio is used because it has short transmission and reception times, which substantially reduces the average power consumption.

The platform has a PIC MCU that operates at 4 MHz allowing execution of 1 MIPS. The MCU includes an ADC that is used to measure battery voltage and read analog sensors. The

Table 4.3 Comparison of existing low-power WSN platforms.

Platform	Microcontroller	Transceiver	On-board sensors	Size	Energy efficiency
Mica	ATmega103L	TR1000	–	1856 mm^2	Moderate
Mica2	ATmega128L	CC1000	–	1856 mm^2	High
Mica2Dot	ATmega128L	CC1000	–	492 mm^2	High
MicaZ	ATmega128L	CC2420	–	1856 mm^2	High
BTnode ver3	ATmega128L	ZV4002+CC1000	–	1890 mm^2	Moderate
Medusa-II	ATmega128L + ARM7	TR1000	T, L, A	4500 mm^{2a}	High
EYES node	MSP430	TR1001	–	2600 mm^{2a}	Very high
ScatterWeb ESB	MSP430	TR1001	L, AC, A, PIR	3000 mm^2 a	Very high
TinyNode	MSP430	XE1205	–	1200 mm^2	Very high
Tmote Sky	MSP430	CC2420	H, T, L	2621 mm^2	Very high
ProSpeckz	CY8C29666	CC2420	–	704 mm^2	Moderate
WiseNET SoC	Cool-RISC	WiseNET	–	12 mm^{2b}	Ultra-high

[a] Estimated.
[b] including the SoC chip only.
Abbreviations of on-board sensors: A – acceleration, AC – acoustic, H – humidity, L – light, PIR – passive infrared, T – temperature.

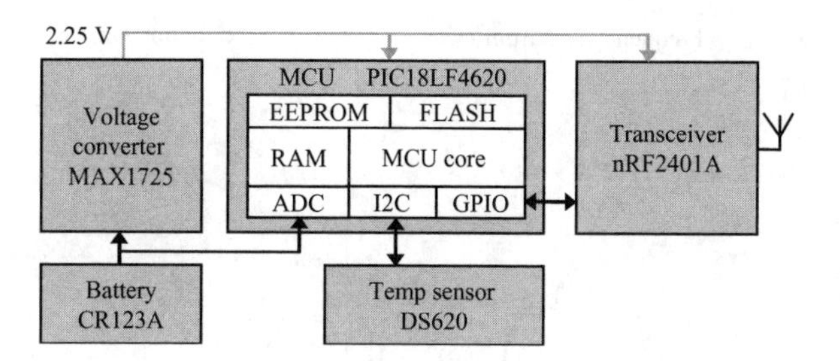

Figure 4.6 Temperature-sensing platform architecture.

implemented nodes are equipped with a Dallas DS620U digital temperature sensor with I2C interface. The sensor guarantees 0.5 °C accuracy over entire −55 °C to 125 °C temperature range. A push-button and a LED are implemented for simple user interface and diagnostics.

Variable environmental conditions set high requirements for batteries, which typically are very sensitive to operating temperature. Thus, a CR123A primary lithium battery is selected. The battery is specified with 3 V / 1600 mAh (milliamperes per hour) capacity and from −40 °C to +60 °C operating temperature. Battery voltage is converted to 2.25 V supply voltage by a MAX1725 linear regulator. The regulator is selected due to very low quiescent current, low electromagnetic interferences, small size and adequate energy efficiency with the required voltage dropping. According to our measurements, linear regulators are well suited to the WSN node's current profile, which consists of very short and high current bursts, while around 99% of the time the node is in low-power sleep mode.

The implemented TUTWSN temperature sensing platform is presented in Figure 4.7 (Kohvakka et al. 2006c). The platform contains two attachable boards: a MCU board and an extension board, which also provides power. The utilization of a detachable extension board allows the change of power source and easy implementation of application-specific sensors and network interfaces.

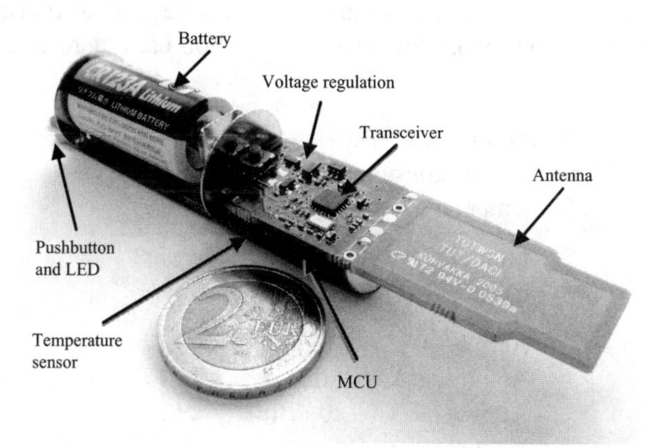

Figure 4.7 Implemented TUTWSN temperature-sensing platform. © 2006 IEEE. Reprinted, with permission, from *Proceedings of 2006 IEEE 17th International Symposium on Personal, Indoor and Mobile Radio Communications.*

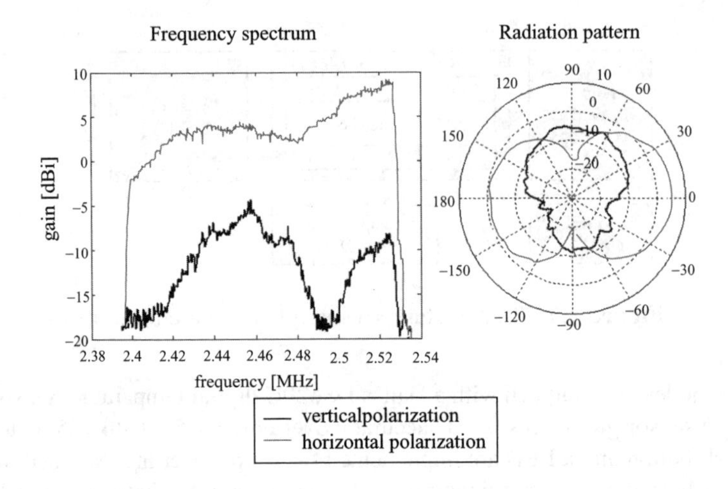

Figure 4.8 Antenna measurement results of the temperature-sensing platform with tapered loop antenna.

The TUTWSN temperature-sensing platform contains a tapered loop Printed Circuit Board (PCB) antenna, which has been developed and fine tuned according to the results of practical test situations. The radiation properties of the antenna are presented in Figure 4.8.

Static power consumption

The static power consumption of the temperature-sensing platform in different modes is presented in Table 4.4. The power consumption was measured with a 3 V DC power supply. The minimum current consumption is 12.3 μA equaling the power of 37 μW when sleep mode is utilized and the radio is turned off. The maximum power consumption is 60.17 mW when the reception is active. Reception requires more energy than transmission because the utilized transmission powers are low and the reception uses more complex circuitry.

Table 4.4 The measured static power consumption of TUTWSN temperature-sensing platform.

MCU mode	Radio mode	Power[a] (mW)
Active	RX	60.17
Active	TX (0 dBm)	42.17
Active	TX (−20 dBm)	29.57
Active	Idle	3.29
Active	Sleep	3.17
Sleep	Sleep	0.037

[a]Measured at 3.0V supply voltage.

4.3.2 SoC Node Prototype*

A very simple SoC node prototype is implemented targeting very low-cost and small-sized TUTWSN subnode implementation. An important requirement is compatibility with other TUTWSN prototypes, which are equipped with larger memory and processing resources for acting as headnodes in the network. This is essential to be able to form a large multihop network.

The prototype is based on a Nordic Semiconductor nRF24E1 chip, which integrates a radio transceiver, MCU, and ADC on a single package. The PHY layer of the radio is similar to the nRF2401A providing good compatibility with the TUTWSN temperature sensing platform (Kohvakka et al. 2005c).

The hardware architecture of the SoC node prototype is presented in Figure 4.9 (Kohvakka et al. 2005c). The chip includes a 8051 compatible MCU core running at 16 MHz. The processor performance is about 2 MIPS and has 4 kB RAM (random-access memory) for a program and 256 B RAM for data. In addition, there is a low-power RC oscillator for a wake-up timer. An external non-volatile Electrically Erasable Programmable Read Only Memory (EEPROM) is used for storing the program code, when supply voltage is switched off.

The RC oscillator provides low, 2 μA sleep mode current consumption, but is very inaccurate. The oscillator frequency is typically between 1 kHz and 5 kHz (kilohertz) depending on temperature and supply, voltage. However, the oscillator frequency can be measured by the help of an external high-frequency crystal, which allows the calibration of the low-power wake-up timer. The prototype utilizes a Maxim MAX6607 temperature sensor and the ADC embedded in the nRF24E1 chip. For simplicity, the prototype uses an internal bandgap voltage reference. ADC has at maximum 100 kHz sample rate and 12-bit resolution.

The SoC node prototype is presented in Figure 4.10 (Kohvakka et al. 2005c). The upper side of the prototype contains the nRF24E1 chip with crystal, antenna, and a programming connector. The underside contains voltage converter, EEPROM, temperature sensor, and a power switch. The size of the prototype is 23 × 13 × 5 mm.

Power analysis

The power consumption of the node is measured while the radio, MCU, ADC, and sensor are programmed to enter different operation modes. The MCU power consumption is measured in an active mode when the MCU core is executing the sensor application, and in a sleep mode when only a wake-up timer is active. The ADC and sensor power consumptions are measured in active and shut down (off) modes. The radio power consumption is measured in sleep, reception, and transmission modes with a minimum and maximum transmission power level.

The results are presented in Table 4.5 (Kohvakka et al. 2005c) and include the power dissipation in the voltage converter, while the prototypes are supplied with 3 V supply voltage. The minimum power consumption for the prototype is 15 μW, when all components are inactive. The maximum power consumptions are achieved when all components are in active mode and the radio is in receive mode. Then, the subnode prototype power consumption is 66.03 mW.

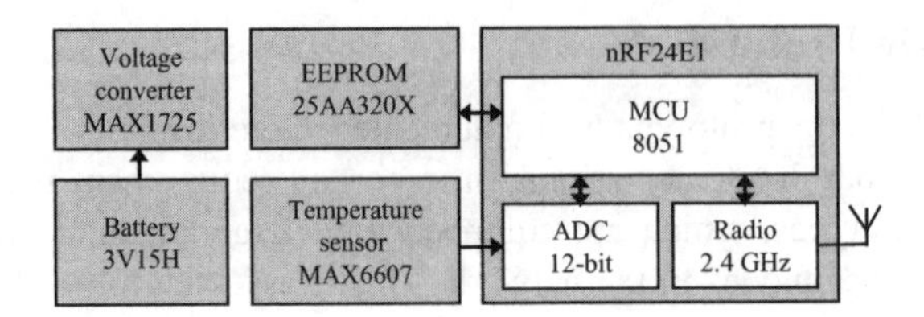

Figure 4.9 SoC node prototype hardware architecture. © 2005 IEEE. Reprinted, with permission, from *Proceedings of 2005 IEEE 16th International Symposium on Personal, Indoor and Mobile Radio Communications*.

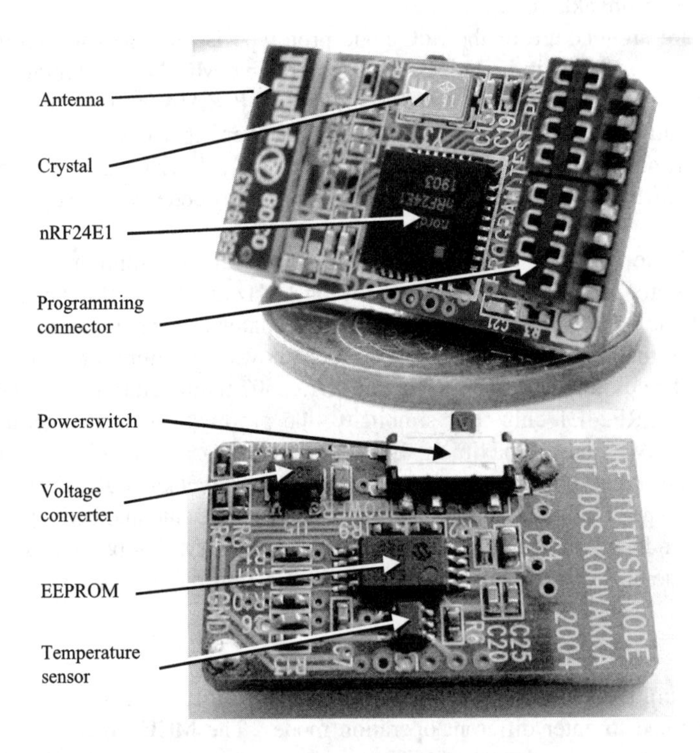

Figure 4.10 The SoC node prototype on a 5-euro-cent coin. © 2005 IEEE. Reprinted, with permission, from *Proceedings of 2005 IEEE 16th International Symposium on Personal, Indoor and Mobile Radio Communications*.

4.3.3 Ethernet Gateway Prototype

This section presents the hardware implementation of a TUTWSN Ethernet gateway designed for high-performance sink nodes.

The hardware architecture of the prototype is presented in Figure 4.11. The prototype consists of a radio transceiver, a MCU, Ethernet interface, a temperature sensor, an illumination sensor, a user interface, and a voltage converter.

Table 4.5 The power analysis of the SoC node prototype. © 2005 IEEE. Reprinted, with permission, from *Proceedings of 2005 IEEE 16th International Symposium on Personal, Indoor and Mobile Radio Communications*.

MCU mode	ACD mode	Sensor mode	Radio mode	Power[a] (mW)
Active	Active	Active	RX	66.03
Active	Active	Active	TX (0 dBm)	44.10
Active	Active	Active	TX (−20 dBm)	32.90
Active	Active	Active	Sleep	11.64
Active	Active	Off	Sleep	11.60
Active	Off	Off	Sleep	10.56
Sleep	Off	Off	Sleep	0.015

[a]measured at 3.0 V supply voltage.

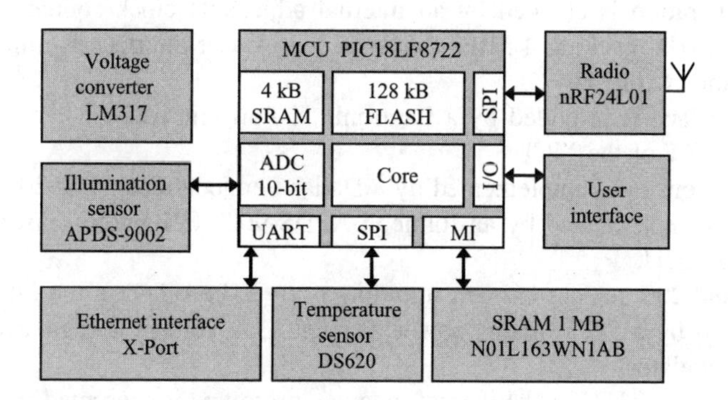

Figure 4.11 TUTWSN Ethernet gateway hardware architecture.

The Ethernet gateway utilizes the Nordic Semiconductor nRF24L01 transceiver operating at 2.4 GHz frequency band. The selected transceiver is an improved model of nRF2401A transceiver, which is used in the TUTWSN temperature-sensing platform. Compared to nRF2401A, the new model doubles the maximum data rate while reducing current consumption resulting in significantly improved energy efficiency. Selectable data rates are 1 and 2 Mbps. When operating at 1 Mbps data rate, the transceiver is compatible with nRF2401A model. Transmission power is adjustable between −18 dBm and 0 dBm. The radio has internal data buffer and Cyclic Redundancy Check (CRC) error detection logic, similar to the nRF2401A transceiver. A loop type antenna is implemented as a trace on PCB.

The operation of the TUTWSN Ethernet gateway is controlled by a Microchip PIC18LF8722 MCU having 128 kB program memory and 4 kB data memory. Internal 1 kB Electrically Erasable Programmable Read-Only Memory (EEPROM) is used for non-volatile configuration data. Additional data memory for data buffering is obtained from an external NanoAmp N01L163WN1AB 1 Mb Static Random Access Memory (SRAM).

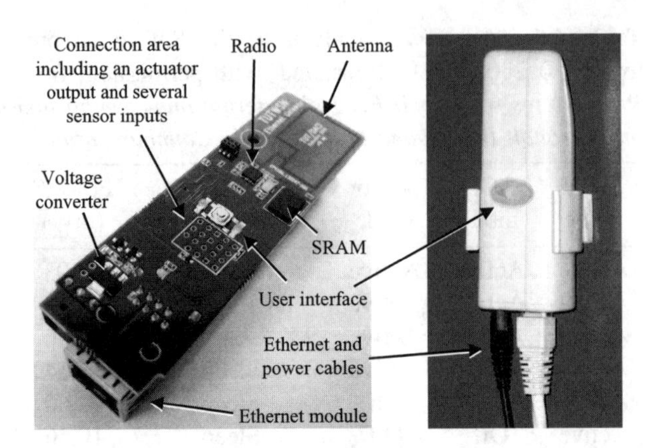

Figure 4.12 TUTWSN Ethernet gateway prototype.

Active mode operation is clocked by an internal adjustable clock source. Utilized clock frequency is 16 MHz resulting 4 MIPS performance. A user interface is implemented with a push button and two LEDs.

Ethernet interface is provided by a Lantronix X-Port embedded ethernet module connected to the UART of the MCU.

Temperature sensing is implemented by a Dallas Semiconductor DS620 sensor. Illumination sensing is implemented by an Infineon APDS-9002-021 photo sensor connected to the ADC of the MCU.

Due to around 200 mA current consumption of the X-Port device, the Ethernet gateway is mains powered by a 5.5 V mains power adapter. This voltage is regulated to 3.3 V by LM317 linear regulator.

The implemented TUTWSN Ethernet gateway prototype is presented in Figure 4.12.

The most important goal in the antenna design process was to reach omnidirectional antenna structure. In addition it needed to be long range and sized to a maximum area of only 2.7×4 cm. Different structures were simulated and finally an asymmetric loop structure was concluded. Simulated and measured properties of the antenna are presented in Figure 4.13.

4.4 Antenna Design

4.4.1 Antenna Design Flow

Figure 4.14 presents the phases of antenna design. First phase captures the requirements, such as frequency range, radiation pattern, and physical size. These are defined together with other design groups. The actual design begins once the requirements have been captured and agreed upon. Electrical and mechanical properties can be designed in parallel. Antenna theory is essential in this phase since it determines antenna dimensions. Dimensions are based on the structure, materials, and manufacturing methods of the antenna.

When the antenna parameters are known, a design phase can be started. First, different possible antenna types are compared according to their suitability to the case in question.

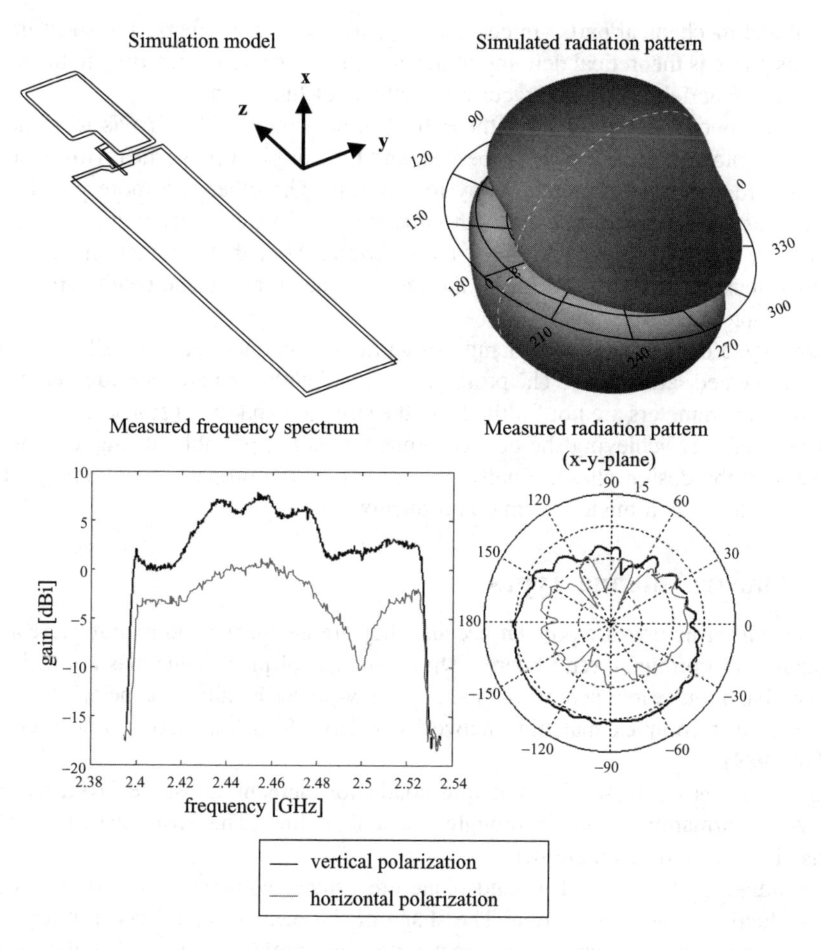

Figure 4.13 Radiation properties of Ethernet gateway prototype with asymmetric loop antenna.

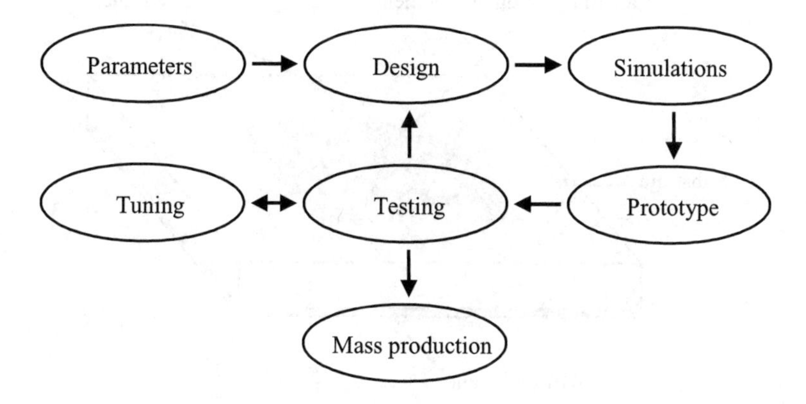

Figure 4.14 Antenna design workflow.

Electrical and mechanical parts are designed in parallel to each others. The most important part of this phase is theoretical defining of the antenna dimensions according to the decisions on structure, materials, and manufacturing methods of the antenna.

There are two possibilities after theoretical dimensioning. The first is to manufacture an antenna prototype, measure its properties, and tune original properties. However, this is usually a hard and time-consuming way to progress. The other and more popular option is to simulate operation of antenna with simulation software. Here, rough theoretical dimensions are given to initial values for the software. After that, simulations are executed and dimensions are tuned according to the results. Therefore, all the results are calculated without complex measurements.

A prototype of the simulated antenna structure is manufactured after all the simulation results are deemed satisfactory. The prototype is tested with accurate measurement methods. Typically, all parameters are not fulfilled and the structure and dimensions need to be tuned. If it is realized that achieving the desired properties is not possible, it might be necessary to go back to the design phase. Finally, the version of the properly functioning prototype can be considered as a model for mass production.

4.4.2 Planar Antenna Types

Planar antenna structures are popular because they are inexpensive to manufacture and easy to integrate in to printed circuit boards. Disadvantages of planar antennas are their narrow frequency band, quite low achievable gain and low-power handling capacity. Power losses can also arise if complex matching networks or large feed lines have to be used (James and Hall 1989).

Planar antennas consist of a radiating conductor element, substrate layer, and ground plane. An information signal is brought via a feed line. The basic structure of planar antennas is presented in Figure 4.15.

The radiating element and ground plane are usually constructed of copper due to its good conductivity and cheap price. The shape of the element could be arbitrary but typically common geometric shapes like rectangles and circles are used. A polarization and

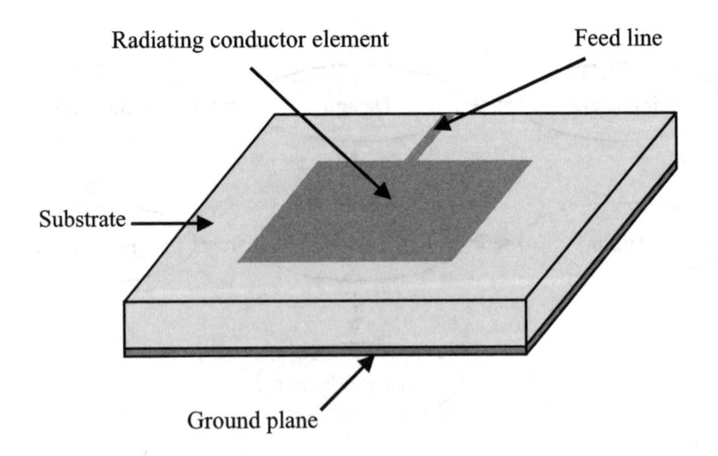

Figure 4.15 Planar antenna structure.

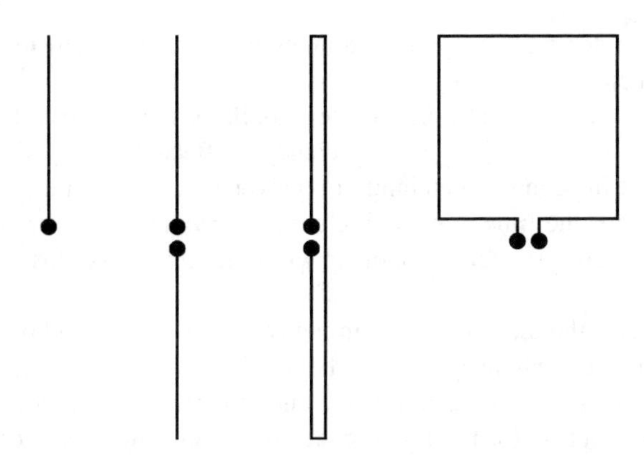

Figure 4.16 Different antenna structures for the same operating frequency. From left to right: monopole, dipole, folded dipole, and loop. Feeding points are marked with black dots.

radiating pattern of the antenna can be affected with the shape of the element. At the same time, resonance frequency of the antenna is determined by the biggest dimension of the radiating element.

The feeding of the element can be implemented with microstrip or coaxial cable. The feed line and transceiver circuit are easy to attach to the same board when a microstrip is used. The coaxial feeding is harder to integrate on to the board due to the complexity of the junctions, but the noise is reduced at the same time. The feed line also affects the impedance of the antenna. The closer the feeding point is moved to the center of the element the smaller the impedance becomes at the resonance frequency (Stutzman and Thiele 1998). In addition, if the feeding point is moved towards the corner of the element, the radiating pattern would also be modified.

Substrate is a dielectric layer between the radiating element and ground plane. The thickness and material of the substrate affects the size, impedance, bandwidth, and efficiency of the antenna. Ground plane can cover the whole board only when the antenna achieves radiating to the semi-space (James and Hall 1989). If the ground plane is chopped, the antenna would become to more omnidirectional. Traditional antenna types can be imitated if ground plane is totally removed. These types are, for example, monopoles, dipoles, folded dipoles, and loops. The different antenna structures are presented in Figure 4.16.

Monopoles are antennas that consist of one straight wire. The length of the wire is usually a quarter of the operating wavelength. The wire is fed along the ground. Alternatively, dipoles are two-wire antennas. The wires are laid one after the other and the total length of the structure is half of the wavelength. A differential feed line is connected to the center point. A folded dipole is similar to the dipole but the open ends of the wires are connected to each other with parasitic element. Finally, the loop antenna consists of one wire, which is formed in to a ring. The length of the loop is often one wavelength, but also other lengths are possible. The feeding is again differential.

4.4.3 Trade-Offs in Antenna Design

Antenna design is challenging because many decisions in the different design phases can affect more than one antenna parameter. Therefore, all decisions should be balanced in order to fulfill all the desirable properties. Sometimes a trade-off has to be made between conflicting requirements.

Energy efficiency is important especially in mobile or maintenance-free radio systems, where a long lifetime is striven. The total efficiency is affected by impedance matching and materials. With good impedance matching any power loss in the radio boundaries will be minimized, but at the same time, a matching network requires a larger area. Good-quality materials and accurate manufacturing methods will minimize power losses to heat but costs can climb significantly.

The antenna is often the most area-consuming component in the radio system. Therefore, by minimizing its size, the whole system can be fitted into a smaller casing. Antenna size is mostly determined by the operating frequency and materials. Frequency bands are subject to license, so the only option for minimizing the size of antenna is to choose the applicable materials. In planar antennas this can be achieved by carefully choosing the printed circuit board substrate. The more dielectric the substrate is, the smaller the antenna dimensions can be for reaching the frequency in question. However, theoretical radiation efficiency declines at the same time. A thinner substrate alleviates this problem but the structure will become more fragile and narrow-band. In addition, when the antenna becomes smaller, the energy efficiency descends and so achieving the desired radiation pattern becomes harder.

The shape of the radiation pattern significantly affects the operation of the antenna. It is often desirable for the antenna to be omnidirectional with an achievable range as long as possible. However, these requirements conflict with one another. Increasing the range of omnidirectional antennas is made possible by utilizing a higher transmit power and better efficiency. The first option decreases the lifetime of mobile systems due to higher power consumption and the second option might raise material and manufacturing costs.

5

Design of WSNs

A universal all-purpose WSN that fulfills the requirements of all possible applications is not a reasonable goal. The diversity of applications creates contradictory objectives that cannot be met with a single design. Instead, all the components of a deployed WSN should be selected depending on the application requirements. Further, commonly accepted solutions for different WSN applications are absent at the moment. Therefore, the design and implementation of a new WSN completely from scratch is also an option. However, armed with minimal historical knowledge and experience, making the correct choices and decisions regarding deployment is extremely difficult.

Constantly emerging new node platforms, protocols, and applications continuously widen the WSN design space. Furthermore, the protocols can typically be configured to different operation modes. Thus, in near future, the WSN design space will be so large that designers will not be able to handle it without a suitable support tool. Therefore, design automation is an important challenge for WSNs.

The unique properties of WSNs make traditional design automation tools poorly suited to their needs. At the moment, the higher abstraction level, the management of dependencies, and automated-code generation are commonly available. Yet, the suitability of such designs to the resource constrained nodes and their exploration in realistic deployment is not achievable with current design methods. Still, it is extremely important to consider all design issues simultaneously, since a single unoptimized part can ultimately ruin the overall energy efficiency or lead to excessive memory usage.

5.1 Design Dimensions

The design dimensions discussed here partly overlap with the general requirements presented in Section 2.3. The main conceptual difference between this and what is presented here is that the general requirements are common for all WSNs and they define abstract qualifiers. Instead, the design dimensions are parameters for specific deployment. These parameters can be either selected from bounded options or defined as the limits of variations for numerical values. For example, considering *lifetime* as a general requirement means that WSNs are expected to be functional for long periods of time. As a design dimension, it defines that the given WSN should operate at least for six months, for example.

The following set of dimensions for structuring WSN design space is proposed in Römer and Mattern (2004). The main effort is to identify the most crucial properties that should be considered in WSN design. The dimensions are selected according to two major guidelines. First, a dimension should vary notably between different applications and secondly, it should have a significant effect on design choices.

- **Deployment**: The actual deployment process can be *random* or nodes can be *manually* set to designated places. Furthermore, the dispersion of nodes may be either *one-time* or *iterative*, in which case battery or node replacements are continuous.

- **Mobility**: In general, WSNs can be categorized to be either *immobile*, *partly*, or *fully* mobile depending on the ratio of moving nodes. Thus, partly mobile means that only a subset of nodes is moving, while in fully mobile network all nodes move. Mobile nodes can be attached to moving objects, influenced by forces of nature (e.g. wind, water, earthquake, avalanche, landslide), or have actuators enabling node mobility. Thus, the mobility of nodes can be either *active* or *passive* depending whether the node can influence to its mobility. Nodes can also travel either *continuously* or only *occasionally*.

- **Cost, size, resources, and energy**: The size of an individual node can be roughly described from largest to smallest as *brick*, *matchbox*, *grain*, or *dust*. The nodes can be either *mains powered*, or energy can be stored in *batteries* or *scavenged* from the environment. The resources of nodes vary from very *limited* and simple node platforms with only a sensor, a simple MCU, and a transceiver to *full-feature* nodes having the resources and computation capacity of workstations. The cost of a single node can vary from a *few cents* to *hundreds of euros* depending on the size, resources, and complexity of the node.

- **Heterogeneity**: Traditionally, all devices in WSN were envisioned to be *homogeneous*. More recently, the pros of *heterogeneous* platforms have been recognized. Complex computation can be centralized to more powerful platforms. Further, it is not practical to populate all physical sensors on every node (e.g. a GPS receiver is not needed in every node in dense deployments). This makes originally similar node platforms heterogeneous.

- **Communication modality**: The communication modality in WSNs is typically *radio* waves. Still, there are several other possibilities that may be more suitable for some applications. Examples of other communication modalities are *light* beams, *inductive* coupling (used e.g. in RFID systems), *capacitive* coupling, *sound* or *ultrasound*.

- **Infrastructure**: The communication architecture in a wireless network can be either *infrastructure* or *ad hoc* or a combination of these two. In general, WSNs are considered to be ad hoc networks but in case there is an existing infrastructure in the area of deployment, its usage is justified.

- **Network topology**: The topology has a major effect on WSN operation. The maximum number of hops affects the latency, robustness, and capacity of the network. The topology can be *single-hop*, where each node can communicate directly with other nodes, *star* or *networked stars*, in which base stations manage the communications, or a multihop arbitrary *graph* or structured *tree*.

- **Coverage**: The coverage of a WSN node means either sensing coverage or communication coverage. Typically with radio communications, the communication coverage is significantly larger than sensing coverage. For applications, the sensing coverage defines how to reliably guarantee that an event can be detected. Therefore, from the application point of view, the sensing coverage is an important factor. The coverage of a network is either *sparse*, if only parts of the area of interest are covered or *dense* when the area is almost completely covered. In case of a *redundant* coverage, multiple sensor nodes are in the same area.

- **Connectivity**: The connectivity of WSN depends on the radio coverage. If there continuously exists a multi-hop connection between any two nodes, the network is *connected*. The connectivity is *intermittent* if WSN is partitioned occasionally, and *sporadic* if the nodes are only occasionally in the communication range of other nodes.

- **Network size**: The network size depends greatly on other dimensions, i.e. coverage and connectivity. Depending on these and the size of the area of interest, the number of nodes in the deployed WSN can vary from *few nodes* to *thousands* or more.

- **Lifetime**: The requirements for the lifetime of a WSN depends on the application. In general, the lifetime ranges from *hours* to several *years*.

- **Other QoS requirements**: The application and target environment set several other QoS requirements to WSNs. These aspects must be considered in the WSN design and they have also an impact on the other dimensions. *Realtime constraints*, *robustness*, *tamper-resistance*, *eavesdropping resistance*, *information security*, and *unobtrusiveness* or *stealth* are examples of the possible application QoS requirements.

An argument against the given list of design dimensions could be that some of the common, measurable performance metrics, such as *throughput*, *delay*, or *reliability*, are absent. These might be partly covered in *other QoS requirements*, but the question arises as to whether they should be merited in their own individual dimensions. The straight answer is no. Unlike traditional computer networks, in WSNs the measurable bandwidth in bits/s or the delay in time units are not important. Instead, the point of interest is the overall application QoS that is realized, e.g. as a realtime event detection or as the quality or accuracy of the data. These are included in the other QoS requirements in the design dimensions. Throughput and other performance metrics are consequences of the design choices that were made based on the application requirements.

Before the design of a new WSN deployment is started, the values and importance of the design dimensions must be defined. This probably results in contrary objectives that cannot be met due to the interrelations of the dimensions. Therefore, trade-offs between conflicting requirements need to be considered before the actual design process begins.

5.2 WSN Design Flow

Compared to traditional HW/SW (hardware/software) embedded system design (Wolf 2001), WSN design has similar characteristics. Yet, the scale and one-time nature of deployments make the system testing and verification more complicated. Therefore, issues that can ease these phases should be considered earlier in the design flow.

An example WSN design flow is depicted in Figure 5.1 (Kuorilehto et al. 2007c). In the following, *WSN application* defines the set of tasks that the network must fulfill. Hence, the WSN application may consist of multiple distinct tasks without interrelations. A *customer* is the entity (an individual or a company) that has ordered and defined the WSN application. The *designer* is one or more persons representing the producer of the network. The designer is responsible for technical specification, design, implementation, and testing. A *component* is a distinctive HW or SW module (e.g. node platform, protocol layer implementation, or Gateway (GW)).

The phases in the design flow are

- **WSN application**: The WSN application defines the functionality, requirements, and target environment for the designed WSN. Depending on the expertise and skills of the customer, the level of detail of the specification can vary from a vague vision to a detailed textual specification.
- **Evaluation of design dimensions**: Based on the WSN application, the designer defines the design dimensions together with the customer. Some of the dimensions might be defined in the application requirements, but the evaluation of some of the dimensions requires the expertise of the designer. The trade-offs between contradictory dimensions are solved before the actual system design is started.
- **System model design**: The design dimensions and the functionality requirements of the application are the input for the system design. The system design consists of several phases. First, the designer defines the technical requirements and specifies the

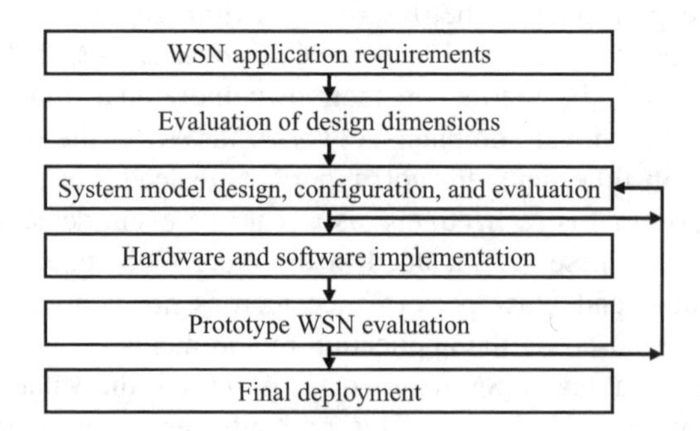

Figure 5.1 An example WSN design flow. With kind permission of Springer Science and Business Media.

system in detail. Second, the components that are needed to fulfill the requirements are selected. This includes node platforms, protocols, GWs, and external databases and User Interfaces (UIs). Finally, the designer either chooses the components from existing ones or gives a detailed specification for implementation.

- **Configuration and evaluation**: Typically, WSN protocols contain configurable parameters that affect, for example, the duty cycle of a MAC protocol or to routing decisions. In the configuration phase, the values of these parameters are set and their effect on the overall network performance estimated. Further, the preset design dimensions (e.g. coverage, size) can be reconfigured if it is necessary to fulfill application requirements.

- **Implementation**: In the implementation phase, new HW and SW components are implemented. This phase also includes module testing and the integration of existing and new components to the system. The integration consists of the compilation of SW to the HW components used, the adaptation of communication, and the modification of external databases and UIs for the application.

- **Prototyping**: After the implementation and configuration, the operation of the components is tested on a small scale. A limited number of nodes are deployed (e.g. to a laboratory environment) and the performance of the network is monitored in various conditions and modes that depend on the application.

- **Final deployment**: After the necessary iterations needed to validate the operation of the network, WSN is deployed to its target environment. Nodes programmed with protocols and applications, GWs, and external components are delivered to the customer, setup, and activated.

In the configuration evaluation and prototyping phases, the operation of the tested WSN is evaluated against the application requirements. If the requirements are not satisfied, a new design iteration is started from the system design phase. The start point of the iteration depends on the nature of detected flaws and the presumable corrections or modifications.

A typical embedded system project also contains a maintenance phase after deployment. This includes possible error corrections, and HW and SW updates. In WSNs the maintenance is significantly more difficult due to the nature of the deployment. Even though it might be possible to reprogram the nodes remotely after deployment, such a procedure adds uncertainty and consumes resources. Therefore, in WSNs the operation reliability of the deployed network should be guaranteed.

Tool support has been proposed for WSN design to overcome the challenges due to the complicated nature of WSN deployments and laborious and iterative design and testing processes. Tools that exploit higher abstraction levels are one option for managing the design space and shortening the design cycle (Baldwin et al. 2004; Kuorilehto et al. 2007b). The major challenge of such tools is the applicability of the generated executables to the resource constrained sensor nodes in terms of processing capacity and code and data memory consumption.

5.3 Related Research on WSN Design

The immaturity of WSN technology and the lack of established practices make the categorization of design tools and methodologies difficult. The methodologies vary in target objectives and domains as well as in the approaches taken. Further, quantitative metrics for the comparison of design methodologies cannot be easily derived, since the emphasis is merely based on preferences that depend on the subjective experiences of a designer (Kangas 2006).

5.3.1 WSN Design Methodologies

A design flow or a design methodology defines a structured approach for system development. In WSN domain, the research considering these approaches is still evolving. The proposals discussed in this section cover different phases in WSN design flow. The approaches can be categorized as *model-driven*, *component-based*, and *platform-based design* methodologies. Integrated development environments are considered as a separate subclass. In general, the proposals provide abstractions, methods, and tools that aid a designer to manage the process. Fundamental issues, such as programming paradigms and programming models for data extraction and network control (as discussed by Culler et al. (2001), Gracanin et al. (2005), and Woo et al. (2004)) are not considered here.

Model-driven design

A model-driven design, or development, uses domain-specific, tailored models and abstractions for aiding the design process (Schmidt 2006). In this context, the model-driven design methodologies include those proposals that support WSN design with existing models or metamodels.

Liu et al. (2003) abstract the internals of WSN operation by collaboration groups, which are collections of nodes or more abstract entities, i.e. agents. An application is designed on top of this abstraction using a state-centric programming model. Application performance can be estimated with high-level simulations.

Yu et al. (2005) propose two simple models that abstract communication. The methodology supports analytical evaluation of application algorithms on top of communication models. The exploration of suitable network parameters is performed manually.

Tinker (Elson and Parker 2006) does not directly fall into the category of model-driven design methodologies but this is closest to the classification used. Tinker focuses on application data and it abstracts communication and node platforms by few simple loss models. The main objective of the tool is the exploration of different data processing algorithms for input data streams.

Component-based design

In WSNs, component-based design is motivated by TinyOS architecture (Hill et al. 2000). TinyOS is the most widely known Operating System (OS) targeted for resource constrained WSN nodes. The main abstraction in the design process is a software component implementing a dedicated function. A system is designed and constructed as a composition of these components. A component is accessed through a well-defined interface (Gay et al.

2003). The reuse of existing designs is directly enabled by the component-based architecture, assuming that the component interfaces and specifications are standardized (Archer et al. 2007; Chu et al. 2006).

All existing proposals in this category exploit or complement TinyOS. A key technology is the nesC (Gay et al. 2003) programming language, which extends C language with a built-in component model as well as concurrency and communication support. While characterized as a language, the nesC is aimed at holistic system design.

In WSN design flow, the nesC language addresses only the prototyping phase. A graphical interface for component design is provided by GRATIS (Völgyesi and Ákos Lédeczi 2002) and Viptos (Cheong et al. 2006). GRATIS offers a graphical interface for assembling the components of a complete system and generates runtime code for connecting the components automatically (Völgyesi and Ákos Lédeczi 2002).

Viptos is built on top of the Ptolemy II design environment (Ptolemy II 2006) and supports all phases in the design flow. Viptos integrates the Model of Computations (MoCs) available in Ptolemy II to nesC, TinyOS, and TOSSIM (Levis et al. 2003). Viptos complements the functionality of VisualSense simulation environment (Baldwin et al. 2004). Graphical designs can be simulated with VisualSense, or generated as executables for TOSSIM simulations or TinyOS nodes (Cheong et al. 2006).

Platform-based design

In platform-based design, an application specification is refined with a set of abstractions and mapped onto potential implementation platforms (Bonivento et al. 2006; Kangas 2006). Typically, this methodology includes an automatic exploration of platform configurations for the given application. In WSN context, the platform includes communication protocols in addition to physical node parameters.

Bonivento et al. (2006) separate the design to distinct models by three abstraction layers for applications, protocols, and node platforms. The key concept is the exploration of the protocol parameters so that application requirements are met within the constraints that are set by node platforms. The requirements are extracted with a Rialto tool that analyzes all possible communication combinations of the application.

Algorithm design methods proposed by Bakshi and Prasanna (2004) possess similar characteristics. Node platform and communication are modeled coarsely by a virtual architecture, on top of which application algorithms can be synthesized to actual programs mapped to the nodes. Each phase is performed manually.

Shen et al. (2006) introduce models for rapid prototyping of WSNs. In a design phase, the methodology abstracts nodes and networking using a set of energy models and connectivity graphs. The design space exploration is based on the simulation of applications with the energy models for obtaining (mainly hardware) parameters. Prototyping is used for validating evaluation accuracy.

Integrated environments

The development environments covered in this section cannot be considered as design methodologies since they do not offer any abstractions for system design. Instead, they provide tools for supporting different phases of the design process.

EmStar (Elson et al. 2004) is a Linux-based development environment for protocol and application software development. The main component in EmStar is a runtime environment, which runs on diverse execution platforms. This allows evaluation of the final executables with simulations, prototyping, or their combination.

A quite similar approach is taken in Worldsens (Fraboulet et al. 2007). Compared to EmStar, Worldsens incorporates two simulators, WSNet for large-scale network simulations and WSim for cycle accurate simulation of a single node. Both simulators run native target platform code. Furthermore, the simulators can be synchronized for cycle accurate large-scale network simulation.

Analysis

The main characteristics of existing design methodologies are outlined in Table 5.1. As stated, the comparison of design methodologies is difficult because of the lack of unambiguous metrics, and contradictory goals and approaches. The parameters defined in the table set a starting point for qualitative comparison of WSN design environments. The table highlights the main properties of the design methodologies and assesses the tool support in different phases of WSN design flow.

The first column in the table identifies the methodology and second defines the main approach. The next one lists the design flow phases that are tackled. Prototyping denotes the implementation on physical node platforms. The abstraction defines the interfaces and models that hide the low-level implementation details. Components considered by the design methodology are listed in the fifth column. Possible components are application (app), communication protocols (comm), and node platforms (node). The evaluation column lists a method or methods that are used for assessing the selected design. The last two columns indicate whether the property is supported by the design methodology. Design space exploration means either automated or manual configuration optimization for the given deployment.

Most of the methodologies support design phase but only a few also cover evaluation and prototyping. There are divergent opinions regarding whether there should be a separate high-abstraction level design and a low-level implementation phase (Fraboulet et al. 2007). While separate environments make the creation of higher level abstractions easier, the porting of the models to the final implementation may introduce additional bugs (Varshney et al. 2007).

In general, model-driven and component-based design methodologies provide suitable abstractions for WSN design. Platform-based design methodologies support automatic exploration of different network or node platform configurations. While integrated development environment introduce a rich set of tools for the prototyping phase, they lack design abstractions.

In WSNs, the most fundamental design choices need to be made typically before a large scale deployment. Thus, the early design flow evaluation has a significant importance in the process. If its results are reasonably accurate, the design choices can be based on a solid foundation.

From the design methodologies, Viptos (Cheong et al. 2006) creates the most extensive framework. It enables graphical design of system models, and the evaluation of these models through simulations. Further, Viptos has an integrated support for code generation on top of TinyOS (Hill et al. 2000).

Table 5.1 Comparison of WSN design methodologies.

Methodology	Approach	Covered phases	Abstraction	Components	Evaluation	Graphical design	Design space exploration
Liu et al. (2003)	Model-driven	Design, evaluation	Collaboration group, state	App	Simulation	–	–
Yu et al. (2005)	Model-driven	Design	Communication model	App	Analytical	–	Manual
Tinker	Data-centric	Design	Application-level data streams	App	Simulation	–	Manual
nesC	Component-based	Prototyping	Component interfaces	App, comm	Code analysis and optimization	–	–
GRATIS	Component-based	Design, evaluation	Component model	App, comm	Component validation	Component assembly	–
Viptos	Component-based	Design, evaluation, prototyping	Ptolemy II MoCs	App, comm, node	Simulation	Ptolemy II	Manual
Bonivento et al.	Platform-based	Design, evaluation	Interface abstractions	App, comm, node	–	–	Automatic
Bakshi et al. 2004 (2006)	Platform-based	Design	Virtual architecture	App	–	–	Manual
Shen et al. (2006)	Platform-based	Design, evaluation, prototyping	Platform models	App, node	Model-based simulation	–	Automatic
EmStar	Integrated development	Evaluation, prototyping	Runtime environment	App, comm	Simulation	–	Manual
Worldens	Integrated development	Evaluation	–	App, comm, node	Simulation	–	Manual

5.4 WSN Evaluation Methods

The concept of WSN evaluation differs from a customers and designers point of view. While a customer evaluates the network subjectively, a designer is merely interested in numerical performance figures. A customer mainly assesses the final outcome of the application tasks, e.g. how fast a critical temperature change was detected or how reliably an animal or a human crossing the state border is identified. While a designer is also interested in these aspects, during the development process metrics such as throughput, delay, and packet reliability are significant for the evaluation of design choices.

The conceptual difference between the evaluation approaches is presumable. A customer has the possibility to monitor and evaluate the operation of the network only in product previews and after the final deployment. During the development a designer may not be able to test the network with the intended application but instead uses different kinds of test and diagnostics SWs. Therefore, numerical metrics are the most reliable, if not the only, way to assess the operation of an evaluated configuration.

Different approaches to the evaluation of protocols and overall network performance have been studied both for WSNs and wireless networks in general. Common approaches are *analysis*, *simulation*, and *prototyping*.

Analysis models the operation of protocols and platforms by mathematical equations. Separate mathematical models are defined for protocol algorithms, communication primitives, and platform energy characteristics. The analysis models form a set of tools that can be used to evaluate the measurable performance metrics for a single node or for a complete network. While analytical models scale to very large deployments, they typically consider only a single point of the network omitting the interactions. Furthermore, the modeling of the dynamic nature of the network and its environment is difficult in analysis.

The network simulation has been used for the protocol evaluation in wired and wireless networks. For WSNs, simulators allow the evaluation of large-scale, long-term WSN operation and performance. Depending on the design methods and tools, the simulations can be performed during the system design, implementation, or large-scale testing phases, or parallel with all these. Pure functional network simulation is not feasible for WSNs but strict resource and environmental constraints together with physical phenomena are fundamental aspects that need to be considered. Furthermore, the accuracy of the simulation tool is very important. Even the smallest and insignificant issue might cause severe changes, for example in network power consumption. Further, an effort should be made to construct extensive and acceptable simulation cases, which may not always be straightforward (Andel and Yasinsac 2006).

Prototyping in a realistic environment is laborious, time-consuming, and always requires HW platforms. In prototyping several design factors, such as processing and memory capacities and radio interfaces, have been fixed when selecting prototype platforms. In WSNs the extensive number of nodes make the construction of a realistic test case even more tedious. For example, the deployment process of a thousand-node test WSN cannot be randomized unless it is not essential to gather up the nodes. In a short-term test of a small-scale prototype network, several crucial aspects cannot be extensively tested. Congestion, data buffering delays and overflows, effects of route selections, channel access failures, and the exceeding of physical processing and storage capacities are issues that are difficult to generate in prototype networks but are typical challenges for large-scale networks.

5.5 WSN Evaluation Tools

Currently, the main approach for WSN design time evaluation is simulation. Analysis is widely used in literature but general frameworks have not been proposed. Prototyping, although also widely used, is application-specific and limited to small scale. Therefore, we concentrate on the tools that implement a framework for large-scale, long-term WSN simulations.

Legacy computer network simulators, such as ns-2 (Ns-2 2006), GloMoSim (Zeng et al. 1998), Qualnet (Scalable Network Technologies 2006), OPNET (OPNET 2006), OMNeT++ (OMNeT++ 2006), Scalable Simulation Framework (SSF) (ssfnet.org 2006), and J-Sim (J-Sim 2006) enable the simulation of wireless network behavior and protocol stack operation but lack accounting for WSN characteristics. This is partially solved by simulators proposed specifically for WSNs, which can be categorized to networking oriented and sensor node simulators. Networking oriented simulators model the transmission medium in detail and are more suitable for large-scale WSN simulations. Sensor node simulators mainly simulate the operation of a single node but still implement a lightweight communication model.

The basic requirements and the properties of the WSN simulators are quite similar. Therefore, the discussion of the simulators is structured so that the implementation of different properties are handled one at a time. After the simulators are presented, they are analyzed against each other in order to establish an overview of the field.

5.5.1 Networking Oriented Simulators for WSN

Most of the proposed networking oriented simulators are based on the legacy computer network simulators. SensorSim (Park et al. 2001) extends ns-2 with general WSN features. senQ (Varshney et al. 2007) and its predecessor sQualnet (Networked & Embedded Systems Laboratory 2006) are built on top of Qualnet. While both possess quite similar features, we focus only on senQ. Simulator for Wireless Ad-hoc Networks (SWAN) (Liu et al. 2001) is based on SSF. SENSIM (Mallanda et al. 2006) and simulation template for EYES (Dulman and Havinga 2003) utilize OmNet++ environment, while J-Sim sensor simulator (Sobeih et al. 2006) adds WSN features to its parent simulator. VisualSense (Baldwin et al. 2004) is an extension to Ptolemy II (Ptolemy II 2006), Prowler (Simon et al. 2003) utilizes MATLAB, and H-MAS (Mochocki and Madey 2003) and SENSE (Chen et al. 2004) implement custom simulation environments.

The most realistic transmission media and protocol stacks are available in SensorSim, senQ, and J-Sim sensor simulator. While the first one relies on the models available in parent simulators, the last two also include a set of WSN protocols. VisualSense has several models for transmission media, which vary in their accuracy. SWAN and Prowler include abstracted transmission media and lowest layer protocol models that estimate the network operation. SENSIM, EYES simulator, H-MAS, and SENSE have error-free transmission medium models, in which signal propagation is dependent only on the transmission range. More simple protocol stacks are available for SENSIM, H-MAS, and SENSE. In VisualSense and EYES simulator, the protocol stack is implemented by the designer.

A separate sensing channel also containing the sensed targets is used for phenomena modeling in SensorSim, senQ, and J-Sim sensor simulator. VisualSense has also a dedicated channel for modeling the propagation of different phenomena. For the other simulators, SWAN models catastrophic plume dispersion, and H-MAS generates random sensor readings. Prowler, SENSIM, EYES simulator, and SENSE do not support phenomena sensing.

Concerning the node capabilities, the power consumption is accounted in several simulators. SensorSim, senQ, SENSE, SENSIM, and J-Sim sensor simulator have detailed power models, which consider battery discharge rate and relaxation. EYES simulator uses a linear battery model, in which maximum energy capacity is available independent on the discharge rate. Prowler and VisualSense estimate power consumption based on activity, whereas in SWAN and H-MAS power modeling is not taken into account. senQ emulates individual clock drifting in nodes.

The simulated code is applicable directly for hardware platforms in SensorSim, senQ, J-Sim sensor simulator, and partly in VisualSense. In SensorSim, simulated SensorWare applications are compatible with custom hardware platforms (Boulis et al. 2003). The other three simulators enable the execution of applications on Berkeley motes (Crossbow Technology 2006) on top of TinyOS (Hill et al. 2000) and senQ on top of SOS (Han et al. 2005).

VisualSense uses a graphical notation for design and supports a combination of different MoCs of Ptolemy II. Abstracted application scripts can be simulated also in Prowler.

5.5.2 Sensor Node Simulators

Most of the proposed sensor node simulators are targeted to TinyOS motes. Complete TinyOS system can be simulated with TinyOS Simulator (TOSSIM) (Levis et al. 2003), Atmel Emulator (ATEMU) (Karir et al. 2004), Avrora (Titzer et al. 2005), WMNet (Wu et al. 2007) and TinyOS Scalable Simulation Framework (TOSSF) (Perrone and Nicol 2002). TOSSF itself is an extension to SWAN. SENS (Sundresh et al. 2004) supports only TinyOS application simulation. Sensor Network Asynchronous Processor (SNAP) (Kelly IV et al. 2003) is a hardware emulator, which connects several processors on a Network-on-Chip (NoC).

VMNet incorporates a realistic transmission medium model that considers path loss and noise. The TOSSIM transmission medium model is directed graphs with individual bit-error rates, whereas in ATEMU and Avrora the transmission medium is error-free accounting only the transmission range. TOSSF utilizes a transmission medium model from SWAN, and in SENS medium and networking protocols are combined into a simple model. In SNAP, NoC models the transmission medium. Protocol stacks in TOSSIM, ATEMU, Avrora, VMNet, and TOSSF depend on simulated configuration. SNAP implements simple protocols for system testing.

Phenomena sensing is modeled in TOSSF by the plume dispersion model of SWAN. TOSSIM, Avrora, and VMNet get sensor readings from an external source. SENS incorporates a separate environment model that supports sensing and actuating. In SNAP and ATEMU phenomena are not modeled. Only VMNet simulates node power consumption.

The applications from all sensor node simulators, and protocols from other than SENS, can be directly mapped to hardware platforms. However, the platform is restricted to a specific one.

5.5.3 Analysis of Evaluation Tools

The comparison of WSN simulators is summarized in Table 5.2 and Table 5.3. The comparison is based on the public information available about each simulator and possible parent simulator engine. If the exact scalability information is not available, the simulator is assessed according to the largest reported simulations.

The scope of input parameters is vital when comparing the configurability of simulators to different kinds of platforms, protocols, and applications. Also, the availability of Graphical User Interfaces (GUIs) and the type of information output by the simulator are accounted in the comparison. The possibility to use simulated protocols and applications for final implementation on physical platforms defines the applicability of a simulator as a complete design and development environment.

For the comparison, the term "accurate results" denotes very close correspondence of the simulation results to real-world measurements with physical WSN prototypes. Accurate results in full-scale simulations need to combine at least realistic models for communications (application, transmission medium, transceiver unit, and low-level communication protocols) and node platform (energy, memory, peripheral I/O, and computation). The node state changes, peripheral activation, and leakage currents contribute to the accuracy of energy consumption. The memory allocation and thread scheduling during simulation depend on the accuracy of the OS model of the simulator.

As shown in the table, most of the simulators are capable of simulating WSN scenarios consisting of thousands of nodes. This is an acceptable limit for current WSNs, but in future the capability to simulate in order of magnitude larger networks is required.

Major differences between the simulators are in input and output parameterization. Although sensor node simulators emulate a single node platform in detail, they do not allow for the testing and evaluation of applications and protocols on other types of sensor nodes. Furthermore, they omit the modeling of node power consumption. Of the simulators, the possibility to define different platforms is only available in senQ.

Most of the simulators visualize network topology and key parameters through GUI. The event information is gathered to trace or log files with varying level of detail. The ns-2 based simulators output extensive trace files.

The code that runs in sensor node simulators is directly applicable for node platforms. In networking oriented simulators, two approaches are taken for the final implementation. Either the simulated protocols and applications are converted to executables for node platforms, or an existing code library is used for emulating a node in large-scale simulation. In the latter, the node implementation already exists, and the final implementation needs only the configuration parameters acquired by simulations.

A rapid protocol evaluation for a specific application is possible only if the simulator protocol stack is modular and its layers interchangeable. Most of the simulators that descend

Table 5.2 Comparison of existing networking oriented WSN simulators.

Simulator	Simulator engine	Scalability (# of nodes)	Simulator input	Simulator output	Final implementation	Benefits	Deficiencies
SensorSim	ns-2	~2000[a]	Power model, protocols (TCL)	ns-2 nam UI, traces (data, energy, errors)	Applications for SensorWare, ns-2 protocols	Accurate results, ns-2 protocols modular	No memory or CPU modeling
senQ	Qualnet	~1000	Nodes, traffic, protocols (scripts)	Qualnet Visualizer statistics (data, energy)	SOS applications, directly	Accurate results, clock skew	No memory or CPU modeling
SWAN	SSF	~10000	Nodes, plume dispersion	GUI, system prints (data, delay)	WiroKit routing protocol	Scalability	Node, sensing, modularity
SENSIM	OmNet++	~5000	Protocols (ini-file for OmNet++)	OmNet++ GUI (data)	–	Modular	Node, sensing
EYES simulator	OmNet++	<1000	Protocols (ini-file for OmNet++)	OmNet++ GUI (data, errors)	–	Modular	Nodes, energy, medium
J-Sim sensor simulator	J-Sim	>1000[b]	Protocols (script)	Text output, GUI (data)[c]	Applications	Modular	Nodes, no GUI
VisualSense	Ptolemy II	~100	MoCs (for Ptolemy II)	Ptolemy II GUI (topology, node info)			
Prowler	MATLAB	<100	Application (script)	GUI (data)	Algorithms to TinyOS[d]	Graphical design, MoCs	Protocols, nodes
H-MAS	Custom	>100	Nodes, protocols (text)	GUI, event files (data)	–	–	Modularity, communication
SENSE	Custom	~1000	Topology, traffic (script)	–[e]	–	Modular	Output, nodes, medium, sensing

[a]See reference (Mallanda et al. 2006).
[b]See "Evaluation of J-Sim", available: `http://www.j-sim.org/comparison.html`
[c]J-Sim outputs simulation data to an "instrument channel", which allows GUI implementation.
[d]See VisualSense homepage, available: `http://ptolemy.eecs.berkeley.edu/visualsense/`
[e]No information about the output of SENSE is given, published results indicate only the performance of the simulator itself.

Table 5.3 Comparison of existing sensor node simulators.

Simulator	Simulator engine	Scalability (# of nodes)	Simulator input	Simulator output	Final implementation	Benefits	Deficiencies
TOSSIM	Custom	~10000	TinyOS code	TinyViz, debug (data, node info)	Directly	Scalability	Homogeneous code, energy model
ATEMU	Custom	~100	TinyOS code nodes (XML)	XATDB debugger (debug data)	Directly	Emulation	Energy, sensing
Avrora	Custom	~10000	Simulated object file	Monitor traces	Directly	Energy model, scalability	Simple sensing and medium
SENS	Custom	~10000	Node profiles (text)	GUI, text files (data, energy, errors)	Applications, directly	Scalability	Node, medium, protocols
TOSSF	SSF	~10000	TinyOS code SWAN parameters	SWAN output[a] (data)	Directly	Scalability	Node, sensing, medium
WMNet	Custom	~500	Simulation case	Log files (statistics)	Directly	Accurate energy and medium	Sensing
SNAP	FPGA emulator	<100[b]	FPGA configuration	FPGA debug interface (node)	Directly	Emulation approach	Medium, sensing, scalability

[a]TOSSF I/O is not specified, but SWAN I/O is assumed due to the relation of the simulators.
[b]SNAP emulator boards can be chained to increase scalability, but the evaluation of the concept is not published.

from a legacy computer network simulator incorporate a modular protocol stack. Graphical design of protocols and applications is possible only in VisualSense, in which the high-abstraction level designs can also be embedded to node platforms.

Generally, the networking oriented simulators are more suitable for network-wide evaluation, whereas the strength of the sensor node simulators lies in testing and optimization of single node operation. SensorSim and senQ implement the most comprehensive simulation environment with accurate battery and sensing models.

Part III

WSN PROTOCOL STACK

6

Protocol Stack Overview

Network protocols are typically divided into several distinct layers according to their responsibilities, which together form a protocol stack. Each layer has precisely defined interfaces, which permits flexible updates and changes in the software and hardware implementations in a modular manner.

An Open Systems Interconnection (OSI) model (Stallings 2004) is a widely used protocol stack that comprises seven layers: application, presentation, session, transport, network, data link, and physical layers. Application layer is the topmost layer that utilizes other layers. Presentation layer converts application data so that both communication ends understand it, a simple example being when end nodes use different byte order. Session layer handles session and connection coordination and hides errors that occur on lower layers from the application. Transport layer manages the reliability of the end-to-end data transfer and implements flow control to avoid congestion. Network layer forwards data through the network by forming a multihop routing path. Data link layer prepares data for physical layer, multiplexes data streams, performs data frame detection, medium access, and error control. Finally, physical layer connects the protocol stack to the transmission medium.

6.1 Outline of WSN Stack

Figure 6.1 compares the OSI (Stallings 2004) model with a widely utilized WLAN computer protocol stack (Kuorilehto et al. 2005b). In a WLAN computer, the Transmission Control Protocol (TCP)/Internet Protocol (IP) stack is used through a socket's API. The WLAN adapter that contains the MAC protocol and the WLAN radio is accessed via a device driver.

Because the requirements for WSN applications differ significantly from desktop applications, WSNs do not utilize all the layers defined in the OSI model. A WSN protocol stack is typically presented by five layers, as presented in Figure 6.1. Most notably, session layer is not usually utilized and a WSN transport protocol rarely utilizes end-to-end flow control. The middleware layer implements API for WSN applications and may contain sophisticated functionality for task allocation and resource sharing. It is often seen as a part of application layer.

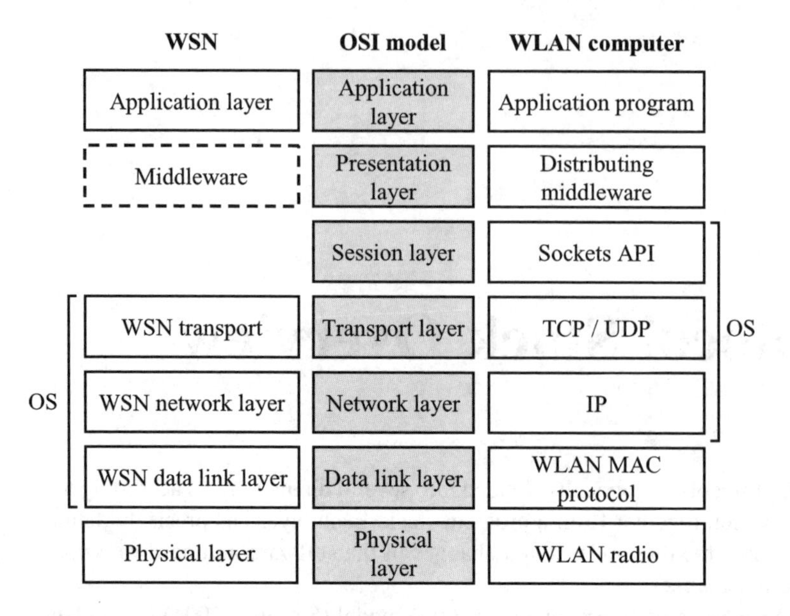

Figure 6.1 OSI model, WSN, and distributed system in WLAN protocol layers (Kuorilehto et al. 2005b).

In WSNs, the most essential protocol layers are the physical layer PHY, the MAC protocol on the Data Link Layer (DLL), and the routing protocol on the network layer. Next, we present the five-layered WSN protocol stack in detail.

6.1.1 Physical Layer

Physical layer (PHY) implements a network communications hardware, which transmits and receives messages one bit or symbol at a time. In practice, the PHY receives analog symbols from the medium and converts them to digital bits for further processing in the higher layers of the protocol stack. The PHY functions available in most transceivers are the selection of a frequency channel and a transmit power, the modulation transmitted and demodulation of received data, symbol synchronization and clock generation for received data.

A PHY transceiver may also include additional functions, which reduce the processing requirements of MCU. For example, an IEEE 802.15.4 compliant PHY includes: data frame synchronization for perceiving the start of an incoming frame; clear channel assessment for detecting ongoing traffic in a frequency channel; Received Signal Strength Indicator (RSSI) and Link Quality Indication (LQI) for measuring signal strength and estimating link quality to neighboring nodes; Cyclic Redundancy Check (CRC) calculation for checking bit errors on received frames; data encryption/decryption for improving network security; and automatic acknowledge transmissions after received frames. Since these features are implemented most efficiently in physical layer, they can improve overall network energy efficiently. Yet, the increased complexity increases hardware costs. In practice, the lowest power COTS transceivers available today include only some of these features.

6.1.2 Data Link Layer

The data link layer (DLL) interfaces with the physical and network layers. It typically consists of a MAC sublayer and a Logical Link Control (LLC) module. The MAC sublayer provides a fair mechanism to share access to the medium among other nodes. The MAC sublayer determines how and when to utilize PHY functions. Hence, MAC plays a key role in the maximization of a node's energy efficiency.

The LLC operates above MAC and is responsible for encapsulating message segments into frames and adding appropriate header information, with destination and source addresses and control and sequencing information and CRC calculation. According to this information, a desired destination node can receive frame, ensure frame integrity, and maintain proper sequencing of frames.

6.1.3 Network Layer

The network layer is responsible for network self-configuration and data routing. For configuring network topology, the network layer selects an appropriate operation mode for a node and determines the most suitable neighbors with which to associate and form communication links. The network topology is updated after link failures or at regular intervals for assuring network connectivity and optimizing network lifetime by balancing energy consumption among other nodes in the network.

A routing protocol executed in the network layer performs end-to-end data routing. The routing protocol decides a suitable next-hop node in which to forward each data frame, such that the frame eventually reaches its desired destination. Hence, the source and destination nodes are connected together by a chain of hops. For maximizing network lifetime, it is important to evaluate an optimal next-hop node for each data frame according to energy budget and delay requirements. Yet, the energy costs of lower protocol layers caused by associations and neighbor discoveries should also be considered.

6.1.4 Transport Layer

The transport layer performs flow control to regulate traffic flow through the network according to observed congestion. The transport layer is also responsible for upper layer error control to detect missing or corrupted frames not perceived by DLL. Due to the low transmission power levels and harsh operation conditions in WSNs, link reliability is much worse than in conventional wired and wireless networks. Thus, it is more feasible to perform flow and error control separately for each hop than from end-to-end, as in conventional networks (Karl and Willig 2005).

In addition, the transport layer performs fragmentation for dividing upper layer application data into small segments suitable for DLL. On the other hand, the transport layer reorders and reassembles received data segments into data packages applicable for application layer.

6.1.5 Application Layer

The application layer offers network services and the actual functionality for the node by network applications. The application layer typically contains several processes executed in parallel, for example sensing applications for various sensors, actuator and node diagnostics applications, and network configuration application. Application layer protocols define the format of exchanged messages and the order of message exchanges between different processes.

7

MAC Protocols

Generally energy efficiency can be achieved at various layers of the communication protocol stack. The bulk of energy-related research has focused on the physical layer, for example (Havinga and Smit 2000). Due to fundamental physical limitations, the focus of WSN research has recently shifted to the communication protocol level.

A MAC sublayer is the lowest part of data link layer and it operates on top of the physical layer. A MAC protocol manages radio transmissions and receptions on a shared wireless medium and provides connections for overlying routing protocol. Hence, it has a very high effect on network performance and energy consumption.

7.1 Requirements

The design requirements of WSN MAC protocols differ completely from the MAC protocols of traditional wireless computer networks, as presented in Table 7.1. While the latter pursue to maximize achieved throughput, low-energy WSN MAC protocols are aiming to maximize the following considerations (Gang Lu et al. 2004):

- **Energy efficiency**: The replacement or recharging of batteries is difficult due to a very large number of nodes, which may be deployed on rough operation environments. Thus, the applicability of WSNs is adequate only if the network lifetime is at least several months.
- **Adaptivity**: Due to harsh and uncertain operation environments, robust and energy-efficient operation should be maintained, while the network size, topology, and RF propagation conditions vary.
- **Scalability**: MAC protocol should tolerate and maintain high energy efficiency and performance in very large and dense networks, as well as in small and sparse networks.
- **Fairness**: MAC protocol should ensure that sinks receive information from all sources equally.

Table 7.1 Opposing MAC requirements for wireless computer networks and low-energy WSNs.

Requirement	Criticality for wireless computer networks	Criticality for low-energy WSNs
Energy efficiency	Lowest	Highest
Adaptivity	Low	High
Scalability	Moderate	High
Fairness	Moderate	Moderate
Latency	High	Low
Throughput	Highest	Lowest

- **Latency**: The required source-to-sink latency depends on the application. In environmental monitoring applications the latency may be even tens of seconds. In surveillance applications a detected event should be reported as soon as possible.

- **Throughput**: Throughput requirement depends on the application, too. In most WSN applications a sufficient throughput may be around few kilobits per a second.

- **Feasibility**: Protocol should be feasible also in practice, as the PHY layer and antenna radiation patterns are nonideal and available resources are very constrained.

Among these important requirements for MACs, energy efficiency is typically the primary goal in WSN. The power consumption can be reduced to some extent by a multihop routing. The reduction of the transmission distance to one half requires only the transmission power of 1/8 to 1/4 depending on the path loss exponent (Seidel and Rappaport 1992) of the operation environment. Thus, data is routed in the network by several quite low energy hops. As the transceiver consumes static power in reception and transmission modes regardless of the used transmission power level, the advantage of multihop routing is limited. Thus, the reduction of a transmission power level to below 10 mW does not bring a significant energy saving with typical low-power transceivers.

To be able to reach adequate energy efficiency, a MAC protocol should be able to minimize the following considerations (Karl and Willig 2005):

- **Idle listening**: Idle listening occurs when a node is actively receiving a channel, but there is no meaningful activity on the channel, which results in wasted energy.

- **Collisions**: When two nodes transmit simultaneously at the same frequency channel their transmissions collide in the overlapping area of their transmission ranges. The data received in this area is more likely to be corrupted resulting in unnecessary receiving costs at the destination node and unnecessary transmitting costs at the source node.

- **Overhearing**: A unicast transmission on a shared wireless broadcast medium may cause nodes other than the intended destination to receive a data packet, which is most probably useless to them and consumes unnecessary energy.

- **Protocol overheads**: As WSNs utilize relatively short packets, protocol headers and trailers may be even longer than the actual transmitted payload data, causing significant energy consumption. In addition, the energy consumption caused by control packet exchange, such as RTS/CTS, may be very significant.

Furthermore, the transient change of operating mode may dissipate a lot of energy, e.g. from sleep mode to transmit mode, which should also be considered, (Benini et al. 2004; Raghunathan et al. 2002; Warneke et al. 2001).

As a principle, the highest energy efficiency is achieved when a source and a destination node are activated and tuned to the correct RF channel simultaneously, while other nodes remain in sleep mode. This is very difficult in large and resource constrained WSNs where the network topology is constantly changing. The design requirements can be reached only by careful cross-layer design, where each protocol layer is designed with an accurate knowledge of the influences to lower protocol layers, and finally to node energy and memory consumption and network performance.

7.2 General MAC Approaches

According to the channel access mechanism, MAC protocols can be categorized into contention and contention-free protocols. A third category of protocols are multichannel protocols that aim to improve network performance by employing multiple frequency channels.

7.2.1 Contention Protocols

In contention protocols, nodes compete for a shared channel avoiding frame collisions by randomizing transmission moments and possibly by carrier sensing for detecting ongoing transmissions. Contention protocols typically allow peer-to-peer network topology, where all nodes are logically homogeneous and can form communication links with any other node in their range.

ALOHA (Roberts 1975) is the simplest contention protocol where nodes operate without any coordination. Nodes continuously listen to the channel for possible incoming data, except their own data transmission moments, which occur on demand. When the traffic is high, risk of collision with other transmitting nodes limits the maximum throughput to below 20% of the available capacity. A slotted ALOHA doubles the achieved throughput by dividing time into slots. When a packet is ready for transmission, a node waits until the start of the next slot. This eliminates collisions caused by partially overlapped frame transmissions but requires clock synchronization.

One of the initial MAC protocols designed for ad hoc multihop wireless networks was the DARPA Packet Radio Network (PRNET) (Kahn et al. 1978). The PRNET combines slotted-ALOHA and Carrier Sense Multiple Access (CSMA) MAC protocols. PRNET introduced many of the functionality utilized in current low-power MAC protocols such as random delays, link quality estimation, and low-duty cycle through node synchronization (Polastre et al. 2004a).

CSMA (Kleinrock and Tobagi 1975) protocol reduces collisions and further improves achievable throughput by checking channel activity prior to transmission. If the channel is

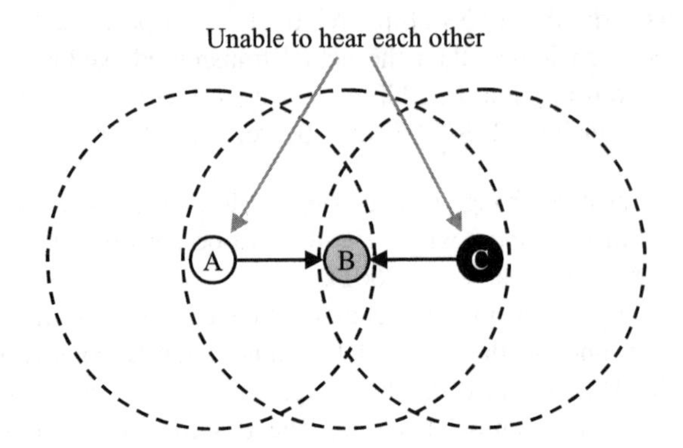

Figure 7.1 Hidden-node problem, where A and C transmit simultaneously and transmissions collide at destination B.

idle, node starts transmission. If the channel is busy, a transmission is delayed and retried later. Although CSMA achieves higher throughput than slotted ALOHA, collisions may still occur due to a hidden-node problem (Tobagi and Kleinrock 1975). Nodes separated by two hops cannot detect each other and their transmissions may collide with the receivers of their intermediate nodes. The hidden-node problem is illustrated in Figure 7.1. Here, node A begins transmitting a packet to node B. However, since node C is out of range of node A, it cannot detect A's transmission. Occasionally, C begins transmitting a packet while A is still sending. This results in collisions at the receiver B. Observe that neither of the two senders is aware of the collision and therefore cannot take preventive measures.

Several authors have developed different solutions to the hidden-terminal problem. Multiple Access with Collision Avoidance (MACA) (Karn 1990) protocol reduces the hidden node problem by performing a Request-To-Send (RTS)/Clear-To-Send (CTS) handshaking prior to data frame transmissions. This is illustrated in Figure 7.2.

Whenever a node wishes to transmit a packet to a neighbor, it first transmits a RTS message. If the desired neighbor is ready for receiving a frame, it responds with a CTS message. Upon receiving the CTS message, the sender begins transmitting the packet. Other nodes that hear the RTS or CTS frame hold off transmitting until the frame has been transmitted

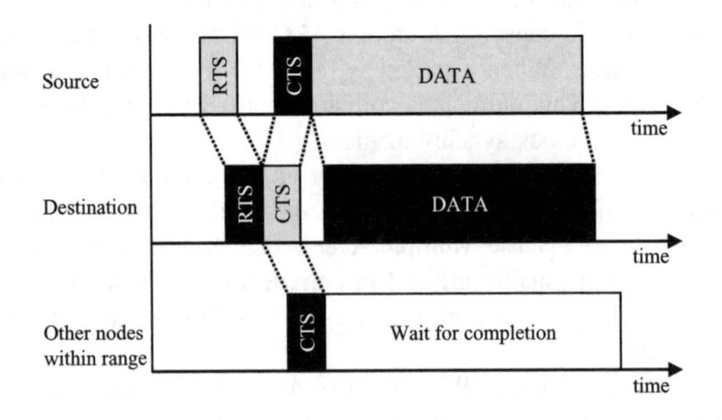

Figure 7.2 Collision avoidance by RTSRTS/CTS exchange.

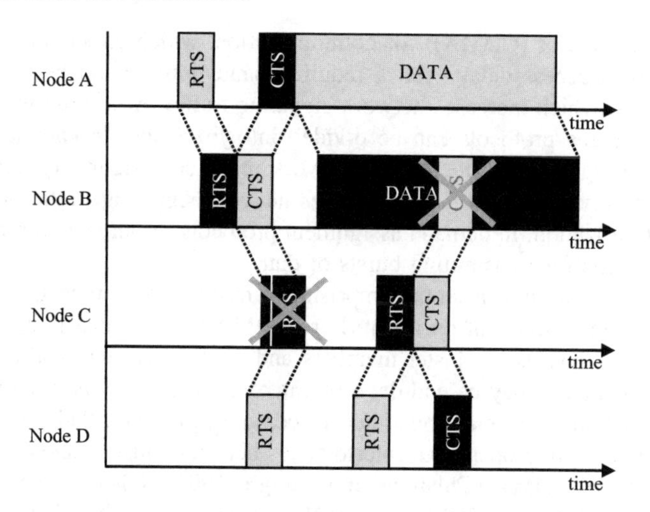

Figure 7.3 Collision problem with MACA protocol.

completely. This is possible since both RTS and CTS frames contain the length of the packet. It is possible that the RTS message or its CTS may suffer a collision. In this case the sender executes a binary exponential backoff algorithm and tries to send a RTS again later.

Although RTS/CTS handshaking reduces collisions, it may also collide with data frames. This problem is illustrated in Figure 7.3. In this example, the nodes can hear only their immediate neighbors. In the figure, node A begins a data transmission by transmitting RTS to node B. Node B responds with a CTS. Almost simultaneously, node D is trying to transmit data to node C by sending a RTS. This RTS frame collides with the CTS of node B, and node C cannot receive either of these frames. However, node A receives CTS correctly and begins transmitting a data frame. As node D does not receive CTS from node C, it retransmits a RTS frame after some time. Since node C does not know anything about the data transmission between nodes A and B, it responds with a CTS, which collides with the data frame at node B. This significantly reduces achieved throughput since the recovery of collisions is performed by a transport layer.

To be able to detect these collisions, Floor Acquisition Multiple Access (FAMA) (Fullmer and Garcia-Luna-Aceves 1997) protocol proposes the use of a longer CTS frame than RTS frame. Thus, at least a part of CTS can be detected. For improving throughput, Media Access protocol for Wireless LANs (MACAW) (Bharghavan et al. 1994) protocol uses link layer ACKs. Hence, the delay from a collision to a retransmission is minimized. The link layer ACKs are also adopted in the IEEE 802.11 standard.

7.2.2 Contention-free Protocols

While contention-based protocols work well under low traffic loads, their performance degrades drastically under higher loads because of collisions and retransmissions. In practice, channel capacity should be over-dimensioned for achieving a required throughput. This problem can be solved by contention-free protocols. In contention-free protocols nodes get separate time slots (Time Division Multiple Access (TDMA)), frequency channels (Frequency Division Multiple Access (FDMA)), or different spreading codes of PHY (Code

Division Multiple Access (CDMA)) for communication, which effectively eliminates collisions and channel access delay. Yet, a required bandwidth should be reserved prior to data transmissions, which increases network signaling traffic. According to the reservation policy, contention-free protocols can be divided into *fixed* and *demand* assignment protocols. In fixed assignment protocols the network resource assignment between nodes is long term and relatively constant. This reduces network signaling traffic but may lead to suboptimal slot utilization. In demand assignment protocols resource assignments are made on a short-term basis for transmitting bursts of data.

TDMA divides time into n slots comprising a frame, which repeats cyclically. Only one node is allowed to transmit on a slot. Pure TDMA requires that nodes form clusters, where each cluster consists of cluster members and a single cluster head acting as a base station. Cluster members may be end-devices, intercluster gateway nodes, or cluster heads of other clusters. End-devices are restricted to communicate with their cluster head only. For assuring collision-free operation, interferences between clusters should be handled by some form of spatial TDMA (Chlamtac and Farag'o 1994; Nelson and Kleinrock 1985), for example using FDMA or CDMA across clusters and then to run independent TDMA schedules within each cluster (Heinzelman et al. 2002). TDMA protocols are attractive for low-power applications since energy is not wasted for collisions and overhearing. Also, nodes may sleep during the slots assigned to other nodes, which reduces idle listening. When new nodes join or old nodes leave the cluster or if the bandwidth requirements for nodes change, then the cluster head must adjust the frame length and slot allocation, which limits scalability and adaptivity. In addition, an accurate synchronization is required increasing the complexity of implementation. Moreover, the maximum number of cluster members in a cluster may be limited. TDMA is very suitable for applications requiring high reliability and guaranteed bandwidth. TDMA is dominating in cellular systems, such as GSM (Rappaport 1996).

7.2.3 Multichannel Protocols

The RTS/CTS exchange results in the *exposed terminal problem* in multihop networks. This is presented in Figure 7.4. In the figure, node B is transmitting data to A. At the same time, node C wants to transmit data to D. However, it detect B's transmission by carrier sensing and must wait until the transmission is complete. However, since D is outside the

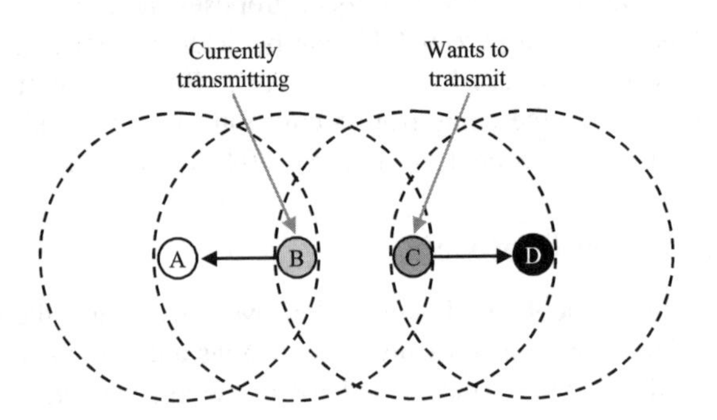

Figure 7.4 Exposed-terminal problem where C unnecessarily suspend its transmission to D until B has completed its transmission to A.

radio range of node B, the transmission would not cause any interference to node D, and D would be able to receive data correctly from C.

In practice, the carrier sensing and RTS/CTS exchange forces most of nodes at two-hops distance from the transmitting node to suspend their data exchanges until the transmission is complete. This reduces network performance unnecessarily since most of the transmission could be carried out without interference.

The utilization of multiple channels for data exchange allows multiple communication to take place in the same region simultaneously. There are many approaches and some of them are briefly discussed next.

Multichannel MAC (MMAC)

MMAC (So and Vaidya 2004) protocol is designed for infrastructure-less ad hoc networks. MMAC enables nodes to dynamically negotiate channels for communication. The main idea is to divide time into fixed-time intervals bounded by beacons, and have a small window at the start of each interval to indicate traffic and negotiate channel usage during the interval using a common signaling channel. After that, nodes switch to their agreed channel and exchange messages on that channel for the rest of the beacon interval. In order to avoid the multichannel hidden terminal problem, nodes need to be synchronized so that every node starts each beacon interval at about the same time. An advantage of MMAC is that it requires only one transceiver for each host.

Hop Reservation Multiple Access (HRMA)

Hop Reservation Multiple Access (Tang and Garcia-Luna-Aceves 1999) is a multichannel protocol for networks using slow FHSS. The hosts hop from one channel to another according to a predefined hopping pattern. When two hosts agree to exchange data by an RTS/CTS handshake, they operate with a fixed frequency for communication. Other hosts continue hopping, and multiple communication can take place on different frequencies.

Receiver-Initiated Channel-Hopping with Dual Polling

Receiver-Initiated Channel-Hopping with Dual Polling (Tzamaloukas and Garcia-Luna-Aceves 2001) takes a similar approach to the HRMA, but the receiver initiates the collision avoidance handshake instead of the sender. These schemes can be implemented using only one transceiver for each host, but they only apply to frequency hopping networks and cannot be used in systems using other mechanisms such as DSSS.

Multichannel CSMA protocol

A multichannel CSMA (Nasipuri et al. 1999) protocol utilizes "soft" channel reservation. If there are N channels the protocol assumes that each host can listen to all N channels concurrently. A host wanting to transmit searches for an idle channel and uses that. Among the idle channels the one that was used for the last successful transmission is preferred. A variation of the protocol is presented by Nasipuri and Das (2000), where the best channel is selected based on signal power observed by the sender. A drawback of these protocols is that they require N transceivers for each host, which is very expensive.

Dynamic Channel Assignment (DCA)

DCA (Wu et al. 2000) is a multichannel protocol that assigns channels dynamically in an on-demand style. Nodes maintain one dedicated network signaling channel for control messages and other channels for data. Each host has two transceivers so that it can listen on the control channel and the data channel simultaneously. RTS/CTS packets are exchanged on the control channel and data packets are transmitted on the data channel. In RTS packet, the sender includes a list of preferred channels. On receiving the RTS, the receiver decides on a channel and includes the channel information in the CTS packet. Then, DATA and ACK packets are exchanged on the agreed data channel. Since one of the two transceivers is always listening on the control channel, the multichannel hidden-terminal problem does not occur. Yet, the energy consumption of constant reception is high. This protocol does not need synchronization and can utilize multiple channels with little control message overheads. But it does not perform well in an environment where all channels have the same bandwidth. When the number of channels is small, having one channel dedicated to control messages can be costly. On the other hand, if the number of channels is large, the control channel can become a bottleneck and prevent data channels from being fully utilized. Jain and Das (2001) propose a protocol that uses a scheme similar to DCA (Wu et al. 2000), which has one control channel and N data channels but selects the best channel according to the channel condition at the receiver side. The protocol achieves throughput improvements by intelligently selecting the data channel but still has the same disadvantages as DCA.

7.3 WSN MAC Protocols

The energy efficiency of the above mentioned approaches is not adequate for WSNs. The main problem of most protocols designed for wireless computer networks is that the receiver must always be on. As the power consumed when listening to an idle channel is the same as the power consumed when receiving data, this method is very power-inefficient (El-Hoiydi 2002a). Further energy saving is achieved by duty cycling.

In low duty-cycle MAC protocols a node's operating time is divided into active–sleep cycles. Each cycle consists of an active duration for regular transmissions, receptions, and a sleep duration during which the radio interface is powered off. The smaller the duty cycle, the lower the energy consumption due to idle listening. The main goal of all protocols in this category is to operate at the smallest possible duty cycle while being able to support application loading requirements. In an ideal case, the sleep state is left only when a node is about to transmit or receive packets (Karl and Willig 2005).

Low duty-cycle MAC protocols utilize virtually the same channel access mechanisms in active periods as the above mentioned approaches. The low duty-cycle MAC protocol can be divided into two categories: synchronized and unsynchronized protocols, according to the synchronization of data exchanges.

7.3.1 Synchronized Low Duty-cycle Protocols

Synchronized low duty-cycle MAC protocols utilize predetermined, periodic wakeup schedules for data exchanges consisting of a sleep period T_{sleep} and active period T_{active} repeated

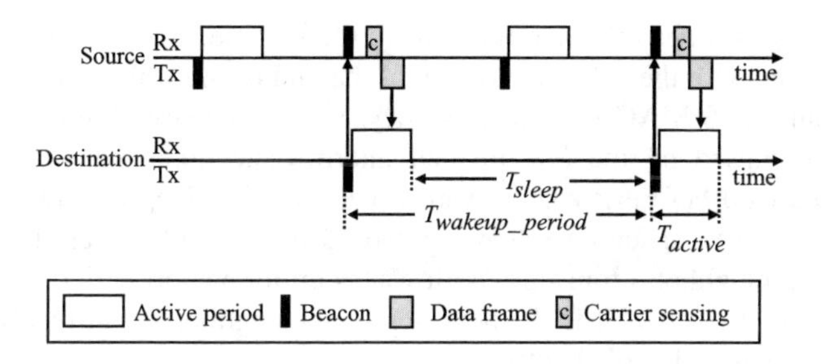

Figure 7.5 Operation of synchronized low duty-cycle protocols.

at $T_{\text{wakeup\_period}}$ intervals, as presented in Figure 7.5. The active period is denoted as a superframe, where a node typically first broadcasts a beacon frame for signaling its schedule and status information, and then listens to the channel for possible incoming data during a channel access period. Due to a synchronized operation, nodes know in advance the exact moment of the active period, and thus, can receive frames with minimum active reception time. As a global synchronization is very difficult in large networks (Lin and Gerla 1997), synchronization is maintained with one or more neighboring nodes by periodically receiving their beacons. To establish a synchronized operation, neighboring nodes are typically discovered by a network scan. The network scan means receiving a long-term reception beacon from neighboring nodes since their schedules and frequency channels are unknown. If all neighbors need to be detected, a node must receive all possible frequency channels for at least one beacon transmission interval. However, the operation in a static network after synchronization is very energy efficient, and 0.1–1 mW average power consumptions are achievable (Kohvakka et al. 2005c; Kohvakka et al. 2006b). In a dynamic network where neighbors and routes change frequently, energy-efficient operation is much more difficult to achieve.

Sensor-MAC (S-MAC)

S-MAC (Ye et al. 2002) is one of the most common synchronized low duty-cycle MAC protocols. The protocol utilizes a fixed 115 ms length active period and an adjustable, long-term (hundreds of milliseconds), network specific wakeup period. Thus, nodes can turn their interfaces off during the sleep duration to save idling energy and they can communicate during the wake periods. By decreasing the active–sleep duty cycle, this protocol can reduce energy consumption with the cost of increased latency and reduced effective channel bandwidth. Active periods are used for transmitting and receiving data to/from neighboring nodes. During the active periods nodes execute an 802.11-like CSMA-CA MAC protocol. Neighboring nodes try to coordinate their active periods to occur simultaneously. This way, nodes form virtual clusters. An active period is divided into three phases, namely Synchronization (SYNC), RTS, and CTS. In SYNC phase, a node receives SYNC frames corresponding beacons from its neighbors containing their time schedules. SYNC frames are transmitted periodically (synchronization period) using CSMA. In RTS phase, the neighboring nodes can send RTS frames using CSMA. A node selects a desired source node according to the received RTS frames and then transmits a CTS frame. The CTS phase is continued by frame exchanges with the selected node until the end of the wakeup period.

A shortcoming of S-MAC is a sleep latency, which increases frame delivery latency. Sleep latency is caused by the fact that an intermediate node may have to wait until the receiver wakes up before it can forward a packet. The sleep latency problem can be reduced by the adaptive listening scheme proposed in S-MAC (Ye et al. 2004). When a node overhears its neighbor's transmission it wakes up for a short period of time at the end of the transmission, so that if it is the next-hop of its neighbor it can receive the message without waiting for its scheduled active time.

Timeout-MAC (T-MAC)

A problem with S-MAC is poor adaptation to changing traffic conditions as the lengths of wake-up and active periods cannot be changed during operation. T-MAC (van Dam and Langendoen 2003) protocol is a variation of the S-MAC, which solves the fixed duty cycle problem by adjusting the active period length according to traffic. T-MAC defines a short (typically 15 milliseconds) time-out window as the maximum idle listening time following CTS phase and each received frame. If no activity occurs during the time-out window, the node returns to sleep. Thus, the length of the active period is adjusted according to traffic. T-MAC saves power at a cost of reduced throughput and additional latency. In variable workloads, T-MAC uses one-fifth of the power of S-MAC (van Dam and Langendoen 2003). In homogeneous workloads, T-MAC and S-MAC perform equally well. A problem of T-MAC is that shortening the active window reduces the ability to snoop on surrounding traffic and adapt to changing network conditions.

DMAC

DMAC (Gang Lu et al. 2004) is an energy-efficient, low-latency MAC protocol for tree-based data gathering in wireless sensor networks. DMAC utilizes this data gathering tree structure to achieve both energy efficiency and low packet-delivery latency. DMAC staggers the active/sleep schedule of the nodes in the data gathering tree according to its depth in the tree. This allows continuous packet forwarding flow in which all nodes on the multihop path can be notified of the data delivery in progress as well as any duty-cycle adjustments.

Data prediction is employed to solve the problem when each single source has a low-traffic rate but the aggregated rate at an intermediate node is larger than what the basic duty cycle can handle. The interference between nodes with different parents could cause a traffic flow to be interrupted because the nodes on the multihop path may not be aware of the interference. The use of a More-to-Send (MTS) packet is proposed to command nodes on the multihop path to remain active when a node fails to send a packet to its parent due to interference.

Self-Organizing Medium Access Control for Sensor Networks (SMACS)

SMACS (Sohrabi et al. 2000) is another proposal for synchronized low duty-cycle protocols. SMACS utilizes frequency and time division between communication links. Nodes perform neighbor discovery at semi-regular intervals by entering a random access BOOTUP mode (Sohrabi and Pottie 1999). In the BOOTUP mode they broadcast invitation messages on a common fixed-signaling channel for inviting new nodes to join the network, and listen to the channel for possible responses and other invitation messages. Since nearly continuous radio reception is required and data exchanges are suspended during the BOOTUP mode, network performance and energy efficiency drop rapidly with increased network dynamics. The operation of mobile nodes in the field of stationary nodes is permitted by an Eavesdrop-And-Register (EAR) algorithm. Mobile nodes register stationary nodes in their range by receiving broadcast invitation messages on the signaling channel. Yet, the algorithm is energy inefficient since the channel must be listened nearly constantly. However, SMACS enables quite high scalability due to frequency division and scheduled data exchanges.

Node Activation Multiple Access (NAMA)

NAMA (Bao and Garcia-Luna-Aceves 2001) uses a distributed election algorithm to achieve collision-free transmissions. For each time slot, NAMA selects only one transmitter per two-hop neighborhood and hence all nodes in the one-hop neighborhood of the transmitter are able to receive data collision-free. However, NAMA does not address energy efficiency, and nodes that are not transmitting switch to receiver mode.

Traffic-Adaptive Medium Access (TRAMA)

TRAMA (Rajendran et al. 2003) is similar to the NAMA protocol (Bao and Garcia-Luna-Aceves 2001) in that the identifiers of the nodes within a two-hop neighborhood are used to give conflict-free access to the channel to a given node during a particular time slot. TRAMA addresses energy efficiency by having nodes going into sleep mode if they are not selected to transmit and are not the intended receivers of traffic during a particular time slot. Furthermore, TRAMA uses traffic information to influence the schedules, which makes TRAMA adaptive to the sensor network application. For instance, an event-tracking application will likely generate data only when an event is detected. On the other hand, monitoring applications may generate data continuously. In either case, TRAMA can adapt its schedules accordingly, delivering adequate performance and energy efficiency.

TRAMA utilizes two assisting protocols: a Neighbor Protocol (NP), which signals a list of node identities from a two-hop neighborhood, and a Schedule Exchange Protocol (SEP), which is used for exchanging node schedules. These protocols are used for commanding a defined set of neighbors to receive a given data frame at a defined time moment. This provides efficient unicast, multicast (i.e. transmitting to only a set of one-hop neighbors), and broadcast (i.e. transmitting to all one-hop neighbors) transmissions. TRAMA updates neighbor information during periodic and relatively long-term random access periods, during which special signaling frames with all neighbors are exchanged. The major advances of TRAMA are that it is inherently collision-free as its medium access control mechanism is schedule-based, and that the duty cycle is adapted dynamically to current traffic patterns.

Low-Energy Adaptive Clustering Hierarchy (LEACH)

LEACH (Heinzelman et al. 2002) is a scheduled MAC protocol with clustered topology. The nodes organize themselves into local clusters with one node acting as the local cluster head. LEACH utilizes a single base station as a root of the network with which all cluster heads employ only direct communications using high-transmission power levels. Thus, a star topology is utilized in two hierarchical levels. Due to the energy consuming communication between cluster heads and the base station, LEACH proposes the use of data fusion in cluster heads to compress the amount of data being sent to the base station.

If the cluster heads were chosen a priori and fixed throughout the system lifetime, as in conventional clustering algorithms, the nodes chosen to be cluster heads would drain their batteries quickly, ending the useful lifetime of all nodes belonging to those clusters. To distribute energy consumption more evenly, LEACH proposes to rotate cluster heads randomly. Nodes elect themselves to be local cluster heads with a certain probability based upon the global knowledge of expected average number of clusters. These cluster heads broadcast their status to the other nodes in the network. Each node determines a cluster to associate with by choosing the cluster head that requires the minimum communication energy. Once all the nodes are organized into clusters each cluster head creates a schedule for the nodes in its cluster. This allows that each cluster member node can switch to a sleep mode at all times except during its transmit time; thus, minimizing its energy consumption.

LEACH assumes that all wireless nodes could reach the base station and that each node can be heard by all other nodes if it uses the maximum transmit power. In practice, this assumption is seldom true. Furthermore, each node has to know a priori the optimum average number of clusters for the entire network, which depends on the global topology and node density. This parameter directly determines the probability of a node becoming a cluster head at each round of the cluster head selection phase (Heinzelman et al. 2000). Since the optimum average number of clusters changes with network topology, LEACH does not support dynamically changing network topology due to node failures, mobility, or addition of new nodes to the network.

Power Aware Clustered TDMA (PACT)

PACT (Pei and Chien 2001) is a clustered MAC protocol, that adapts the duty cycles of the nodes to user traffic where radios are turned off during inactive periods. PACT uses passive clustering (Gerla et al. 2000) that allows nodes to take turn as the communication backbone nodes. Each cluster consists of a cluster head, a few intercluster gateway nodes, and ordinary nodes. Active gateway nodes relay the data traffic between clusters. Thus, the data is multihop-routed using short-range communication. The other gateway nodes will stand by to preserve energy for future use. The number of active gateways is limited by using a simple selection scheme, where the node having the highest number of distinct cluster heads is selected as the active gateway. The role of the cluster heads and gateways are rotated to conserve energy. Unlike LEACH, PACT changes the clustering status of a node in a distributed fashion based on its battery energy level. PACT employs a modular architecture, which minimizes the requirements for utilized network topology or routing protocol.

Self-Organizing Slot Allocation (SRSA)

SRSA (Wu and Biswas 2005) protocol is a TDMA-based MAC protocol that can minimize slot allocation interferences across overlapping clusters without the use of CDMA or FDMA, or any synchronization between clusters. SRSA operates at the cluster level and does not rely on any information beyond a cluster.

The network topology of SRSA is similar to LEACH (Heinzelman et al. 2000), where cluster heads are in direct communication with a base station. An exception is the number of base stations. As LEACH utilizes a single base station, SRSA assumes multiple base stations interconnected by a wired network.

SRSA specifies only intracluster communication, while a separate mechanism is necessary for the base station to cluster head communication. After the election of cluster heads each cluster maintains a local TDMA MAC frame. Each regular sensor node is allocated TDMA slots for uplink transmissions as well as for downlink receptions. After a node receives its slot allocation, it remains active during the allocated slots and sleeps during all other slots. The main idea is to initiate communication with a random initial TDMA allocation and then adaptively change the slot allocation schedule locally based on feedback derived from collisions experienced by the local nodes within a cluster. Reliance only on the local information ensures the scalability of SRSA over very large networks.

7.3.2 Unsynchronized Low Duty-cycle Protocols

Unsynchronized protocols utilize frequent channel sampling for detecting possible starting transmissions, as presented in Figure 7.6. Each transmitted frame typically begins with a long preample (T_{preample}) lasting longer than the channel sampling interval (T_{interval}). Hence, the periodic channel sampling can detect the preample of each transmission prior to actual data transmission. After the detection of a preample the node simply listens to the channel until data is received. Transmissions may be on a random and on-demand basis. Since nodes do not need to remember the schedules of their neighbors, unsynchronized protocols are relatively simple and require a small amount of memory compared to synchronized protocols. Common to unsynchronized low duty-cycle MAC protocols is that transmitting a data frame consumes more energy than receiving it. This approach is well suited for WSNs, which require low data rates and have burst type traffic. A drawback is that energy consumption is higher than in synchronized protocols due to the frequent channel sampling, high overhearing and the transmission and reception energies of long preamples. With the current technology, the power consumption of unsynchronized low duty-cycle MAC protocols is in the order of 5–10 mW (Wong and Arvind 2006).

RF wakeup scheme

RF wakeup scheme by Hill and Culler (2002) is the first preample sampling protocol targeting to reduce idle listening cost. In the RF wakeup scheme the analog baseband of the radio

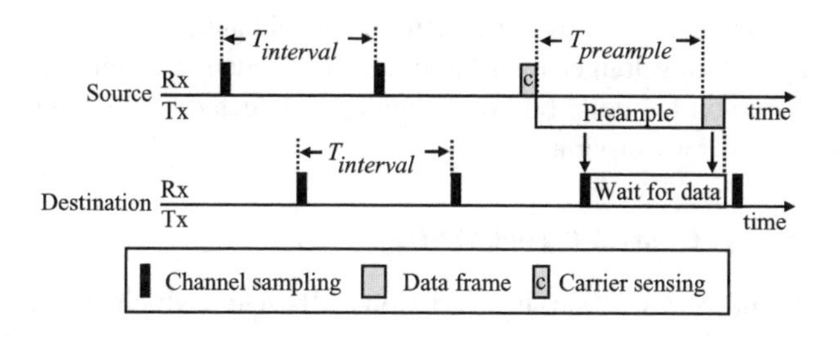

Figure 7.6 Operation of unsynchronized low duty-cycle protocols.

is quickly sampled for energy every 4 seconds. The performance of the protocol is demonstrated on an 800 node multihop network, where below 1% radio duty cycle is achieved.

ALOHA with preample sampling

ALOHA with preample sampling (El-Hoiydi 2002a) is a related technique that works on more complex radios. ALOHA with preamble sampling presents a low-power technique similar to that used in paging systems (Mangione-Smith 1995). To allow the receiver to sleep for most of the time when the channel is idle, nodes periodically wake up and check for activity on the channel. If the channel is idle the receiver goes back to sleep. Otherwise, the receiver stays on and continues to listen until the packet is received. A received frame is replied immediately by an acknowledgment frame. Nodes send frames randomly as in ALOHA but with long preamples to match the channel check period. A disadvantage of the preample sensing is increased transmission and reception lengths, and an increased probability of collision due to the longer transmissions. Collisions can be reduced by using the long preambles for first frames only for initial synchronization of nodes. Afterwards the nodes transmit and receive on a schedule with normal-sized packets, as in synchronized low duty-cycle protocols. Collisions can also be reduced by replacing ALOHA with CSMA, as first presented in (El-Hoiydi 2002b).

Wireless Sensor MAC (WiseMAC)

WiseMAC (El-Hoiydi et al. 2004) protocol is an iteration of ALOHA with preample sampling mechanism. The network consists of an access point and numerous sensor nodes and is formed in a star-topology. The WiseMAC protocol minimizes the transmission and reception energy consumption by letting the access point learn the sampling schedules of each sensor node. By knowing the schedules, the access point can start the preample transmission just prior to the channel sampling moment of a destination sensor, and hence minimize the preample transmission time. As the access point initiates each data transaction, collisions can be avoided and ALOHA channel access mechanism works well. The disadvantage of the protocol is star network topology, which reduces its suitability in WSNs.

Berkeley Media Access Control (B-MAC)

B-MAC (Polastre et al. 2004a) is a variant of CSMA with preample sampling mechanism. Channel sampling interval is application specific parameter. If the channel is sensed to be busy, node turns on its receiver until a data packet is received or a time-out is occurred. Prior to a data transmission, a source node performs clear channel assessment for CSMA and transmits a preample lasting longer than the duration of channel sampling interval. Hence, each data transmission on the neighborhood should be detected by channel sampling before the start of actual data transmission.

Speck Medium Access Control (SpeckMAC)

SpeckMAC (Wong and Arvind 2006) is a variation of B-MAC, which reduces the destination node channel reception time by transmitting redundant retransmissions to fulfill the role of the long preample of B-MAC. There are two variations of the protocol: Speck Medium Access

Control Backoff (SpeckMAC-B) and Speck Medium Access Control Data (SpeckMAC-D). The former variation sends successive short wakeup packets instead of the preample containing information about the destination address and an exact time to the beginning of the actual data packet transmission. After receiving one wakeup packet a node checks if the destination address matches its own address or a broadcast address, returns to sleep mode, and if needed, turns its radio on again at the beginning of the data transmission.

The SpeckMAC-D operates quite similar to the SpeckMAC-B, but sends the actual data packet repetitively with a short preample instead of the wakeup packet. Hence, a destination node needs only to receive until one data packet is received equaling averagely the time of one and a half packet lengths. This is typically significantly less than the channel sampling interval.

Zebra MAC (Z-MAC)

Z-MAC (Rhee et al. 2005) protocol is targeted to combine the strengths of TDMA and CSMA. Z-MAC adjusts its behavior according to the level of contention. Under low contention it behaves like CSMA, and under high contention more like TDMA.

The network is formed as a flat multihop topology. Z-MAC utilizes time-slotted channel access, as in TDMA. Time slot assignments are performed at start-up. Thus, the entire network should be deployed at the same time, and most nodes should have fixed locations during the entire network lifetime. At start-up, nodes perform a simple neighbor discovery protocol and exchange their neighbor lists with all neighbors, resulting in each node obtains a list of all two-hop neighbors. Slots are then assigned according to the known two-hop neighbors, such that each slot is assigned only to a single node in the two hops neighborhood, quite similarly to that in SRSA.

In contrast to TDMA, each node may transmit during any time slot. Z-MAC solves contention by performing CSMA in each slot. However, the owner of each slot gets the highest priority to access the channel by getting a shorter contention window size than other nodes. Thus, the slot owner always has an earlier chance to transmit in the slot. Other nodes can steal the slot if the slot owner does not have data to transmit. As a result, Z-MAC performs like a CSMA protocol, which uses a TDMA schedule as a hint to enhance contention resolution. In the worst case, protocol falls back to CSMA.

Since each node may transmit at any slot, all nodes should be ready for frame reception at every slot. Normally, this necessitates constant radio reception for all nodes sacrificing energy-efficiency. However, Z-MAC improves energy efficiency by utilizing the low-power listening scheme of B-MAC below the Z-MAC protocol. Thus, each frame is preceded by a long preample. Z-MAC improves the performance of B-MAC by reducing collisions and improving fairness.

7.3.3 Wake-up Radio Protocols

As presented in Guo et al. (2001), the utilization of a separate wake-up radio can effectively minimize idle listening time. The wake-up radio scheme is based on the assumption that the listen mode of the wake-up radio is ultra-low power and it can be active constantly. At the same time, the normal data radio is in power-down mode as long as there are no packet transmissions or receptions required. If a neighbor node wants to transmit a packet, it first sends a wake-up beacon over a wake-up channel to trigger the power-up of the normal radio and then sends the data packet over the data radio. This protocol is successful in avoiding overhearing and idle listening problems in the data radio, but it is unable to solve the collision problem. Moreover, the difference in the transmission range between data and wake-up radio may pose significant problems.

Power Aware Multi-Access protocol with Signaling (PAMAS)

PAMAS (Singh and Raghavendra 1998) protocol combines the MACA protocol with a wake-up radio on a separate signaling channel (Shong Wu and Li 1987; Tobagi and Kleinrock 1975). In PAMAS, the RTS/CTS message exchange takes place over a signaling channel that is separate from the channel used for packet transmissions. A node normally sleeps, but when it has data to transmit, it first sends a signal through another channel to wake up the receiver node. Since the only function of the radio for wake-up channel is to wake up sleeping nodes, the hardware for that interface can be simpler and have significantly lower data rate; and thus, be less energy hungry. For reducing energy consumption and overhearing, PAMAS utilizes a forced-interface sleep mechanism allowing a primary transceiver to sleep when there are no ongoing frame transmissions or receptions for the node. A drawback is that an extra transceiver is required, increasing the cost and complexity of the PHY layer.

Sparse Topology and Energy Management (STEM)

STEM (Schurgers et al. 2002) protocol trades energy savings for latency through listen/sleep modes as in Ye et al. (2002) but by using a separate radio. When a node wants to send a packet, it polls the target node by sending wake-up messages over a signaling channel. Upon receiving a wake-up message, the target node turns on its primary radio for regular data transmissions. The purpose of using a separate signaling channel is to prevent polling messages from colliding with ongoing data transmissions. This scheme is effective only for scenarios where the network spend most of its time waiting for events to happen. Otherwise, the polling through a stream of wake-up messages, collisions, and overhearing may cancel out the energy savings obtained by sleep modes.

A variation of STEM utilizes a signaling channel for transmitting “busy tones” using a separate transceiver. Thus, when a node wants to transmit a packet, it transmits the preamble of its packet (this contains the receiver’s address). The receiver responds by transmitting a busy tone in the signaling channel. On hearing the busy tone on the signaling channel, the sender continues sending the packet using the data radio. It is easy to see that the hidden terminal problems described previously are handled with this solution. Nevertheless, network throughput is reduced due to the exposed terminal problem.

7.3.4 Summary

The low-power WSN MAC protocols presented in this section are summarized in the Table 7.2.

Table 7.2 Comparison of low-power WSN MAC protocols.

Protocol	Network topology	No. of channels	Channel access	Idle listening avoidance	Synchronization	Major contributions
S-MAC	Flat[a]	1	CSMA	Periodic sleep	Yes	Active/sleep schedules
T-MAC	Flat[a]	1	CSMA	Periodic sleep, adaptive duty cycle	Yes	Adaptive duty cycle
DMAC	Flat	1	CSMA	Periodic sleep	Yes	Staggered active/sleep schedules
SMACS	Flat	Many	TDMA/FDMA	TDMA	Yes	Hybrid TDMA/FDMA
NAMA	Flat	1	TDMA	No	Yes	Distributed algorithm for slot assignments
TRAMA	Flat	1	TDMA	TDMA with adaptive duty cycle	Yes	Energy saving to NAMA
LEACH	Clustered	1	TDMA	TDMA	Yes	Low-energy clustering, rotation of cluster heads
PACT	Clustered	1	TDMA	TDMA with adaptive duty cycle	Yes	LEACH with multihop routing
SRSA	Clustered	1	TDMA	TDMA	Inside clusters	Adaptive slot assignments for reducing interferences
Z-MAC	Flat	1	TDMA/CSMA	Periodic sleep	Yes	Hybrid TDMA/CSMA above B-MAC
WiseMAC	Star	1	Random access	Periodic sleep	No	Learning of preample sampling schedules
B-MAC	Flat	1	CSMA	Periodic sleep	No	Preample sampling with CSMA.
SpeckMAC	Flat	1	CSMA	Periodic sleep	No	Reduction of preample reception time
PAMAS	Flat	2	CSMA	Wakeup radio	No	MACA with a wake-up radio scheme
STEM	Flat	2	CSMA	Wake-up radio with periodic sleep	No	Wake-up radio with active/sleep scheduling

[a] Virtual clusters.

8

Routing Protocols

A routing layer operates on top of the MAC layer and is responsible for delivering data packets via multiple hops from a source to a target. As a packet can often be routed through several paths, the routing decision has a great impact on load balancing between nodes, experienced reliability, and delay. The selected route construction and maintenance method dictates the energy efficiency of the routing and how well the mobility is supported.

8.1 Requirements

Routing in a WSN is fundamentally different from other wireless networks and has several unique requirements.

- **Data-centric communication**: Traditional networks use nodecentric communications where nodes exchange data with each other by using unique addressing. However, the routing in WSNs is typically data-centric. The data-centric communications are based on sensed data, while knowing exact node identifiers is not important.
- **Asymmetric traffic**: In WSNs, communication between arbitrary nodes is seldom required. Instead, the traffic is usually highly asymmetric as the majority of the data is routed from nodes acting as data sources to the nodes acting as data consumers (sinks). Typically, the number of sinks is much lower than the number of data sources. While most of the traffic is directed to the sink, the data sent to the nodes might actually be more important. For example, periodic environmental measurements sent by nodes can tolerate high delays and packet losses, as the lost data is replaced later with another measurement. However, it is imperative that control data sent by a sink is not lost as it may not be repeated and might contain important configuration information.
- **Resource constraints**: As sensor nodes are very resource constrained, the routing cannot use computationally heavy operations or store extensive routing-state information.

Ultra-Low Energy Wireless Sensor Networks in Practice: Theory, Realization and Deployment

- **Network lifetime**: Many uses of WSN require long-term deployment, thus emphasizing the importance of network lifetime. As each transmission and reception consumes energy, the routing protocol must keep messaging overheads to a minimum. A routing protocol must not only minimize route maintenance overheads but also try to deliver packets via the most energy-efficient route. However, simple routing schemes, such as shortest-path routing, are not desirable in WSNs. Instead, several aspects must be considered when choosing a route. Low-reliability links are likely to cause packet loss, thus potentially causing retransmissions and unnecessary energy loss. Since transmission energy is proportional to the square of the distance, sending data via an intermediate node might actually reduce the required transmission energy. In addition, minimizing the energy consumed by a single transmission is not enough, but a routing protocol must equal the total network energy consumption. Otherwise, the nodes that forward most traffic die first, which may partition the network and thus prevent nodes from communicating with the sink.

- **Scalability**: Routing protocol must be able to cope with networks consisting thousands of nodes. Due to memory limitations and large number of nodes the routing tables may not be used to maintain routing information. Therefore, distributed routing is required, preferably using nodes knowing only their immediate neighborhood.

- **Aggregation**: The data-centric nature allows the reduction of transmitted packets by aggregating data. In aggregation, the amount of data is either reduced with an aggregation function (such as average, sum, maximum, or minimum) or by concatenating several samples into single packet, thus reducing the overheads per sample.

- **Robustness**: WSNs operate on very error-prone environments. Transmission errors are common because a typical deployment contains obstacles and changing environmental conditions. In addition, a node can die unexpectedly due to low energy. Since the nodes are typically deployed randomly, some areas may be densely populated causing interference, while the nodes in sparsely populated areas have to utilize low-quality links. Therefore, a WSN routing protocol cannot rely on a single link and must use redundancy to achieve reliable routing.

- **Mobility support**: In a typical WSN network, the majority of the nodes are stationary. However, node or sink mobility may be required depending on application. In target tracking a node may be attached to the monitored target that moves around the network. Sink mobility is required when a mobile user collects data from the surrounding sensor network.

- **Application-specific behaviour**: As a WSN is highly application specific, the routing protocol must be tailored for different types of applications. For example, mobility tracking differs greatly from static environmental monitoring.

8.2 Classifications

WSN routing protocols can be classified based on their operation as nodecentric, data-centric, location based, negotiation based, query based, cost field based, or multipath based.

These classes are not mutually exclusive as a routing protocol may be both data-centric and query based while having features seen in location-based protocols.

The route construction is either *proactive*, *reactive*, or *hybrid*. In proactive routing routes are constructed before they are actually needed, while reactive routing creates routes only on a on-demand basis. Hybrid protocols use a combination of these techniques. Reactive routing is preferred in dynamic networks as precomputed routes may become invalid, thus requiring costly route maintenance. Reactive routing also requires less memory on nodes as all possible routes may not be active simultaneously. The drawback in reactive routing is the delay when routing the first packet caused by route construction. Also, it is usually more efficient to compute all routes at once rather than creating an individual route.

A routing protocol may utilize either *sender-decided* or *receiver-decided* packet forwarding. In sender-decided forwarding a node explicitly selects a next-hop neighbor that a packet is forwarded to. In receiver-decided forwarding a node broadcasts a packet to all its neighbors and neighbors determine independently whether they forward the packet or drop it. As several neighbors may decide to forward the packet, the packet may be unnecessarily duplicated. The benefit is that a node does not have to know its neighbors, thus reducing the need for memory-consuming routing tables.

Receiver-decided protocols use *flat* routing topology where a node does not know its neighbors, while sender-decided protocols usually form *tree* or *mesh* routing topology. In tree topology, a node knows its neighbors that are closer to the root of the tree, thus having a limited set of next-hop neighbors to choose from. The tree is rooted at the sink. In mesh topology, a node knows most or all of its neighbors and can make more advanced routing decisions.

8.3 Operation Principles

In this section, we describe WSN routing protocols according to their operation principles. Related work is presented on each category.

8.3.1 Nodecentric Routing

The nodecentric approach is used in wired and wireless ad hoc networks, thus allowing compatibility with the existing protocols and possibility to bridge data from other networks. However, the use of required global unique addressing scheme in WSNs is challenging. The reason that a decentralized address maintenance is preferred is because of the network size and error-prone nature of the sensor nodes. Separate network partitions may join the network, which makes maintaining a consistent addressing system rather energy consuming.

Related work

Dynamic Source Routing (DSR) (Maltz et al. 1999) is a reactive routing protocol targeted at wireless ad hoc networks. Route discovery is performed by broadcasting a route discovery packet to the network. As the packet is forwarded, each node adds its address to the packet. Thus, when the destination receives the discovery packet it can send a reply containing the exact path. DSR is unsuitable for WSNs because the packets should be as small as possible for energy reasons. Also, in large-scale WSNs a route may consist of tens or hundreds of hops.

8.3.2 Data-centric Routing

In data-centric routing the data is routed depending on its content rather than using sender or receiver identifiers. Data-aggregation can be naturally performed as the data-centric routing is already content aware.

Related work

Directed Diffusion (DD) (Intanangonwiwat et al. 2003) names data with attribute-value pairs. Initially, a data-collecting node (sink) requests a large data transmission interval, thus generating only light traffic. The interest is injected into the network where it gradually disseminates to each node. The sink refreshes its interest periodically to recover from unreliable interest propagation. When a node receives an interest it establishes a gradient towards the sender node. Once a source node obtains an interest, it generates the requested data and broadcasts it to the sink via multiple gradients as shown in Figure 8.1.

Loops are prevented by maintaining a data cache that contains recently received items and data rates for each gradient. After the sink receives the requested data, it reinforces one or a small number of the gradients by redefining the interest with a smaller interval. Reinforcement continues towards source nodes as each node reinforces the hop from which the data was received. Once the gradients have been established, DD offers energy efficient node-to-sink data delivery (Bokareva et al. 2004).

The basic DD can be optimized to fit into a particular purpose by adding directional flooding to interest propagation, choosing how data is propagated (single or multipath), and selecting reinforcement policy. In self-stabilizing DD protocol (Bein and Datta 2004) nodes can query cached interests from their neighbors. This allows better error recovery and reduces the need for refreshing interests. Source-initiated DD (Bush et al. 2005) increases reliability by sending a packet via multiple paths.

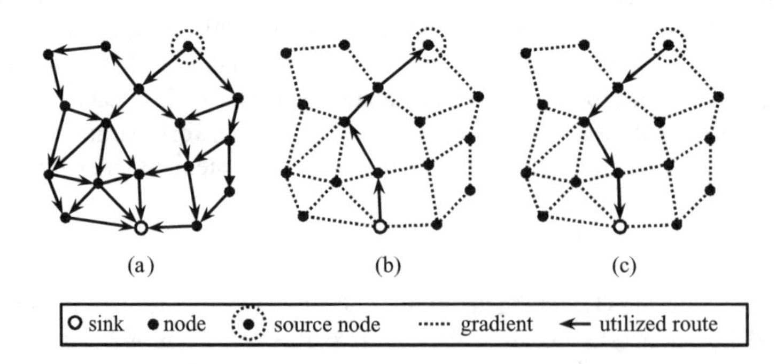

Figure 8.1 Directed diffusion: (a) low traffic broadcast through the network after receiving an interest; (b) reinforcement of a gradient; and (c) high traffic along a reinforced gradient.

8.3.3 Location-based Routing

Location-based (geographic) routing uses position information when making routing decision. Sensor measurements are usually related to a specific location, which makes the approach natural to WSN. Location-based routing scales as well as routing tables or a global knowledge of the network are not required. However, each node must know its position and a source node must know the destination location. This can be a problem as the use of positioning chips (such as GPS) increases the price and energy consumption, and manual configuration during deployment is not suitable for large-scale networks. The problem can be alleviated by using positioning chips only in part of the nodes, while the other nodes calculate position with the assistance of their neighbors. It should be noted though, that the requirement to know the destination location is similar to the nodecentric protocols where the destination address must be known. The challenges in location-based routing include finding an efficient route around a hole in the network. Also, a routing protocol must be able to handle possible inaccuracies in the location information.

Related work

Cartesian routing (Finn 1987) uses greedy selection by choosing the next-hop neighbor that is nearest to the target. While the protocol is simple and finds the shortest path, it does not tolerate holes in network.

Greedy Perimeter Stateless Routing (GPSR) (Karp and Kung 2000) operates normally in a greedy forwarding mode that is similar to the Cartesian routing. However, if a node x does have a neighbor that is closer to the target D, a hole is detected and packet is marked into a perimeter mode. GPSR forwards a packet in the perimeter mode around a hole according to a right-hand rule, in which the most forwarding node that is on the right side of the $\overline{xD}$ line is selected. The forwarding around a hole in GPSR is show in Figure 8.2(a). Eventually, the hole perimeter crosses routing path $\overline{xD}$, after which the routing continues with the greedy algorithm. The difficulty with GPSR is to detect when the greedy search has become trapped, for example, when the target is mobile.

Routing packets directly from target to the destination is not desired or possible in many situations. These situations include holes and avoiding nodes under certain conditions, such

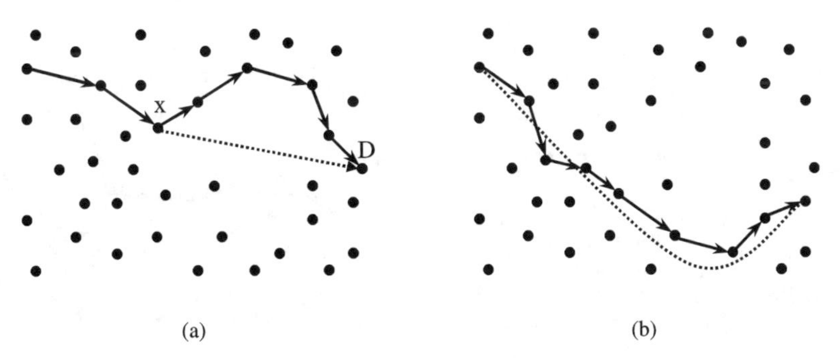

Figure 8.2 Hole avoidance in location-based routing protocols: (a) GPSR follows the perimeter of a hole; (b) trajectory-based forwarding defines a curve to avoid undesired areas.

as low-energy or network congestion. Trajectory-Based Forwarding (TBF) (Niculescu and Nath 2003) routes the packets along a predefined curve (trajectory) that can be used to avoid undesired areas. The operation of the TBF is presented in Figure 8.2(b). A source node calculates the trajectory and assigns a description of the trajectory into a packet. The trajectory is expressed as a formula, therefore causing computation on each node but taking only a little space from a packet. Energy consumption and any forwarding delay can be controlled by choosing the next-hop near the curve with different policies. These policies include a minimum deviation from the curve, selecting the most forwarding node, selecting a node that has the most battery left, or using random selection. With different trajectory definitions, TBF can be used in unicast, multipath, broadcast, or multicast routing. The benefits of TBF is that it is immune to source and intermediate node mobility because the trajectory does not encode intermediate member identifiers of the path. However, the actual trajectory generation may be complicated and requires global knowledge of network to avoid obstacles.

Trajectory- and Energy-Based Data Dissemination (TEDD) (Goussevskaia et al. 2005) uses the concept of TBF but generates trajectories based on the global knowledge of the remaining energies in sensor nodes (energy map). Thus, the protocol increases network lifetime by routing through the nodes that have high energy reserves and by avoiding low-energy nodes. The energy map is constructed on a monitoring node, which receives energy estimates from the network. The protocol uses receiver-decided forwarding instead of sender-decided forwarding, as used in TBF, to eliminate the need for neighbor tables and to allow better support for network dynamics.

Mobicast (Huang et al. 2003) defines the target as a delivery zone instead of a single destination. The delivery zone can change over time as shown in Figure 8.3(a). Therefore, the protocol is well suited for target tracking where the delivery zone moves with the target object. A Mobicast session is defined as $(m, Z[t], T_s, T)$, where m is a message, $Z[t]$ is delivery area that changes over time t, T_s is sending time, and T is the duration of the session. While the protocol requires only a small network-wide storage footprint for the information being delivered, it is not efficient when the delivery zone has to move over a hole.

Face Aware Routing (FAR) (Huang et al. 2005) is based on Mobicast but differs from it by addressing situations better, in which the delivery zone must be moved over a hole. The protocol defines the node's spatial neighborhood to consist of neighbors that are located in the faces adjacent to the node. From the definition, it follows that nodes on the opposite site

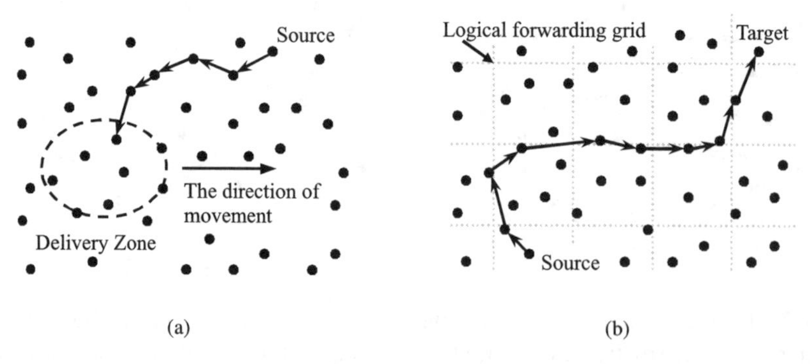

Figure 8.3 Location-based routing protocols: a) Mobicast; b) TTDD.

of a hole belong to each other's spatial neighborhood. A node that has a spatial neighbor in a delivery zone forwards the packet, thereby forwarding succeeds over holes. As each node may have nodes belonging to its neighborhood that are several hops away, the neighborhood must be determined in a separate neighborhood-discovery procedure.

TTDD (Luo et al. 2005) proactively constructs a forwarding grid structure by using the greedy algorithm (holes are not considered). Data travels in two tiers as shown in Figure 8.3(b), first through dissemination nodes on the grid, later within the local grid. A sink connects to the nearest dissemination point located at the grid points. A query is flooded on the grid, while grid nodes update soft-state information about the sink. The information is used to direct reply data streams back to the sink. TTDD expects that sensor nodes are stationary but supports multiple, mobile sinks.

8.3.4 Multipath Routing

In multipath routing a packet traverses from a source node to a target node via several paths. Thus, the target may receive several duplicates of the packet. Duplicating packets adds robustness to the data delivery, but increases network load and energy usage as extra transmissions are required.

Related work

Classic *flooding* is the simplest multipath routing algorithm, where a node forwards a packet to all of its neighbors except the node that the packet was received from. For preventing the retransmission of duplicates, flooding typically uses a simple routing table that contains a source node and last-received packet sequence number entries. Still, flooding is extremely inefficient as a node receives the same packet from each of its neighbors.

Many WSN routing protocols use flooding during route the set-up phase, which has inspired research on optimizing flooding. In *Gossiping*, each node does not automatically forward a message but uses certain probability instead, which greatly reduces the routing overheads. However, the probability must be chosen carefully since too small a probability may prevent the packet reaching all destination nodes.

Gradient Broadcast (GRAB) (Ye et al. 2005) is a receiver-decided protocol that forwards data along an interleaved mesh. Each packet is assigned with a budget that is initialized by the source node. The budget consists of the minimum cost to send a packet to the sink and an additional credit. When a node receives the packet, it compares the remaining budget against the cost required to forward the packet to the sink. If the cost is smaller or equal to the budget, the node forwards the packet. As the credit increases the budget, it allows the packet to be forwarded along paths other than the minimum-cost paths. Thus, the credit determines the amount of redundancy for the packet and has a trade-off between used energy and reliability. If the credit is zero, the packet must be forwarded along the minimum-cost path. To suppress duplicates each node maintains a cache that contains identifiers of recently forwarded packets.

8.3.5 Negotiation-based Routing

Negotiation-based routing protocols exchange negotiation messages before actual data transmission takes place. This methods saves energy as a node can determine during the negotiation whether the actual data is needed or not. High-level data descriptors are used to distinguish different data, which helps to detect and eliminate redundant messages. For example, a data descriptor might describe sensor type and area. For negotiation protocols to be useful the negotiation overheads and data-descriptor sizes must be smaller than the actual data.

Related work

Sensor Protocols for Information via Negotiation (SPIN) (Kulik et al. 2002) is a negotiation-based protocol family that contains separate protocols optimized for point-to-point and broadcast networks. SPIN uses a resource manager that allows nodes to cut back their activities if their energy is low. Resource awareness allows nodes to send data without negotiation, where the cost of sending the data is smaller than the negotiation cost.

8.3.6 Query-based Routing

Query-based routing protocols request specific information from the network. A query might be expressed as high level (such as Structured Query Language (SQL)) or natural language. For example, a sensor network might be queried for data as "average temperature around area *x, y* during the last hour".

Related work

Active Query forwarding In Sensor Networks (ACQUIRE) (Sadagopan et al. 2003) uses cached, local information to partially resolve a query. When a node receives the query, it performs an update from neighbors within lookahead of *d* hops. The query progresses either via a random or guided path until it has been fully resolved. Then, a complete response is sent back to the querying node. A query with a one-hop lookahead is illustrated in Figure 8.4(a). To be beneficial ACQUIRE requires that data is replicated and is useful mainly for complex, one-shot queries. Still, ACQUIRE can be used in conjunction with sensor databases where data is distributed to several nodes.

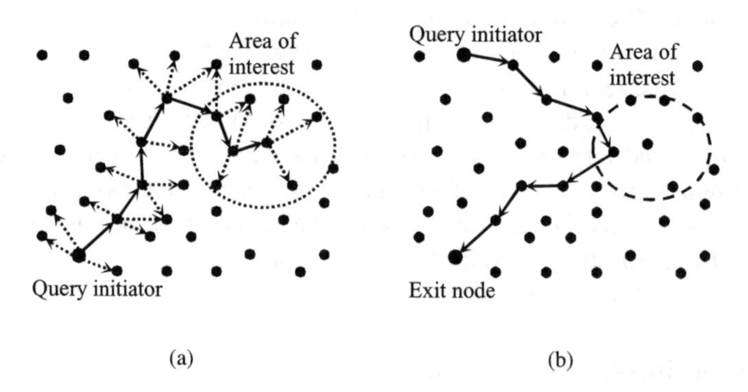

Figure 8.4 Query-based routing: (a) ACQUIRE with one-hop lookahead; (b) information-directed routing with M equal to 1.

Information-directed routing in ad hoc sensor networks (Liu et al. 2005) is a reactive, location-based routing protocol that is targeted at tracking applications. It supports queries to moving targets and allows returning the query to a prespecified exit node. The exit node can be the same or a different node to the one originally performing the query. An example of a protocol operation with different entry and exit points is presented in Figure 8.4(b). The protocol selects the next-hop among neighbors with M hops distance. The hop that minimizes communication overheads (e.g. minimizes hops) while maximizing the gathered information is selected. The protocol requires prior knowledge of the network for setting the optimal M value. A too small M will not route around holes, while a too large value has unnecessarily high overheads.

8.3.7 Cost Field-based Routing

A cost field-based routing assigns each node a cost value that is determined based on the distance between the node and a sink. The actual cost may be calculated from an arbitrary chosen metric, e.g. the number of hops to the sink or a cumulative energy metric. The only requirement is that the cost is increased on each hop, otherwise a loop could occur.

The routing begins typically with a set-up phase in which a sink broadcasts a route advertisement containing an initial cost. This method is presented in Figure 8.5. When a node receives an advertisement message it compares the cost included in the message against its own cost. If the cost decreases, it accepts the new cost, increases the cost contained in the advertisement, and resends the advertisement to its neighbors. Thus, the nodes that are further away from the sink have a higher cost and data can be sent to a neighbor that has a lower cost.

The advantages of the cost field-based approach is that the knowledge of forwarding path states is not required. Instead, a node needs only to know its cost to the sink. The disadvantage is that the routes must be created proactively. While data to the sink is forwarded efficiently, flooding might be required for the other direction. However, this is usually acceptable since traffic is often asymmetric.

Related work

Minimum-Cost Forwarding (Ye et al. 2001) uses a backoff-based algorithm to reduce the message overheads during the set-up phase. A node defers route advertisement until a

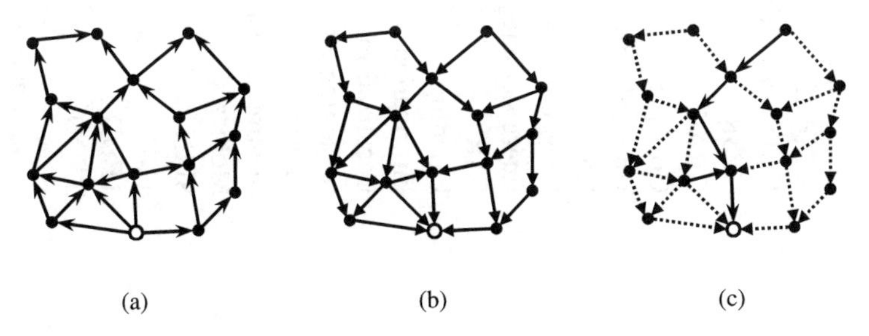

Figure 8.5 Cost field-based routing: (a) set-up phase; (b) established gradients; and (c) traffic to the sink.

Table 8.1 Summary of WSN routing protocols.

Routing protocol	Classification	Topology	Mobility support	Scalability
ACQUIRE	Data-centric/query-based	Flat	Poor	Medium
Cartesian routing	Location-based	Flat	Partial[a]	Good
GPSR	Location-based	Flat	Partial[a]	Good
GRAB	Multipath/cost field-based	Flat	Poor	Good
Gossiping	Multipath	Flat	Poor[b]	Good
Directed diffusion	Data-centric	Mesh	Poor	Medium
DSR	Nodecentric	Flat	–	Poor
Energy-aware routing	Cost field-based	Tree	–	Medium
Energy-efficient routing	Cost field-based	Tree	–	Medium
FAR	Location-based	Flat	Full	Good
Flooding	Multipath	Flat	Full	Good
Information-directed routing	Locationcentric/query-based	Flat	Partial[a]	Good
Localized max-min remaining energy routing	Cost-field-based	Tree	–	Medium
Mobicast	Location-based	Flat	Full	Good
Minimum-cost forwarding	Cost field-based	Tree	–	Medium
Mobicast	Location-based	Flat	Full	Good
SPIN	Data-centric/negotiation	Flat	Poor	Good
TBF	Location-based	Flat	Partial[a]	Medium
TEDD	Location-based	Flat	Partial[a]	Medium
TTDD	Location-based	Grid	Partial[c]	Good

[a] Target node must be stationary.
[b] May be unreliable.
[c] Only source and target nodes may be mobile (intermediate nodes are stationary).

node no longer receives advertisements with a lower cost. Therefore, the optimal cost-field can be determined in a single pass. Once the costs are resolved, it uses receiver-decided forwarding. Each packet contains a minimum cost from source to sink and a consumed cost (initially zero). A node forwards a packet if the sum of the consumed cost and the cost at the node matches the source's cost.

The cost value can be used to balance the energy consumption between nodes. Energy-efficient routing presented in Schurgers and Srivasta (2001) selects the next-hop randomly between routes having the same cost. If the energy remaining in a node is low, the node discourages other nodes from routing through it by increasing its cost. Shah and Rabaey (2002) describes a cost field-based routing, in which a forwarding node is selected with a probability that depends on the energy metric of the route.

Localized max-min remaining energy routing (Bachir and Barthel 2005) avoids updating the cost value when the energy metrics change. Instead, it minimizes the exchange of routing information between nodes by using a localized approach to detect energy levels in a route. When a packet is forwarded a node introduces an extra delay that is inversely proportional to the remaining energy. Thus, the route with the lowest delay optimizes the energy usage.

8.4 Summary

The discussed routing protocols are summarized and classified in Table 8.1. As routing protocols may use hybrid techniques to achieve their intended target application, some of the protocols are classified to several categories.

Mobility support is evaluated according to the amount of nodes that can move and the effort required for route reconstruction. A protocol does not have mobility support if the movement of a node requires expensive route reconstruction. Poor mobility support allows light mobility, and a node is able to fix its routes independently. In partial mobility support only certain types of nodes, such as target nodes, may be mobile. When the mobility is fully supported, any node can move without significant effect on the routing process.

Scalability is rated as poor, medium, and good based on required memory and route maintenance overhead. The scalability is denoted as poor when routes must be maintained globally, or when the required memory usage is directly proportional to the network size. In the medium scalability rating, routes are constructed locally, and only a subset of the network nodes can increase the memory usage. For example, the memory usage may be proportional to the amount of sinks in the network. In the good scalability rating, routes are determined locally and the memory usage is (near) static.

As can be expected, the location-based protocols have the best mobility support and scalability. However, when selecting a WSN routing protocol, there are also several other criteria that should be considered, such as complexity (required processing power and memory), energy usage on intended deployment scenario, QoS support, and provided data-centric features. Due to the unique requirements in WSN routing, an all-purpose routing protocol, like IP in traditional networks, does not exist. For example, tracking applications have quite different requirements than a static, environmental measurement network.

9

Middleware and Application Layer

This chapter focuses on the frameworks and interfaces that ease the application development, runtime configuration, and management of WSNs. Pure application implementations are not presented here but discuss and examples can be found in Section 2.2. Furthermore, frameworks that are targeted to a single application or a very small domain of applications are not discussed because of the limited possibilities for generalization. Nonetheless, such frameworks can be extremely beneficial in their selected scope and their contributions should not be underestimated.

Application and middleware layers are discussed together, because the main functionality of these layers (apart from pure application tasks) is the communication abstraction between different applications, either network-wide or locally (Stallings 2004). In WSNs, a strict hierarchical layering is impractical due to the limited resources, and therefore the functionalities of these layers are often combined.

In the widest scope, the definition of term *middleware* refers to software that connects other software components or applications so that they can exchange data. More specifically, it can be defined as a layer that lies between the operating system and applications in a distributed system. Hadim and Mohamed (2006) define WSN middleware *as a software infrastructure that glues together the network hardware, operating systems, network stacks, and applications*.

An overview of the SW architecture of a node with a middleware is depicted in Figure 9.1. At its upper interface, a WSN middleware implements an API for applications. A middleware operates on top of the OS and network protocol stack . However, in WSNs it can be considered to glue all the layers down to Hardware (HW) level.

9.1 Motivation and Requirements

In computer networks, the networking stack (TCP/IP) and the interface through which it is accessed (sockets) are widely spread and commonly accepted. Accordingly, the main

Ultra-Low Energy Wireless Sensor Networks in Practice: Theory, Realization and Deployment

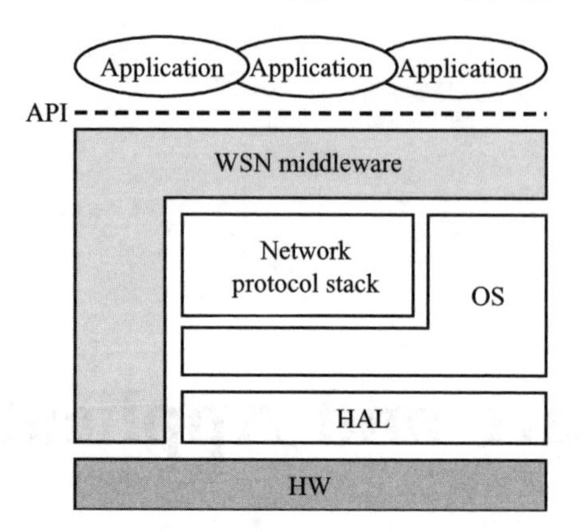

Figure 9.1 WSN middleware and its architectural relation to other SW components.

reason for an additional layer between applications and the networking stack interface is the need for more sophisticated functionality, e.g. support for distributed processing. Computer network middlewares, such as Common Object Request Broker Architecture (CORBA), implement an abstraction layer between network protocols and applications (Vinoski 1997). The main objective is to allow the execution of distributed systems on top of heterogeneous hardware and software architectures. For example in CORBA, Object Request Brokers (ORBs) provide the means for platform independent communication and activation of objects on demand.

Common OS conventions, networking protocols, and interfaces are not yet defined for WSNs. Therefore, a single starting point and basis for middleware design is absent. As a consequence, the requirements, architectures, and proposed solutions for WSN middlewares vary. The main purpose of middlewares is to support development, maintenance, deployment, and execution of WSN applications (not WSNs as a whole, refer to Chapter 5) (Hadim and Mohamed 2006; Römer et al. 2002). The main functionalities of WSN middlewares are

- **Abstraction**: As in computer networks, one of the most important functions for WSN middlewares is the abstraction of underlying hardware and software architectures. This is crucial in WSNs as the heterogeneity of platforms is one of the network characteristics. Thus, a middleware must provide API that is independent of underlying OS architecture and network protocols.

- **Task formulation**: The tasks of WSN are defined before the deployment, but they may change constantly. Therefore, an expressive formulation is required to define the tasks of the network. This formulation should be able to define simple request/response interactions, asynchronous event notifications, and complex sensing tasks.

- **Task allocation**: Middlewares are responsible for splitting the tasks to atomic operations and distribute the execution and control for these operations to individual sensor nodes.

- **Aggregation, data fusion, and in-network processing**: The middleware layer is responsible for the processing of data in the network because it possess the required information for making QoS, relevancy, and coverage-related decisions. The aggregation of application task data (e.g. min/max/average/assembly) is the simplest form of the WSN middleware data processing. In data fusion, outputs of application tasks are merged to higher level results. In-network processing combines data fusion to reactive decisions through actuators.

- **Access interface to external networks**: In addition to an internal interface to the node protocols and OS services, middlewares should also implement the interface for the external networks and devices that access WSN. Thus, a middleware should abstract the internals of WSN and provide the tools and service descriptions that allow interaction between different types of networks.

- **Application QoS**: Typically, application level QoS requirements cannot be directly mapped to the QoS classes and parameters of the network. Therefore, a middleware adapts application QoS to the protocols, makes necessary trade-offs, and monitors realized QoS. Further, the integration of real-world aspects, such as time and location, should be integrated into middlewares.

The benefits of a middleware layer in WSNs are not evident. In networks with a limited number of simple, distinct applications, which do not require complex processing and communication, a middleware consumes only limited node resources. While the network and application complexity, heterogeneity of platforms, and diversity of protocols have increased, a need for middlewares have emerged. Yet, due to the resource constraints, the WSN middlewares are not as comprehensive as middlewares for computer networks (Kuorilehto et al. 2005b; Stankovic et al. 2003).

For the realization of ubiquitous computing paradigm (Weiser 1993) middlewares are essential. Middlewares implement an interface that allows external users to discover and access WSNs in an ad hoc manner without predetermined knowledge of the type of services the network can provide (Juntunen et al. 2006).

9.2 WSN Middleware Approaches

The approaches for WSN middlewares vary greatly due to the immature technology. Therefore, a unified classification for WSN middlewares is absent. Hadim and Mohamed (2006) propose a taxonomy of programming models for WSNs. The following classification partly follows the presented models but adds more consideration for application layer support.

The classification of WSN middlewares according to their main purpose is presented in Figure 9.2. WSN middlewares are first categorized to *programming abstractions*, *runtime support*, and *interfaces* depending on the main objective. The first creates an abstracted view of the network and sensor data, while the second focuses on runtime systems and mechanisms for supporting application execution (Gummadi et al. 2005; Hadim and Mohamed 2006). The interfaces offer an external data access interface without any additional functionality. However, the classification is not unambiguous since typically all middlewares offer some sort of an external data access interface, and runtime environments provide some

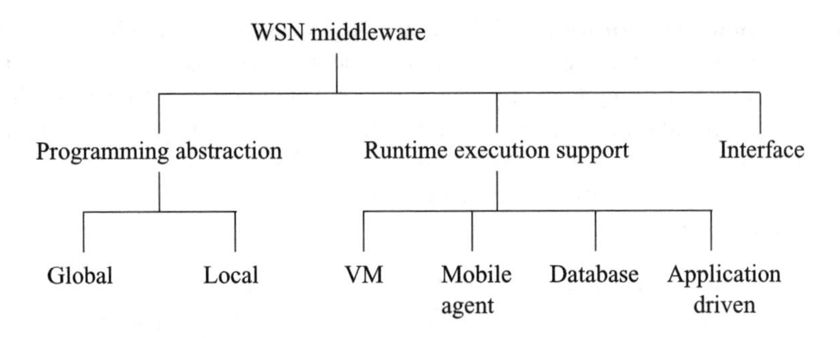

Figure 9.2 WSN middleware classification.

sort of abstractions for WSN. On the other hand, programming abstractions also requires runtime support for realization.

The programming abstractions are subcategorized into global and local, depending whether the network behavior is viewed as a single entity or divided into local groups defined by, for example, data content, location, or network topology. Nevertheless, these should not be confused with programming languages, such as nesC (Gay et al. 2003) as discussed in Chapter 5. While programming languages are abstractions of the capabilities of computer hardware, these are abstractions of network capabilities.

The runtime execution support is divided into four subcategories depending on the basic system architecture. Again, the classification to Virtual Machines (VMs), databases, mobile agents, and application-driven middleware may partly overlap. For example, mobile agents typically execute on top of VM. VMs abstract underlying HW and SW through a virtualized environment, while database middleware view WSN as a data storage that can be queried for data (Stankovic et al. 2003). Mobile agents realize mobile code paradigm, where instead of data, code is moved to data origins for local processing (Fuggetta et al. 1998). Application-driven middleware control and allocate tasks to nodes based on the QoS requirements and guidelines set by the application (Hadim and Mohamed 2006).

9.3 WSN Middleware Proposals

Several different kinds of middlewares have been proposed for WSNs. Proposals can be classified according to the categories presented above depending on their main functional purpose.

9.3.1 Interfaces

The standardization efforts for smart transducer interfaces are often considered in the context of WSNs. For example, IEEE 1451.5 Work Group (WG) (IEEE 2006) and Object Management Group (OMG) (STI 2003) specify interfaces and conventions for the data gathering from smart transducers.

A pure functional API is specified by Java Specification Request (JSR) 256 Mobile Sensor API (JCP 2006) that defines an interface, which allows Java ME applications to fetch data from sensor nodes. However, it does not support any methods for the controlling or commanding of sensor nodes. Similarly, service platform and interfaces for WSNs are proposed in Sgroi et al. (2005). The basic API is implemented by query and command service interfaces and supplemented by timing, location, and concept repository services.

9.3.2 Virtual Machines

Maté (Levis and Culler 2002) is a simple custom bytecode interpreter for very resource-constrained nodes. Maté instructions are WSN specific that enables very small code size. Maté applications are divided into 24 B capsules that can be efficiently injected to WSN. Yet, these capsules does not possess features of mobile agents but instead they disperse and infect the whole network.

9.3.3 Database Middlewares

Database middlewares abstract the data gathering from WSN by database queries issued to a distributed database implemented by the network. The basic approach is SQL type query interface, which may be supplemented with support mechanisms for more complex processing.

In Sensor Information and Networking Architecture (SINA) (Shen et al. 2001), database queries are injected into the network as Sensor Qucrying and Tasking Language (SQTL) (Jaikaeo et al. 2000) scripts. These scripts migrate from node to node depending on their parameters. The allocation of queries to individual nodes is implemented by a Sensor Execution Environment (SEE) that compares SQTL script parameters to node attributes and executes script only if these match. The expressivity of SQTL scripts also allows more complex tasks, e.g. by timer utilization for execution triggering.

In Cougar (Yao and Gehrke 2002), a query optimizer at the gateway node determines energy-efficient query routes. The query plans generated by the query optimizer are parsed in the nodes by a query proxy. Local query proxies make sensing, aggregation, and communication decisions, based on the data flow and computation plan specifications defined in query plans.

TinyDB (Madden et al. 2003) takes a quite similar approach to Cougar. In TinyDB, a query processor in a node supports basic SQL-type query operations and data aggregation for improving network energy efficiency. In addition, TinyDB supports event-based queries that are initiated in-network after the occurrence of a specified event is detected.

A database approach is also taken in DSWare (Li et al. 2004). DSWare improves data availability by distributing frequently queried data to the network and increases robustness by grouping nodes with similar objectives for management. In addition, DSWare implements an event detection service that signals condition changes on predefined events, which improves the reactivity of the network.

TinyLIME (Curino et al. 2005) extends LIME middleware (Murphy and Picco 2001) to WSNs. In LIME, data and communication use tuple space (Gelernter 1985) for sharing information. TinyLIME extends LIME middleware to include sensor data provided by WSN nodes. When a LIME client (PC or similar) accesses a sensor tuple in a tuple space, a base station node queries data from the node relating to the tuple. The scalability of TinyLIME is limited to nodes directly connected to network base stations.

The approach taken in TCMote Middleware (Diaz et al. 2005) is similar to that of TinyLIME. Unlike TinyLIME, TCMote middleware does not extend the tuple spase used by the middleware outside WSN. The queries and predefined alerts utilize tuple channels for information exchange. The TCMote middleware architecture is limited to clusters having a star topology.

9.3.4 Mobile Agent Middlewares

A mobile agent is an object that in addition to the code carries its state and data. A mobile agent makes its migration and processing decisions autonomously. In order to obtain platform independency and relatively small object size, mobile agents are typically implemented on top of VMs for platform independency (Fuggetta et al. 1998; Qi et al. 2003).

In Sensorware (Boulis et al. 2003) the tasks are implemented by Tool Command Language (TCL) scripts. A user queries the WSN data by injecting a script into the network. In addition to the functionality of the task, each script contains an algorithm that controls the script migration. The TCL scripts are typically very small sized, which allows their efficient migration.

Mobile agents in Agilla (Fok et al. 2005) are implemented with custom bytecode, quite similar to Maté (Levis and Culler 2002). High-level instructions result to small-sized agents, which migrate efficiently between nodes. The agents communicate through a local tuple space (Gelernter 1985). Reactions are used for notifying an agent of the events that it is interested in.

In MagnetOS (Barr et al. 2002) application tasks are implemented as Java objects. MagnetOS utilizes automatic object placements algorithms that attempt to minimize communication by moving Java objects nearer to the data source. MagnetOS is implemented on top of a Distributed Virtual Machine (DVM) (Sirer et al. 1999), which abstracts the network as a single Java Virtual Machine (JVM). The resource consumption of the DVM makes MagnetOS suitable only for resource rich sensor nodes.

Mobile Agent Runtime Environment (MARE) is merely targeted to mobile ad hoc networks, but its resource requirements are comparable to MagnetOS. In MARE, mobile agents move towards data sources and perform communication locally. A tuple space (Gelernter 1985) is used for resource discovery and communication between agents.

9.3.5 Application-driven Middlewares

Here, application-driven middleware covers all proposals that perform networking and execution control based on application initiated guidelines (Hadim and Mohamed 2006). Typically, this involves a task-allocation functionality for distributing application processing to nodes.

Middleware Linking Applications and Networks (MiLAN) (Heinzelman et al. 2004) takes an application QoS specification as an input and adapts the network operation during runtime to meet application QoS and to lengthen the network lifetime. Multiple applications are interleaved according to their importance. MiLAN is tightly integrated to the underlying protocol stack, which enables networking adaptation according to the application needs.

Like MiLAN, a cluster-based middleware proposed by Yu et al. (2004) uses application QoS specification for runtime control of network operations. The middleware manages underlying network topology by forming clusters, within which it controls resources and task allocation. Tasks are allocated by a heuristic algorithm that attempts to minimize computation and communication energy costs.

WSN node middleware (Kuorilehto et al. 2005a) takes a quite similar approach to MiLAN and cluster-based middleware. The main objectives for the runtime task allocation are the fulfilling of application QoS and maximizing the network lifetime. The communication and computation load caused by tasks and node roles are divided evenly between nodes. Further, an option for task binary transfers is included for more robust application operation.

An application task graph defines the guidelines for task and role assignments in DFuse (Kumar et al. 2003). Application tasks are implemented as fusion functions that are allocated to the nodes by a distributed algorithm. While the operation is initiated by a root node, the allocations are dynamically re-evaluated in order to find a more optimal configuration.

Park and Srivastava (2003) combine application-driven task allocation and mobile agents. Communication and dependency graphs are extracted from application at design time. The graphs are modified for minimizing the communication costs by link and schedule optimization. During runtime, the execution is adapted by mobile agents.

An application adaptor of Impala (Liu and Martonosi 2003) changes active tasks, and the nodes executing them, based on the application parameters. The changes are coordinated by a state machine that describes the application functions and the conditions that change them. Application tasks can be updated during runtime by a centralized control entity.

Mires (Souto et al. 2005) is merely a message-oriented middleware but its actions are controlled by the application. Mires uses a publish/subscribe architecture in which nodes advertise their tasks (topics) and external user applications select desired tasks for execution.

9.3.6 Programming Abstractions

A global programming abstraction is also referred to as macroprogramming since it allows the programming of WSN as a whole and hides the individual nodes from the application (Hadim and Mohamed 2006). Local abstractions focus on the nature of sensed data within a local context (Hadim and Mohamed 2006).

Kairos (Gummadi et al. 2005) creates a global programming environment with node, neighbor, and data access abstractions. Nodes and neighbors are accessible independent of the underlying topology. A distributed program implemented with the abstractions is converted in compile time to individual instances and executed in nodes on top of Kairos runtime. The runtime environment manages global and shared variables and communication between program instances.

Abstract regions (Welsh and Mainland 2004) are a communication abstraction among local groups of nodes. A runtime environment supports neighbor querying and data sharing that can be utilized for distributed application programming. The algorithms for region construction can be dynamically changed.

The main abstractions in FACTS (Terfloth et al. 2006) are facts and rules. Facts abstract data representation and communication, while rules define the means for data processing. A rule is executed on top of an interpreter, known as a rule engine. The execution is fired by an event with a prerequisite that the conditions (e.g. availability of data) of the rule are satisfied. A function is an abstraction that can be used for accessing low-level resources and for efficient algorithm implementation using machine code.

A local abstraction is provided by a Global Sensor Network (GSN) middleware (Aberer et al. 2006) that allows the programming of virtual sensors using Extensible Markup Language (XML). Virtual sensors abstract both the sensor data access and communication from the application programmers. Runtime support is implemented by Java-based containers that host and manage virtual sensors.

There are few other programming abstractions, such as Protothreads (Dunkels et al. 2006) and Token Machine Language (TML) (Newton et al. 2005), that convert the abstractions during compile time. However, since these do not include runtime support, there are not discussed further.

9.3.7 WSN Middleware Analysis

A fair comparison of WSN middlewares is difficult due to the differing objectives and target platforms, as well as their completely orthogonal implementation approaches. An evaluation framework, together with the categorization of most of the WSN middlewares according to that framework, is presented by Hadim and Mohamed (2006). The analysis described here extends that framework with the requirements given in Section 9.1.

The comparison of VM, database, and mobile agent middlewares is presented in Table 9.1, while application-driven middlewares and programming abstractions are compared in Table 9.2. The purpose of these comparisons is not to order the middlewares according to their superiority but to give the main features of each middleware. This information should ease the evaluation of middlewares for the different kinds of application and node platform domains.

Both tables have similar dimensions for the comparison. If not stated otherwise, the values in a column may be either *full*, *partial*, or *none*. The first value means that the middleware supports the given dimension, while the second denotes that some aspects of that feature are not considered. The last value states that the dimension is not handled by the middleware but it is left to application developer.

The second column in the tables defines the requirements set to the implementation platform. *Tiny* denotes a Mote-class device, *small*, e.g. Intel XScale class devices, while *large* means PC or PDA-type devices. The level of heterogeneity abstraction defines the platform independency provided by the middleware. The fourth column lists the method, that is used to describe the task details for the middleware. This is typically a textual definition that uses, for example, XML or a custom format. The fifth column specifies how application tasks are allocated to nodes. The allocation can be done either statically at design time or dynamically, and it can be performed by a centralized entity or by distributed collaboration. The sixth column depicts how the middleware supports aggregation of data within the network. The next one defines the interface that the middleware offers for external WSN data access. Term *custom interface* is used to denote a highly implementation-dependent function or message interface. The last column lists the other key features of the middleware.

Table 9.1 Comparison of VM, database, and mobile agent middleware proposals for WSNs .

Middleware	Target platform	Heterogeneity abstraction	Task formulation	Task allocation	Aggregation	External data access	Miscellaneous
Maté	Tiny	Partial	–	–	–	Code capsules	Limited expressivity
SINA	Large	Full	SQTL script	Dynamic, distributed	Partial	SQTL scripts	
Cougar	Tiny	Partial	Query plan	Dynamic centralized	Full	Queries	
TinyDB	Tiny	Partial	SQL-type queries	Dynamic centralized	Full	Basic and event-based SQL-type queries	
DSWare	–	Partial	–	Dynamic, centralized	Partial	SQL-type queries, event detection	Data caching
TinyLIME	Tiny	Partial	Data template	Dynamic, centralized	–	Tuple read	
TCMote	Tiny	Partial	Attributes	Dynamic, centralized	–	Tuple get	
SensorWare	Small	Full	TCL script	Dynamic, distributed	–	TCL scripts	
Agilla	Tiny	Partial	Agent code	Dynamic, distributed	–	Mobile agent	Local tuple space for data sharing
MagnetOS	Large	Full	–	Dynamic, distributed	–	–	Single system image
MARE	Large	Full	Task descriptor	Dynamic, distributed	–	Tuple access	Geared for mobile environments

Table 9.2 Comparison of task allocation middleware proposals for WSNs .

Middleware	Target platform	Heterogeneity abstraction	Task formulation	Task allocation	Aggregation	External data access	Miscellaneous
MiLAN	–	Partial	State and QoS graphs	Dynamic, distributed	–	Custom interface	
Yu et al. (2004)	–	Partial	QoS specification	Dynamic, distributed	–	Custom interface	
Kuorilehto et al. (2005a)	Tiny	Partial	QoS levels	Dynamic, distributed	–	Custom interface	
DFuse	Large	Partial	Task graph, fusion function	Dynamic, distributed	Full	Fusion API	
Park et al. 2003	Small	Partial	Task graph	Static, centralized	Partial	Custom interface	Mobile agents for runtime migration
Impala	Large	Partial	Parameter table	Dynamic, distributed	–	Custom interface	Supports runtime reprogramming
Mires	Tiny	Partial	Topic	–	Partial	Publish / subscribe	
Kairos	Small	Full	Centralized programs	Static, centralized	–	Remote access interface	Preprocessor code modification
Abstract regions	Tiny	Full	Region abstractions	–	Full	–	
FACTS	Small	Full	Rule	–	Partial	–	Functions for low-level optimization
GSN	Large	Full	Virtual sensor	Dynamic, centralized	Partial	SQL-type queries	

Making further conclusions about middleware performance is unfair. In general, database middlewares are most suitable to static networks with structured data queries. VMs are suitable for applications that require task mobility. On the other hand, application-driven middlewares may be suitable for both of the mentioned application domains, but their main benefits are in the balancing of network loading and resource consumption usage.

10

Operating Systems

In general, the main objectives of OSs are convenience of computer usage, efficiency of system resource usage, and the ability to develop new system features without interfering with the service. In computer systems, the OS's main target is to make the use of computer programs and resources easier for a human user. The main services of OSs include the concurrent execution and communication of multiple programs, access and management of I/O devices and permanent data storage (i.e. file system), and control and protection of system access between multiple users (Stallings 2005).

Clearly, all of these aspects are not significant for OSs in WSN domain. Most of the listed properties facilitate the use of a computer in the perspective of a human user. Yet, the basic ideology that can be adopted from the list above is also evident in WSNs. The main objective of OSs in WSNs is to make the application and system development easier.

Until recently, the architecture of OSs in computer systems has been a large monolithic kernel that implements all OS features inside the kernel. Modern OSs have adapted a microkernel architecture, which includes only the most essential features of OS, while other OS services are implemented in external processes or servers (Stallings 2005).

10.1 Motivation and Requirements

As with embedded systems in their first evolution phases, most of the initially proposed WSNs lacked *systems software* that controls the execution of applications and the access to resources. Instead, the applications and networking protocols were implemented with a single control loop, in which typically only one application was tightly integrated to the networking protocols. Clearly, with WSNs servicing only one task this is beneficial. Unnecessary system resources are not consumed, the application can be implemented with awareness of network operation, and the system complexity and therefore error robustness is kept to a moderate level.

In spite of the constant evolution of sensor node platforms, the node resources have not increased considerable. Mainly, the physical node size has become smaller and the power consumption of components diminished. Still, the number of applications in WSNs has

Ultra-Low Energy Wireless Sensor Networks in Practice: Theory, Realization and Deployment

increased, their requirements become more complex, and required tasks and their functionality advanced (Akyildiz et al. 2002a).

The WSN protocols together with multiple applications constitute an extremely complex system for a resource-constrained sensor node. Further, due to the tight relation to the real world, realtime requirements are strict. The management of the system resource usage, timeliness, and peripheral access of multiple simultaneously active tasks at a single node level need systems software support for realtime coordination (Stankovic et al. 2003).

10.1.1 OS Services and Requirements

The limited resources in WSN nodes prevent the utilization of full-feature, general-purpose OSs. Instead, WSN OSs are required to have an extremely small memory footprint but still provide the basic OS services for applications and network protocols. Furthermore, the orthogonal requirements of different applications require a support for modularity in OSs (Hill et al. 2000).

All the sophisticated features of modern, general-purpose OSs are not required in WSNs but still a variety of services are needed. The following lists the most important services for WSN OSs:

- **Small memory footprint**: The limited node resources are the main driver for OS, application, and network protocol development. In order to justify its usage in systems, the code and data memory usages of OSs must be very limited.

- **Energy management**: Due to the battery-powered nature and requirements for lifetime, a low energy consumption is crucial. Thus, energy management in terms of Dynamic Voltage Scaling (DVS) and shutting down of unused peripherals is a key requirement for WSN OSs.

- **Concurrency of tasks**: WSNs are concurrent by their basic nature. The network consists of concurrent data flows that must be served in timed manner. In addition, concurrent programming is needed in order to support multiple applications and for serving the different physical sensors within a node. Thus, basic services allowing concurrency (scheduling, Inter-Process Communication (IPC), and synchronization mechanisms) are essential in OSs.

- **Realtime operation**: The tight relation to the real world inherently sets tight timing constraints on WSN operation. Further, the operation of network protocols may be time-sensitive. Thus, realtime coordination is required to control the actions of a single node.

- **Memory management**: The limited memory resources also affect the applications, not only the OS. A memory management service needs to be implemented by OS in order to control the utilization of the especially scarce data memory.

- **Peripheral access**: WSN nodes have several types of peripherals such as sensors, radios, and other I/O devices. OS control is needed to share these resources with tasks and to implement the lowest level access.

- **Robustness**: WSNs may be deployed to environments that need robust and reliable operation. Therefore, OS need to be robust, but in addition provide an environment that facilitates the development of robust applications.

- **Hardware abstraction**: The WSN node platforms are heterogeneous. In order to provide a coherent interfaces for applications, OSs need to abstract the lowest level hardware access and offer a well-defined API to application developers.

- **Modularity**: The limited resources of nodes and diverse applications require configurability from OSs. All OS components are not necessary in all configurations. Therefore, it is beneficial to include only those components that are needed in a deployed OS configuration.

10.1.2 Implementation Approaches

The requirements for OS services are extensive. Still, one of the main objectives is the need for a very limited-memory footprint. Traditional general-purpose OSs or even those targeted to embedded system domains are not suitable for most resource-limited WSN nodes. Therefore, completely new approaches are required for the OS implementation in WSNs.

Both of the traditional implementation architectures, *monolithic* and *microkernel*, can be justified for WSNs. A monolithic OS allows a tight integration of the OS components, which diminishes the size of OS. A microkernel architecture may not be as small in size due to the additional interfacing of the external services, but it achieves a considerably higher level of modularity and support for tailoring to different application cases.

The application-specific nature of WSNs necessitates also application-specific OSs. A general-purpose OS for WSNs is as infeasible as a general-purpose WSN itself. Therefore, in addition to the modularity, application-specific configuration and parameterizing are crucial for WSN OSs (Hill et al. 2000).

Another aspect that varies in the OS implementations for WSNs is the networking support in OS level. It may be either integrated into OS or implemented as a separate service that is executed in parallel with the application. The integrated network stack eases the application development due to the defined networking API, improves networking service predictability, and allows direct access to other OS resources. When the protocol stack is implemented as a separate service, it competes for Central Processing Unit (CPU) time with application tasks, which increases the risk of malfunction and starvation. Yet, such an approach keeps the kernel itself simpler and smaller and also allows easier change or configuration of protocols (Kuorilehto et al. 2005b).

A traditional sequential programming model adopted in the first WSNs is not sufficient in applications that require reliable and realtime responsiveness. Two main architectural alternatives for implementing concurrent processing in terms of OS have evolved in WSNs. A process-based *preemptive multithreading* is adapted from computer systems also to WSNs. Another approach is *event-based*, which adapts to the event-initiated nature of WSNs.

Preemptive multithreading

Preemptive multithreading offers a common programming model to application developers. The main concepts of in preemptive multi threading are illustrated in Figure 10.1 (Stallings

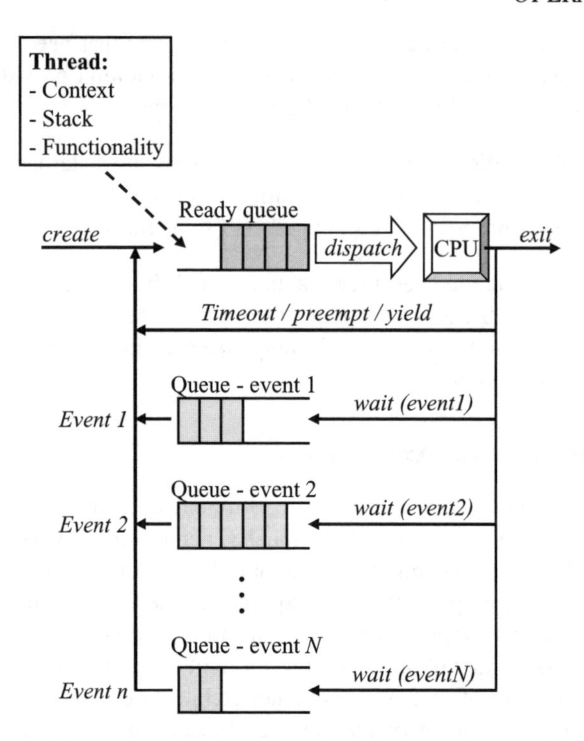

Figure 10.1 Thread management in preemptive multithreading OSs.

2005). Each process or thread (thread is used from now on) is an independent functional unit that is executed in its own context. When a new thread is selected for execution the context of the currently running thread is saved and the context of the new thread is restored to CPU.

When a new thread is created it is put to the ready queue. A scheduler dispatches a thread from the ready queue for execution according to the utilized scheduling policy. A running thread may be preempted from CPU due to a time-window expiration or a wait operation initiated by an OS routine. In the former case, the thread is put back to the ready queue, and in the latter, to an event-specific wait queue. When the event occurs, the thread is put to the ready queue. In WSNs, the events are for example timer, peripheral, or IPC events.

Preemptive multithreading offers a seemingly parallel execution of threads and enables realtime systems, in which the tasks with the highest importance need to be scheduled at defined timestamps. However, in WSNs the context switching adds considerable overheads to the processing of typically very simple application tasks. Furthermore, the enabling of preemption requires that each thread has a dedicated memory for its stack and context.

Event-based kernels

Event-based kernels adapt the event-driven programming model to WSNs. These kernels rely on event handlers, as depicted in Figure 10.2. An event handler reacts to an event, which is initiated by a hardware interrupt that indicates, for example, a reception of data from radio, a timer event, or available data from a sensor.

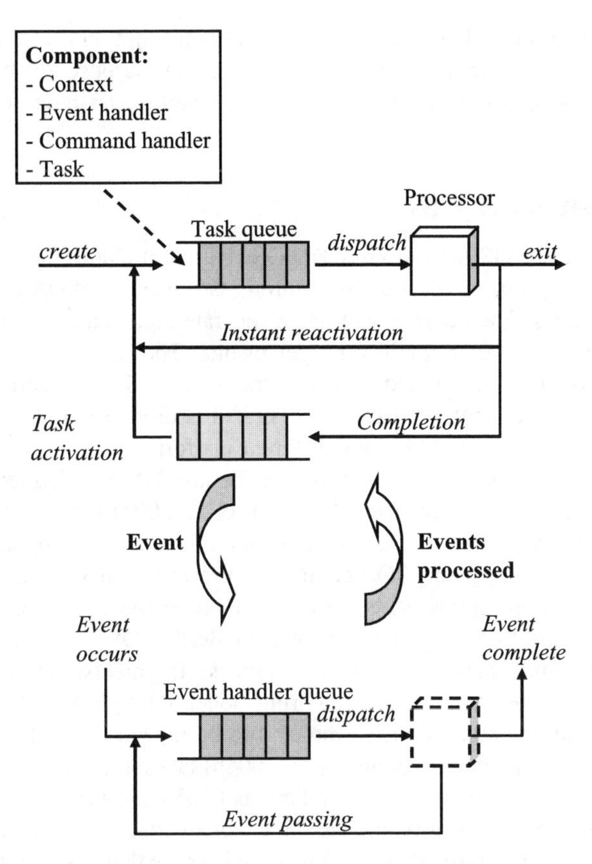

Figure 10.2 Scheduling of events and tasks in event-based OSs.

The architecture of event-handler kernels is typically layered, in which an event is propagated upwards towards the topmost event handler. The actual processing of data is done in tasks that perform regular processing. When a task needs services from lower layers, e.g. for sending data, it issues a command that is similarly propagated downwards through a layered hierarchy. A task can be activated by the task itself, by other tasks, and by event handlers.

Each event handler, command handler, and tasks run to completion, i.e. context-saving is not required. Yet, reactiveness is obtained through event handlers that can preempt the currently running task or command processing (Hill et al. 2000). As a consequence, the data memory usage of such OSs is especially low. On the other hand, the event-driven programming model is somewhat difficult to comprehend. Further, as each task runs to completion, the managing of long-term processing needs to be done at the application level.

10.2 Existing OSs

Embedded Realtime Operating Systems (RTOSs), such as OSE (Enea 2006), QNX Neutrino (Systems 2006), and VxWorks (Wind Driver 2006) are widely used in industrial and telecommunication systems. However, their memory consumption is too large for resource-constrained sensor nodes. Small-memory footprint, general-purpose RTOSs, such as FreeRTOS (FreeRTOS 2006), do not meet the strict timing and power-mode utilization requirements of WSNs.

10.2.1 Event-handler OSs

The most widely known OS for WSNs is TinyOS (Hill et al. 2000) that uses a component-based event-driven approach for task scheduling. Software is divided into components encapsulated in frames. Each component has a separate command handler for upper-layer requests and an event handler for lower-layer events. The processing is done in atomic tasks. The frame defines the context in which the event and command handlers, and the tasks related to the component, are executed. TinyOS is originally implemented for motes, but thereafter it has been ported to many different platforms.

In TinyOS, components and their communication are statically defined at compile time. A more dynamic approach is taken in SOS (Han et al. 2005) that adopts the component model from TinyOS but allows runtime loading and unloading of components. To facilitate runtime component management, SOS contains a kernel that maintains modules, handles memory management, and implements communication between modules.

Similar architecture to SOS is implemented in Bertha, OS for Pushpin nodes (Lifton et al. 2002). The components, referred to as process fragments, can communicate with process fragments in the same and neighboring nodes through a bulletin board system. Runtime loading and unloading allows a process to be moved to another node.

Lightweight OS for BTnodes (Beutel et al. 2004b) takes a more simplified approach. It also relies on events processed by event handlers, but tasks and events are defined statically during compile time. All application processing is done within the event handlers. BTnodes can share their sensory data through a distributed tuple space that abstract the origins of data.

Similar to BTnodes, the basic architecture in CORMOS (Yannakopoulos and Bilas 2005) consists of events and event handlers that are organized into modules. The communication between modules is performed with events that are signaled through event paths. An event path can exist between local or remote modules, which makes the communication between distributed application modules possible.

An event-driven approach that adopts the a Finite State Machine (FSM) programming model is taken with SenOS (Hong and Kim 2003). An event triggers a state transition in FSM, whichs ends to another state. SenOS maintains state information and event queues for FSMs, and transition tables describing application functionality. Multiple applications can coexist as SenOS supports preemption between FSMs.

Contiki (Dunkels et al. 2004) also has an event-handler-based kernel with a support for dynamic loading. An application program module that contains required relocation information can be linked to the OS kernel during runtime. In Contiki, an interrupt handler does not generate an event that traverses to the topmost layer, but only sets a polling flag that is checked by the polling mechanism. This allows Contiki to be run on top of an underlying realtime executive. In addition, Contiki implements a support for preemptive multithreading through a library on top of the event-handler kernel.

10.2.2 Preemptive Multithreading OSs

Preemptive multithreading for sensor nodes with Portable Operating System Interface (POSIX) style API is implemented in Mantis Operating System (MOS) (Bhatti et al. 2005). MOS implements priority-based scheduling, synchronization between threads, and power-saving features. Network stack is implemented as an OS service. The data memory consumption of MOS kernel is only 500 B but this does not include application thread stacks.

A similar approach to MOS is taken in nano-RK (Eswaran et al. 2005), which implements also a fixed-priority preemptive scheduling. In addition to the features also implemented in MOS, nano-RK introduces a resource reservation policy, through which application threads can allocate CPU time, network bandwidth, and sensor resources.

RETOS (Cha et al. 2007) extends MOS and nano-RK features by supporting a dual mode operation that separates kernel and user modes. The RETOS functionality can be reconfigured by runtime loading of modules. OS does not include full-feature network stack, but instead it offers layered interfaces for networking.

Like RETOS, *t-kernel* (Gu and Stankovic 2006) incorporates OS protection. The protection is implemented by modifying the application code during runtime. The same approach is used also for virtual memory abstraction. Yet, runtime code modification may incur unexpected delays and increases code size (Cha et al. 2007).

10.2.3 Analysis

As previously stated, in WSNs the OS and its parameters also need to be configured application specifically. The suitability of OS depends on the resources of node platforms and the requirements of applications and protocols. Thus, even a commercial embedded RTOS, such as OSE, may be the best selection if WSN nodes have enough resources and the application requires the performance and services of such an OS.

The comparison of existing OS proposals for WSNs is presented in Table 10.1. The table lists the main features of OSs. The second column defines the programming paradigm used by the OS and the following two columns show its resource consumption. The task instance specifies the name of the entity implementing an application task. The sixth column defines whether the OS has no realtime support or if it guarantees hard or soft deadlines. Runtime reprogramming is typically supported either by allowing bootloader type reprogramming of the whole code image, or by enabling dynamic loading of tasks. The last column lists additional services provided by the OS.

The values given in the Table 10.1 are based on the information available in cited reference publications, unless stated otherwise. Thus, for example the code and data memory sizes given in the table may differ considerably from those of an application-specific configuration of the OS. Some of the given information may not be valid for the latest version of the OS. Since these may change constantly, their inclusion in the comparison is not practical. Furthermore, as detailed information about new products is not available, e.g. BTnodes have a new lightweight OS, BTnut (Beutel 2006) it is omitted from the comparison.

Table 10.1 Comparison of existing WSN OSs.

OS	Main approach	Code memory (KB)	Data memory (B)	Task instance	Realtime	Runtime reprogramming	Other services
TinyOS	Event-based	3.4[a]	226[a]	Component	–	–	Active messages allowing RPC
SOS	Event-based	20.0	1163	Component	–	Dynamic loading of components	Dynamic memory pool, runtime integrity check
Bertha	Event-based	~10[b]	~1500[b]	Process fragment	–	Tasks	Bulletin board system for RPC
BTnodes OS	Event-based	34.7[c]	1029[c]	Event handler	–	Full-system reprogramming	Bluetooth stack, tuple space, Java smoblets
CORMOS	Event-based	5.5	130	Module	–	–	Transparent RPC
SenOS	State machine	N/A	N/A	FSM	–	Reloading of state transition tables	
Contiki	Event-based / preemptive	3.8	>230[d]	Process / thread	–	Dynamic loading of processes	
MOS	Preemptive	~14[b]	~500[b]	Thread	Soft	–	Network stack, remote command shell
nano-RK	Preemptive	~10[e]	~2000[e]	Task	Hard	–	Network stack
RETOS	Preemptive	23.7	1125	Thread	Soft	Module relocation	Network interface, memory protection
t-kernel	Preemptive	28.2	~2000	Thread	–	–	Virtual memory

[a] Includes a simple multihop protocol and temperature-sensing application.
[b] Accurate value not given.
[c] See Beutel et al. (2003).
[d] Data memory depends on number of processes, maximum event queue length, and number of multithreaded tasks.
[e] Includes eight tasks, eight mutexes, and four 16 B network queues.

For resource-constrained WSNs, the memory resources of individual nodes are typically the main driver for OS selection. In general, low-cost MCUs are equipped with very limited data memories. Thus, preemptive OSs that require a separate data memory area for thread stacks may not be suitable for applications with large number of tasks.

On the other hand, event handler OSs perform poorly in applications with lengthy computation, such as cryptographical algorithms. Furthermore, compared to preemptive kernels, the event-driven programming paradigm can be difficult to understand. This may slow down the development or even cause faulty operation due to incorrect programming concepts.

Contiki attempts to tackle the difficulties of both approaches by offering support for multithreading through a library on top of the event-handler kernel. This may solve the processing of lengthy applications tasks, but for context saving in preemption Contiki does not offer any new solutions. Yet, the main benefit in Contiki is that a designer can freely map the tasks for the event handler and for preemptive multithreading according to the task requirements and available resources.

11

QoS Issues in WSN

QoS has various meanings depending on the context to which it is applied. Generally, it is used to assess whether the service satisfies the expectations. In communication networks, QoS is usually understood as a set of performance requirements to be met for transferring a data flow. QoS is commonly expressed by defining throughput, delay, jitter (variation of transfer delays), and error rates (Gozdecki et al. 2003).

11.1 Traditional QoS

QoS has been extensively studied in wireless LANs and wired computer networks. IP and Asynchronous Transfer Mode (ATM) provide extensive QoS support ranging from best-effort service to guaranteed service. The QoS models in IP can be divided into the following categories: best-effort, relative priority marking, service marking, label switching, Differentiated Services (DiffServ), and Integrated Services (IntServ). The best-effort service is the simplest and means that QoS is not specifically addressed. Relative priority marking and service marking flag the desired service within the IP header. The priority describes the importance of a packet and affects delay and drop probabilities. The service marking allows selecting a routing path that prefers delay, throughput, reliability, or (monetary) cost. Label switching, DiffServ, and IntServ operate on traffic aggregates instead of marking a single packet, which allows more accurate QoS control. In DiffServ, the traffic entering the network is classified and each class is assigned with different a behavior. IntServ provides end-to-end service guarantees by performing resource reservation on each step of the route.

11.2 Unique Requirements in WSNs

A WSN node may consist of different sensor applications, each requiring a certain type of service. For example, it may be acceptable to lose some measurements in repeatedly transmitted environmental data, but transmitted events and one-shot queries must be reliable. In addition, critical alert messages from possibly life-threatening conditions require not only high reliability, but low delay.

The simplest method for providing a sufficient service is to make sure that a network has enough resources for each application. Thus, the capacity is fitted for the worst-case network usage. While this approach can be used in traditional networks, it is not applicable in energy and resource constrained WSNs. While the ideas from traditional QoS can be applied to the WSNs, a WSN differs significantly from traditional computer networks.

In computer networks, only a fraction of the total energy consumption is caused by communications. As wired devices are AC powered and wireless devices (such as laptops) can be easily recharged, the energy is not of significant concern. In a WSN, the majority of energy consumption is caused by wireless communications, therefore requiring *energy awareness*.

Computer networks route most of their data in a wired-backbone network, while only end nodes may be wireless (such as in wireless LANs or cellular networks). In sensor networks, data may be routed via tenths of wireless hops. Each of the wireless links may fail due to radio interference, changing environmental conditions, node mobility, or unexpected failure of a sensor node due to low energy. Even if a link has a small probability of failure, the probability accumulates on each link, which makes end-to-end communication unreliable. Due to the *unreliability*, the connection oriented approach commonly used in traditional networks is impractical in WSNs.

A WSN node has significant *resource constraints* as memory and processing power are limited. This significantly limits the available approaches for QoS. For example, storing states of each flow that passes a node is impossible. The memory constraints and the unreliability prevent using many protocols that rely on end-to-end resource reservation mechanisms, such as IntServ.

11.3 Parameters Defining WSN QoS

In addition to the traditional *throughput*, *reliability*, and *latency* parameters, the level of service can be defined with *security*, *mobility*, and *data accuracy*. The parameters are presented in Figure 11.1. Security and mobility are essential in any wireless network,

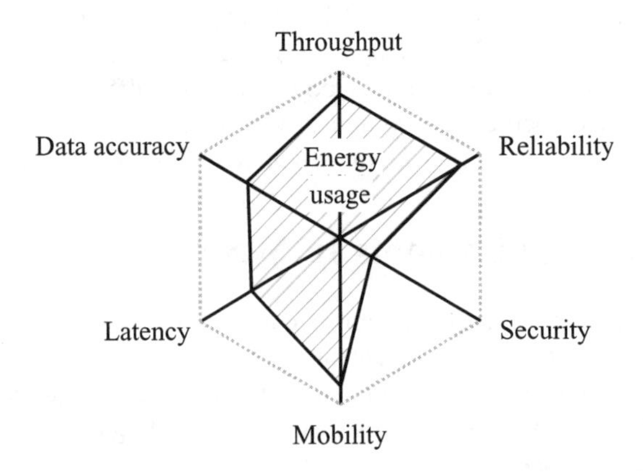

Figure 11.1 Parameters defining QoS in WSNs. Increasing a QoS-parameter also increases energy usage, therefore decreasing network lifetime.

while data accuracy is especially relevant to the sensor networks. Decreasing latency or increasing any of the other parameters usually affects *energy usage*, thus decreasing the network lifetime.

The QoS parameters are listed in the following:

- **Throughput**: Throughput in a sensor network is not usually as significant as other parameters, because in a typical use case a sensor node sends small packets relatively seldom. Still, the use of acoustic and imaging sensors requires significant throughput, as data must be streamed through the network. Thus, certain WSN applications require maximizing throughput and possibly throughput guarantees.

- **Reliability**: Common methods for increasing reliability in communications networks are using acknowledgments and error correction. Also, adding redundancy increases reliability as the network is able to recover from the loss of a single packet, but this method increases energy usage.

- **Latency**: Latency is the time taken for the network to transfer a packet from a source node to the destination node. For critical messages, networks may need to provide delivery guarantees. As sensor networks rarely use realtime streaming applications (e.g. video), the variation of the latency is less important.

- **Security**: Security is achieved by encrypting messages and verifying that a message is authentic. However, these may require significant processing power. In addition, encryption may widen data size and authentication requires additional messaging, thus causing more communicational overheads.

- **Mobility**: The mobility support may range from partial mobility to full mobility support. In partial mobility support, only a part of the nodes can be moved. The maximum degree of mobility may be limited to a maximum amount of mobile nodes. Also, the protocol stack and the utilized transceiver may limit mobility speed, as the communication range is limited and a node may move outside the range before having a chance to send or receive data.

- **Data accuracy**: A node detects a physical phenomenon within certain sensing coverage that is affected by the physical sensor and environmental obstacles. As a network may have redundant sensors and as the measurements in all areas are not equally important, energy efficiency can be improved by switching off some of the nodes. For example, the majority of the nodes in an intruder detection WSN are initially on power-save mode (sleep and low-sensing interval), while border nodes may be more active than the other nodes. When an intruder is detected, nodes are switched on for tracking movement and to determine the type of intruder.

- **Energy usage**: As computation is often much less energy consuming than transmitting, some of the communications may be traded against computation. For example, data may be preprocessed to fit into smaller packets or by performing data aggregation. However, the aggregation has a trade-off between energy usage and reliability, as a large amount of data may be lost on a missed packet. In addition, when the aggregation is performed by combining non-redundant data, the data accuracy decreases.

11.4 QoS Support in Protocol Layers

In this section, we define QoS parameters that each protocol layer controls. Related research is presented on each layer.

Each protocol layer builds on the foundation provided by the lower layers, as show in Figure 11.2. Upper protocol layers define the requirements that ultimately come from the application layer. Based on the available capabilities, the upper layer controls the operation of the lower layer by selecting the level of service. The lower layer provides feedback of its operation, such as link failures, which allows the upper layer to adjust its operation accordingly.

11.4.1 Application Layer

The application layer has the best knowledge regarding the importance of the data. Therefore, an application associates a generated packet with its QoS requirements. The network aims to fulfill these requirements, while minimizing the energy consumption.

The application layer is also responsible for making the sensor measurements and controlling sensors. As such, an application may configure sensors based on measurement accuracy versus time interval, thus making a trade-off between the data accuracy and energy usage.

11.4.2 Transport Layer

The two main tasks for the transport layer are congestion control and reliable transmission. Typically, congestion control limitations the sending of traffic to reduce the utilized bandwidth. As the congestion is reduced, the overall reliability in the network is increased because the data link layer does not have to drop frames. However, throughput limitations may increase delays as the source node must hold onto the generated packets that much longer. Therefore, QoS awareness is required to make a decision regarding which traffic is being more limited than other traffic.

The energy efficiency of a transport protocol depends on the number of transmissions required to deliver a message. This is greatly affected by the acknowledgment scheme being utilized. With the positive acknowledgment scheme, the receiver acknowledges each packet. If an acknowledgment is not received within a certain time interval, the packet is resent. With the negative acknowledgment scheme, missing packets are requested by the receiver node. This reduces messaging as acknowledgments are sent only when required.

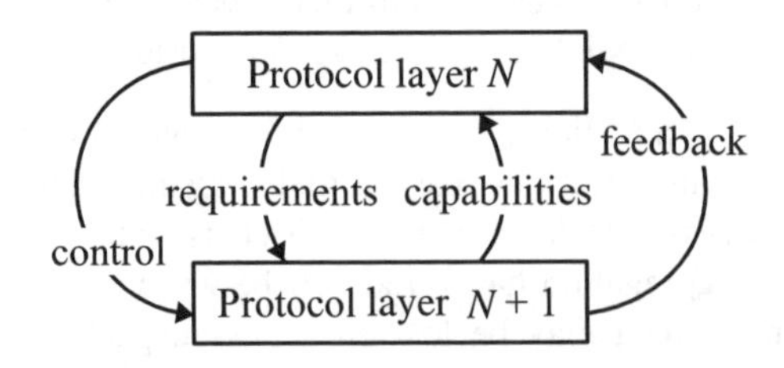

Figure 11.2 Layered approach for providing QoS.

However, the drawback of the scheme is that the sender does not know when a packet is successfully transmitted and must therefore buffer sent packets longer than with positive acknowledgments.

Related work

Due to resource constrain, WSNs rarely implement the transport layer. Instead, the transport layer functionality is often combined with the other layers.

Pump-Slowly, Fetch-Quickly (PSFQ) (Wan et al. 2005) is a WSN communication protocol that allows reliable data transfer with low communication cost. To allow operation on resource-constrained sensor nodes, PSFQ combines the functionality of the transport and network layers. In PSFQ, data is transmitted with relatively slow speed by delaying data forwarding with two configured time values $T_{\min}$ and $T_{\max}$. In broadcast networks, the $T_{\min}$ parameter allows a node to receive a frame multiple times. A node then evaluates the necessity to forward the frame based on how many times it was received. If a sequence number gap in a received frame is detected, PSFQ uses a negative acknowledgment to request all missed frames. The frame is requested in less than $T_{\min}$, which allows reducing latency on error situations.

11.4.3 Network Layer

The network layer controls QoS with traffic shaping and routing protocol. The traffic shaping performs congestion control by classifying packets and providing queuing disciplines that provide per class QoS and fairness. For example, a node may drop low-priority traffic to ensure enough resources for higher priority data. The routing protocol is responsible for selecting an end-to-end routing path fulfilling the desired QoS characteristics. As a route that maximizes one QoS metric may not be optimal on others, the route selection has to make a trade-off between different QoS metrics.

For maximizing network lifetime, a routing protocol not only tries to minimize the energy used for routing the packet, but also perform load balancing between nodes to prevent highly loaded nodes from dying prematurely. However, the shortest path route is not always the most energy efficient. Because the transmission power requirement is proportional to the square of the distance, it might be more energy efficient to forward data through two short hops than through one long hop. As longer routes usually have higher latency, the route selection therefore has to make a trade-off between energy and delay.

The routing protocol design can make the selection between maintenance and routing energy consumption, thus determining how dynamic the protocol is. Generally, the location-based routing techniques provide the best mobility, while the table-based routing techniques are more energy efficient in static networks.

Related work

SPEED (Tian He et al. 2003) uses stateless nondeterministic geographic forwarding and guarantees a certain delivery speed for a packet. In SPEED, the next hop is selected randomly among the neighbors with the probability that is proportional to the link speed. Only the nodes that advance towards the target and meet the delivery time are included in the selection. The link speed is calculated by dividing the distance between nodes (obtained with the geographic location information) by measured link delay. The next-hop selection is combined with feedback received from neighbors. If a node cannot forward a packet due to congestion or a hole in the network, it sends a backpressure beacon. The received backpressure beacons reduce the forwarding probability. SPEED is especially suitable for realtime data delivery, but it does not include any other QoS metrics in its routing decisions.

Sequential Assignment Routing (SAR) (Sohrabi et al. 2000) is a sender-decided routing protocol that forms a multipath tree rooted at a sink node. Next-hop is selected based on the energy cost, QoS metric, and priority level of a packet. The QoS metric may be defined as required, for example, it may be based on link delay. On each hop, SAR calculates weighted cost from the link cost and priority level assigned for a packet. Thus, a higher priority packet results in a lower weighted cost and can traverse through low energy but higher QoS links. The drawback of the SAR is that the changes in QoS metrics, energy, or topology require recomputation of routes. SAR recovers from these changes by performing periodic updates initiated by the sink node.

11.4.4 Data Link Layer

The data link layer achieves energy efficiency by the design choices of the channel access scheme, the selection between synchronized and unsynchronized data exchanges, and the usage of the low duty-cycle operation. These features are discussed more extensively in Chapter 7.

Low duty-cycle operation and access cycle length have a significant impact on available QoS. While a smaller duty cycle requires less energy, it also decreases available throughput, as there are less transmission opportunities per access cycle. In addition, duty cycling also increases latency because a node must wait until the receiver wakes up before it can send a frame. Longer access cycles increase the waiting time, thus further increasing the delay. Therefore, low duty-cycle operation has a trade-off between latency, throughput, and energy usage.

The data link layer is responsible for dividing bandwidth to the traffic based on their priority and QoS requirements. TDMA-based MAC protocols can naturally implement per flow throughput guarantees, because the slot usage of each node is negotiated separately. On the other hand, CSMA can provide per class bandwidth differentiation by assigning each priority with different backoff times, similar to the IEEE 802.11e. The flexibility of the bandwidth usage is another factor. In this respect, CSMA is better as nodes may use bandwidth as required.

CSMA has the further benefit of having smaller delays on lightly loaded networks as a node does not have to wait for its slot to transmit. However, when the network congests, delays and aggregate throughput on CSMA get worse due to the increased number of collisions.

The data link layer mainly controls reliability with the used retransmission scheme, but also by avoiding the collisions and hidden node problems. In addition, the adjustment of transmission power affects reliability. High transmission power enables more reliable transmission, but on the other hand, might cause additional interference within a network.

The mobility support on the MAC layer is largely dependent on how often a node can communicate with its neighbors. Low duty-cycle operation and long access cycles are bad for mobility, as a node may have already moved outside the communication range before it has the chance to communicate. In addition, the requirement for extensive association hand-shaking or transmission slot reservation may also limit mobility.

Related work

Most of the sensor MAC protocols concentrate on minimizing energy. These protocols are presented in detail in Chapter 7.

Q-MAC (Yang Liu et al. 2005) is a CSMA-based sensor MAC that priorizes frames so that higher priority packets experience lower latencies. The traffic is classified into five internal queues, from which the next transmit frame is obtained with the internal priority-based fair scheduling algorithm. The distributed scheduling between nodes is performed by adjusting the contention window according to the priority, thus giving higher priority frames faster channel access. The priority is calculated from an application-defined importance rating, the number of transmitted hops to give packets that have traveled more hops a higher priority, residual energy to increase the importance of low-energy nodes, and the proportional load of the queue to avoid overload on a queue. In simulations, Q-MAC is able to differentiate latency between priorities while using the same amount of energy and having comparable throughput as S-MAC.

11.4.5 Physical Layer

The physical layer comprises not only the transceiver, but also MCU, sensors, and the energy source. Therefore, the physical layer determines limits for the other layers.

While the transceiver causes most of the energy usage, it also imposes several other limitations to the communication protocols. The data rate limits maximum achievable throughput, whereas the used coding scheme affects reliability. As the communication range is limited, the transceiver determines the minimum network density that is needed to route data. MCU puts limits to computational capabilities and available memory, therefore preventing complex protocols and applications. Also, its energy efficiency in sleep and active modes have a significant impact on energy usage.

Physical sensors have certain accuracy and acquisition time-limiting sampling intervals. To overcome these limitations, the network may need to sample data in several nodes on the same region and combine this data to get more detailed values, thus consuming more energy. Still, if a sensor supports selecting sensing accuracy, the accuracy may be purposefully reduced to make a trade-off against energy.

11.5 Summary

While WSNs and traditional networks share the same QoS concepts, the WSN QoS is significantly different. A WSN does not consider some traditional QoS parameters, such as jitter, but defines new parameters. Energy-efficiency is crucial in WSNs, which require a long network lifetime and the avoidance of maintenance. In addition, due to the data centric nature of a sensor network, data accuracy should also be considered as a QoS parameter.

The design choices of the network protocol layers affects the obtainable QoS. The QoS parameters that can be controlled on each protocol layer are summarized in Table 11.1. The operation on separate layers can be complementary. For example, a radio on the physical

Table 11.1 The QoS parameters on each layer.

Layers	Throughput	Reliability	Latency	Security	Mobility	Data accuracy	Energy usage
Application	○	○	○	●	○	●	●
Transport	●	●	●	●	○	○	●
Network	●	●	●	●	●	○	●
Data link	●	●	●	●	●	○	●
Physical	●	●	●	●	●	●	●

layer affects the reliability of a single transmission; the MAC layer is responsible for the reliability of a link; and the routing and transport layers are responsible for true end-to-end reliability. Also, for a real QoS, cooperation between layers is required. Otherwise, each layer may try to maximize different QoS metrics, which will have unpredictable and possibly undesirable results.

12

Security in WSNs

A large part of WSN applications require protection for the data transfer as well as for the nodes themselves. For instance, unauthorized parties should not be able to access private patient information, suppress burglar alarms, or tamper with heating systems. Compared to other wireless technologies, such as WLANs, processing resources and power supplies are significantly more stringent in WSNs (Chapter 2). This calls for highly efficient security designs and implementations. Furthermore, ad hoc networking, multihop routing, as well as node capturing and Denial-of-Service (DoS) threats place new challenges on the security of WSNs, specifically on key management, authentication, routing, and physical protection. This chapter surveys various aspects, alternatives, and proposals for realizing security in WSNs, concentrating on the solutions providing security through cryptographic mechanisms (Menezes et al. 1996).

12.1 WSN Security Threats and Countermeasures*

Security can be defined as a state of defense against willful acts of smart adversaries – people. Security implicitly also covers *safety* to some extent. Safety is defined as the defense against random events, such as accidents and failures. Security design involves specifying a selection of procedures for providing security, such as algorithms, protocols, and their usage. Another key aspect of security development, specifically in WSNs, is to design an efficient implementation of the selected procedures, consisting of their components and interaction. A security attack is defined as any action that attempts to compromise the security of a system. A security threat is the possibility of executing an attack.

WSNs share the security threats of other communication networks, consisting of message interception, modification, and fabrication as well as interruption of communications and operation (Stallings 1995). However, the threats are specifically inherent in WSNs due their wireless operation environment and special constraints, which enable new forms and combinations of attacks.

*With kind permission of Springer Science and Business Media.

Even though WSNs themselves typically have limited capabilities, an attacker can possess powerful tools, e.g. a laptop and a sensitive antenna, for making attacking more effective (Karlof and Wagner 2003). Attackers can be divided into outsiders and malicious insiders (Karlof and Wagner 2003; Shi and Perrig 2004). Whereas an outsider is not an authorized participant of a WSN, an insider may have the knowledge of all the secret parameters of a WSN, such as cryptographic keys, and thus is able to perform more severe attacks. An outsider can become an insider by compromising a WSN node. It is desired that a secure WSN blocks outsider attacks and that security only *gracefully degrades* (or, is *resilient*) in case of an insider attack (Shi and Perrig 2004). This is also related to the fault tolerance requirement specified in Section 2.3, Chapter 2.

12.1.1 Passive Attacks

Interception attacks form the group of passive attacks on WSNs. A passive attack can either result in the disclosure of message contents (eavesdropping) or successful *traffic analysis* (Stallings 1995). Traffic analysis refers to a situation in which the attacker is not able to learn the actual message contents but is able to find out useful information through analyzing message headers, message sizes, transmission frequencies, etc. In WSNs, interception attacks can be performed by gathering information exchanged between nodes, particularly at data aggregation points. Besides regular data transfer, routing information can be exploited for traffic analysis.

Discovering message contents, including exchanged routing tables, can be thwarted by encrypting transmissions as long as encryption keys remain unknown to attackers. Analysis of traffic patterns can classically be deterred by maintaining a constant flow of encrypted traffic, even when there is nothing to transmit (Avancha et al. 2004). However, the solution is typically not suited for WSNs as they should minimize the radio usage for power conservation. Another solution for hindering traffic analysis is to tunnel messages so that their final destinations addresses are encrypted.

12.1.2 Active Attacks

Active attacks on WSNs consists of modification and fabrication of information and interruption in its various forms (Stallings 1995). Interruption is also known as DoS. Impersonation and message replay are the two instances of fabrication. Modification includes changing, delaying, deleting, and reordering messages and stored data. Various modifications and fabrications can be prevented with cryptographic procedures. However, instead of directly tampering with the data itself, in some WSN applications modifications can also be performed by affecting the sensed phenomenon, which cannot be restrained by using cryptography. For example, in a temperature-monitoring application a sensor node can be relocated and heated, which implies false readings. Of the methods for performing active attacks, WSNs are particularly vulnerable to node capturing, resource exhaustion, and tampered routing information.

Node capturing

As WSN nodes are often deployed in publicly accessible locations, they are susceptible to capturing attacks (Perrig et al. 2004). After capturing a node, the attacker can attempt to discover its cryptographic keys or other sensitive information, to destroy it, or to reverse-engineer and modify its functionalities (Wood and Stankovic 2002). Compromising a single node can jeopardize even the whole WSN by allowing insider attacks. Regular node failures can also randomly cause similar effects to capturing attacks (Avancha et al. 2004; Shi and Perrig 2004; Wood and Stankovic 2002).

Routing attacks

Despite that routing is an important aspect in WSNs, their routing protocols are often simple, and thus more susceptible to attacks than general-purpose ac hoc routing protocols (Karlof and Wagner 2003). A number of routing attacks for realizing DoS, performing selective forwarding of advantageous messages, and attracting traffic to a malicious destination for interception have been identified for WSNs (Karlof and Wagner 2003). The methods include routing information modifications, HELLO flooding, acknowledgment spoofing, as well as sinkhole, Sybil, and wormhole attacks. Routing information modifications can be used for creating routing loops, luring traffic, extending routes, and partitioning WSNs. HELLO flooding and spoofed acknowledgments allow advertising non-existent or weak links. Sinkhole, Sybil, and wormhole attacks facilitate attracting traffic to a compromised node or to chosen parts of a WSN.

The countermeasures against routing attacks include link layer encryption and authentication for unicast, multicast, and broadcast transmissions, multipath routing, bidirectional link verification, and geographical routing (Karlof and Wagner 2003). Bidirectional link verification ensures that a link can equally be used for both directions. Geographical routing integrates location information to routing decisions.

12.2 Security Architectures for WSNs

As discussed, a large part of the attacks on WSNs can be prevented by means of cryptography. In this section, selected cryptographic security architectures designed for WSNs are reviewed, namely TinySec (Karlof et al. 2004), SPINS (Perrig et al. 2002), the IEEE 802.15.4 standard (IEE 2003b), and the ZigBee specification (Zig 2004). Also, the Bluetooth security is reviewed as Bluetooth is the most widely used low-cost and low-power wireless technology.

12.2.1 TinySec*

TinySec (Karlof et al. 2004) is a security architecture for protecting the link layer of WSNs developed at the University of California, Berkley. The design goal has been to provide an adequate level of security with the limited resources of WSNs. TinySec supports data authentication by protecting transmissions with Message Integrity Codes (MICs) and confidentiality by encrypting transmissions. Encryption is performed with a block cipher in the Cipher Block Chaining (CBC) mode and MICs are computed with the cipher in the CBC-MIC mode. Applications can configure TinySec to apply only MICs or both MICs and encryption to transmissions. Freshness protection of messages is consciously excluded

*With kind permission of Springer Science and Business Media.

as it is consider too resource demanding. Key distribution or entity authentication schemes have not been specified.

12.2.2 SPINS*

Another academic design called SPINS (Perrig et al. 2002) is a suite of WSN security protocols. It consists of two main components, Secure Network Encryption Protocol (SNEP) and μTESLA. Whereas SNEP supports the data authentication and confidentiality of two-way communications, μTESLA is a protocol for the data authentication of broadcast messages. SPINS includes centralized key distribution, as discussed in Section 12.3.3. It has also been used for building authenticated routing mechanisms.

SNEP encrypts messages and protects them with MICs. Different keys, derived from a shared master key between the two communicating nodes, are used for each different purpose and communication direction. A counter value is included into messages for freshness. Encryption is performed with a block cipher in the Counter (CTR) mode and data authentication in the CBC-MIC mode.

Instead of computationally expensive public-key algorithms, μTESLA uses symmetric-key cryptography. The required asymmetry, i.e. an authentic message can only be generated by a single party but its authenticity can be verified by others, is based on the delayed disclosure of symmetric keys. A broadcast message is protected with a MIC, for which the sender discloses the verification key at a predetermined time instant. In order to work, it is required that the clocks of nodes are loosely synchronized and the sender has enough space for storing the chain of verification keys.

12.2.3 IEEE 802.15.4 Security*

The MAC layer of the IEEE 802.15.4 standard supports three security-related modes: *unsecured*, Access Control List (ACL), and *secured*. The unsecured mode does not include security mechanisms. The ACL mode corresponds to non-cryptographic access filtering using lists of authorized MAC addresses.

The secured mode uses cryptography for providing confidentiality and/or data authentication for the wireless links through seven security suites with varying protection levels. As the schemes are based on the work of the IEEE 802.11i WLAN task group, they also utilize AES in WLAN-related modes of operation. Confidentiality is supported through encryption using the 128-bit-key AES (NIS 2001) in the CTR mode and data authentication through MICs in the CBC-MIC mode. The combination is offered with AES in the CTR with CBC-MIC (CCM) mode.

The ACL entries of a device define the agreed security suites and link keys for each peer address. The 128-bit link key is directly used in the confidentiality and data authentication procedures. The same link key is only allowed to be used for one purpose. Each ACL entry contains sequence counters that are used as inputs to the cryptographic procedures and optionally for freshness (replay) protection. A single default entry is used when a node operating in a network is not explicitly listed but the secured mode is still desired. The default key can be used as a group key as well.

---

*With kind permission of Springer Science and Business Media.

*With kind permission of Springer Science and Business Media.

The standard does not define mechanisms for key distribution or entity authentication. Instead, keys are assumed to be pre-shared secrets or provided by higher protocol layers. Other weaknesses, such as shortcomings in the freshness protection of the CTR suite, have also been identified for the 802.15.4 security design (Sastry and Wagner 2004).

12.2.4 ZigBee Security

The commercial ZigBee specification (Zig 2004) builds on the top of the security design of 802.15.4 by slightly modifying it and specifying key management and entity authentication procedures.

Entity authentication and key management in ZigBee

The Zigbee security is based on three types of secret keys: *pairwise* keys, *network keys*, and *master keys*. The keys can be either dynamically distributed or pre-installed into nodes. Pairwise keys are used for protecting communications at the APS or MAC layer, shown in Figure 3.4. The network key is typically a network-wide key, which can be used at the APS, NWK, and MAC layers. ZigBee nodes can contain several network keys for supporting group keying but only one of the network keys is used for protecting outgoing frames, referred to as the *active network key*. Master keys are long-term keys optionally utilized for establishing the other types of keys.

The ZigBee specification introduces a concept of Trust Center (TC), which is an entity trusted by the other nodes within a ZigBee network. Typically, TC is the node coordinating the network. TC utilizes the APS services for distributing network keys and pairwise keys to enable network-wide and end-to-end security. In order to protect the distribution process, the other nodes must contain a pre-shared master key with TC. Also, a preconfigured network key can be used for protecting pairwise key distributions from outsiders. If pre-shared keys do not exist, keys can still be distributed but the phase is completely open to outsiders.

ZigBee supports two types of configurations related to the role of TC. In the *commercial mode*, TC maintains a list of devices, master keys, pairwise keys, and network keys and enforces security policies as well as controls the network access. In the *residential mode*, TC maintains a network key and controls the network access but lists of devices, master keys, or link keys are not required. TC can be seen to correspond to an authentication server of a WLAN operating in the infrastructure mode.

Entity authentication in ZigBee is carried out with a four-message challenge-response scheme called Symmetric-Key Key Exchange (SKKE). The procedure uses a master key, which the nodes request from TC. The computations utilize AES with 128-bit keys. The authentication procedure does not require direct one-hop connections. Instead, messages can be conveyed through one or more intermediate nodes. Before the authentication procedure begins, the initiator learns the address of the peer, e.g. through service discovery. The nodes can also directly request a pairwise key from TC, in which case they do not need to execute the SKKE protocol before starting a protected communication session.

A same type of an authentication procedure is performed between TC and a new node requesting access to the WSN in the commercial mode. In the residential mode, TC simply decides whether to accept the node based on its address. If required, the network key is transmitted unprotected to the new node. In addition to the time of authentication, TC can periodically run key updates for the nodes of its network.

Confidentiality and integrity in Zigbee

Instead of supporting the separate MIC and CTR suites of 802.15.4, ZigBee specifies a new mode of operation called CCM* for confidentiality and data authentication. Also, CCM* uses AES but in addition to the standard CCM operation, it offers encryption-only and MIC-only configurations. The different configurations are referred to as security levels in the specification. The processing in these configurations is equal to CCM but either the encrypted data or the MIC are discarded. The new mode was defined in order to enable the use of a single key throughout the security levels as well as on the different layers of the ZigBee stack.

When security is turned on, by default the MAC layer is configured to use CCM* in the same security level and with the same sequence counters as NWK. The active network key is used in the default 802.15.4 ACL entry. The pairwise ACL entries contain the APS pairwise keys, sequence counters, and security levels of neighboring nodes. The usage of the pairwise entries is preferred but if they do not exist, the default entry is used. The MAC layer protects only the frames originating from itself. Otherwise the security processing at the MAC layer corresponds to that of the 802.15.4 standard.

Correspondingly to the MAC layer, the APS and NWK layers operate on frames. NWK invokes protection when its frames require protection or when the layer is configured to protect all the traffic of the higher protocol layers. Outgoing frames are processed in CCM* using the active network key and the default NWK security level provided by the higher protocol layers. For incoming frames, NWK maintains a list of neighboring nodes and their security material. Upon the reception of a frame, NWK fetches the corresponding network key and other parameters for the source address, and processes in CCM* using the same security level as for all the outgoing frames.

APS provides security for the frames originating from the application layer. The frames can be protected with either pairwise keys or network keys. Similarly to the lower protocol layers, APS uses CCM* with the NWK security level configured by an application or ZDO. Equally to NWK, outgoing frames are protected with the active network key if network protection is requested. However, if NWK is also applying security, APS does not invoke protection. Otherwise, a pairwise key with the corresponding security material and the NWK security level is used.

For incoming frames APS uses the network keys in the same way as NWK. If a pairwise key has been used for the protection, APS fetches the security material corresponding to the source address from its list of peer nodes and processes in CCM* using the NWK security level. Pairwise keys are preferred for APS command frames and for them the security level is always CCM* with 128-bit MICs. Special keys derived from pairwise keys are utilized for protecting commands used for key establishments and transports.

As a summary, the utilization of a network key provides protection at the network level and the utilization of pairwise keys protection at the node level. The MAC layer can only protect its own transmissions between neighboring nodes whereas the APS layer can provide application layer transmissions with end-to-end protection.

12.2.5 Bluetooth Security

A Bluetooth device can operate in three security-related modes (Blu 2004). In the *non-secure* mode, the device does not initiate any security procedures. In the *service-level enforced security* mode, the device does not initiate security procedures before a channel establishment at the L2CAP layer. In the *link-level enforced security* mode, security procedures are initiated before a Bluetooth link has been established. Whereas the service-level enforced security mode supports different security policies for parallel applications, the link-level enforced security mode enforces the same MAC-layer security level for all connections. In addition to the operating modes, security levels for devices and services have been specified (Blu 2004; Muller 1999).

When security is applied, Bluetooth implements key management, entity authentication, confidentiality, and integrity protection. Security processing is carried out at the baseband and controlled by LM according to the requirements of the protocol layers above the MAC layer. Due to the large number of adjustable parameters, Bluetooth SIG has published additional recommendations for configuring the security procedures in different profiles (Gehrmann 2002). Furthermore, the ambiguities of the Bluetooth specification related to the encryption of piconet broadcasts have been clarified in a separate publication (Morris 2002).

Due to the popularity of the technology, the Bluetooth security has been extensively researched and various vulnerabilities have been identified. A summary of the weaknesses with a comprehensive reference list can be found in (Hämäläinen 2006).

Entity authentication and key management in Bluetooth

The Bluetooth security is based on three types of link keys, *initialization keys*, *combination keys*, and *master keys*. The earlier versions of the Bluetooth specification included a fourth type of a link key called *unit key* but its usage has been deprecated in the newest specification (Blu 2004) due to security problems (Jakobsson and Wetzel 2001). The 128-bit link key is utilized in entity authentication. Depending on its type and the desired level of protection, the link key is also used for generating an encryption key.

An initialization key is typically only used when two Bluetooth devices establish a connection for the first time. The key is generated from a Personal Identification Number (PIN) code, manually supplied to both the devices. A combination key is generated from information shared between two Bluetooth devices. The sharing of the generation information is protected with the effective link key. A master key is a temporary group key distributed by the piconet master and used for protecting broadcast transmissions.

After exchanging a new link key, the devices verify its correctness and each other's identities by subsequently running a challenge-response authentication procedure. The devices are identified by unique Bluetooth addresses. The procedure involving the creation of an initialization key, using it to protect the exchange of a new link key, and running the authentication procedure with the new link key is called *pairing*.

Key generations and the authentication procedure utilize algorithms referred to as E_1, E_{21}, E_{22}, and E_3. They are all based on the SAFER+ block cipher. Despite of the similarities, compared to the shared key authentication method of Wired Equivalent Privacy (WEP) in IEEE 802.11 (IEE 1997), the Bluetooth authentication scheme is more robust. It is believed that an attacker cannot recover useful information from the exchanged authentication messages when proper parameters are used.

Confidentiality and integrity in Bluetooth

Bluetooth optionally provides confidentiality through encrypting frame payloads. Encryption is performed with a proprietary stream cipher called E_0, which is based on four parallel Linear Feedback Shift Registers (LFSRs). Before proceeding with encryption, devices agree on the size of the encryption key, which can vary between 8 and 128 bits. The encryption key is derived from the link key and parameters provided by the master device. E_0 is initialized with the encryption key and the realtime clock of the master, which ensures that each stream produced with the same key is different. For integrity, a CRC checksum is computed and appended to the frame payload prior to encryption.

12.3 Key Distribution in WSNs*

Due to the multihop communications, large number of nodes, ac hoc networking, and resource constraints, distribution of secret keys becomes one of the most challenging security components of WSNs. In order to be applicable and secure, each of the security architectures described above requires a key distribution method. Various distribution solutions and keying mechanisms exist that are suited to different WSN applications and deployment scenarios.

12.3.1 Public-key Cryptography

Public-key cryptography is commonly used for key distribution in traditional computer systems, e.g. on the Internet. For WSNs the benefits of public-key mechanisms are the resiliency against node capturing, the possibility to revoke compromised keys, and scalability (Chan et al. 2004). However, public-key algorithms are computationally intensive and public-key protocols require exchanging large messages, consisting of keys and their certificates. Therefore, the mechanisms are often considered poorly suited for WSNs. For example, the expensive computations and message exchanges can be exploited for DoS. Nevertheless, according to recent research results, Elliptic Curve Cryptography (ECC) can be a feasible solution for some WSN applications (Gupta et al. 2005; Gura et al. 2004). Still, symmetric-key algorithms can be computed considerably more efficiently than ECC algorithms in typical WSN nodes (Hämäläinen et al. 2006).

12.3.2 Pre-distributed Keys

The simplest keying mechanism is to use a single, pre-installed symmetric key for the whole WSN, such as the network key in ZigBee. The solution results in the lowest resource and management requirements as well as enables nodes to create protected connections with all the other nodes in the network (Chan et al. 2004; Karlof et al. 2004). A drawback is that when a single node is compromised, the security of the whole WSN is lost. The

*With kind permission of Springer Science and Business Media.

solution also allows full-scale insider attacks. The effects of node compromises can be decreased, e.g. by using the network-wide key only for setting up pairwise keys during the establishment of the WSN, erasing it afterwards (Chan et al. 2004). However, this solution prevents adding nodes after the initial WSN deployment.

The other extreme is to pre-distribute unique keys for each pair of nodes (Chan et al. 2004), e.g. the way the PIN code is used in Bluetooth. The technique is resilient to node capturing as only the keys of compromised nodes are leaked. Also, key revocation can be supported (Chan et al. 2004). The drawbacks are poor scalability and high storage requirements since in a WSN of n nodes each node has to store $n - 1$ keys, totalling $n(n - 1)/2$ keys for the network.

A solution between the network-wide and pairwise keys is to use group keying (Karlof et al. 2004), as in the case of the default key in IEEE 802.15.4. In this scheme, a node shares a symmetric key with its neighbors. A group-key compromise allows only accessing the communications within the group. The solution does not support creating protected connections between all the nodes in the network.

Depending on the WSN application and available resources, the described keying mechanisms can also be combined (Zhu et al. 2003). Each node maintains a certain number of pairwise and group keys and a network-wide key, which facilitate protected pairwise and group communications within a subset of nodes as well as local and global broadcasts.

Random key pre-distribution

WSN key distribution based on randomly chosen pools of symmetric keys has aroused interest in the research community. In the original scheme (Eschenauer and Gligor 2002) a large set of keys is generated and distributed to nodes during the WSN setup. Each node is allocated a subset of the generated keys such that with a certain probability each pair of nodes ends up sharing at least one key. After the allocation the nodes perform key discovery in order to find out which of their neighbors they share keys with. The links protected with the discovered keys can be further used for agreeing on new keys, called *path keys*. The path keys enable direct communications with neighbors the nodes initially did not share a key with. The scheme has been developed further to be better resilient to node capturing, e.g. by requiring larger number of overlapping keys (Chan et al. 2004) and exploiting WSN deployment information in advance (Du et al. 2006).

The benefit of the random key pre-distribution over unique pairwise keys is the decreased amount of key storage and over network-wide keys the resiliency to node capturing. The mechanism also supports key revocation and re-keying (Eschenauer and Gligor 2002). However, in order to be applicable the node density has to be high and uniform in the scheme. A drawback of the basic scheme is that the randomly shared keys cannot unambiguously be used for entity authentication as the same keys can be shared by more than a single pair of nodes (Chan et al. 2004).

12.3.3 Centralized Key Distribution

Along with public-key cryptography, Key Distribution Centers (KDCs) are utilized for providing authentication and key distribution services in communication systems, e.g. in WLANs. A KDC can be used in WSNs as well (Chan et al. 2004; Hämäläinen et al. 2006; Perrig et al. 2002). Nodes authenticate to a KDC, which generates a symmetric key and securely communicates it to the nodes. For example, TC of a ZigBee network operates as a KDC. It is required that nodes can establish secure channels to the KDC, e.g. through pre-shared symmetric keys, and that the KDC is trusted and has capacity for storing the channel establishment information of all nodes.

The storage requirements in WSN nodes are low as each node has to store permanently only a single key (the node–KDC key). Furthermore, since authentications and key establishments of a WSN are controlled in a centralized location, the scheme is resilient to node capturing, supports simple node revocation, and protects against node replication attacks (Chan et al. 2004). The drawbacks of the scheme are that nodes establishing a connection must always communicate with the KDC first and that the KDC becomes an appealing target for attacks. If the KDC is accessed through multiple hops, the authentication latencies increase, the energy consumption of the nodes close to the KDC grows, and these nodes become targets of DoS attacks. The situation can be alleviated by locating the KDC outside the hostile operation environment of the rest of the WSN and distributing and/or replicating its functionalities (Hämäläinen et al. 2006).

12.4 Summary of WSN Security Considerations

A large part of WSN applications require protection against malicious parties. WSNs are prone to the same attacks as all the other communication networks and their devices. However, the wireless operation environment, self-organization, and the special constraints enable new forms and combinations of attacks. Hence, designing and implementing a reliable security solution for WSNs becomes even more challenging. Most of the protection can be realized with the combination of cryptographic mechanisms and physical protection but settling for trade-offs between the level of security, resource requirements, and cost is often needed. A key aspect in the security design is the level of resiliency in the case of an attack. Similar to the design of other aspects of WSNs, instead of attempting to develop a solution suited for all possible usage scenarios, for cost efficiency the WSN security design should be developed considering the threat model of the target application.

Part IV

TUTWSN

13

TUTWSN MAC Protocol

This section presents the design options and rationalizes the implementation choices for the MAC protocol of TUTWSN. We will focus on network topology, channel access and slot reservation mechanisms, and frequency channel utilization. In addition, we will present algorithms, protocols, and parameter optimizations, which further improve network energy efficiency and performance in dynamic environments.

13.1 Network Topology

A network can be formed using flat or clustered topology (Baker 1981). In a flat topology, all nodes are logically homogenous and they can form communication links with any other node in their range. In a clustered topology, network is formed as clusters of nodes, where each cluster consists of a cluster head and several leaf nodes. Cluster members may consist of low-energy leaf nodes, cluster heads of neighboring clusters, and possibly intercluster gateway nodes (Baker 1981). The leaf nodes can communicate only with other nodes within the cluster. The communication with nodes outside the cluster must be routed through the cluster head.

The network topology significantly affects network energy consumption and bandwidth utilization. In the flat topology, all nodes manage their data exchanges with neighbors similarly. This is inefficient from an energy and scalability point of view. As an example, in synchronized low duty-cycle MACs, each node must transmit and receive beacons, and manage its own active data exchange period (superframe) for communication. Hence, energy is consumed even without actual data exchanges. In addition, each node periodically reserves a frequency channel during which other transmissions in the same frequency channel within two radio ranges (interference range) cause collisions. This limits the network scalability in dense networks, where potentially hundreds of nodes are within the interference range.

In the clustered topology, only the cluster heads are responsible for the management of data exchanges. Most of the nodes are leaf nodes that transmit data to cluster heads on a on-demand basis and under the control of cluster heads. This maximizes the energy efficiency and minimizes the network resource utilization for the leaf nodes. As the number

Ultra-Low Energy Wireless Sensor Networks in Practice: Theory, Realization and Deployment

of leaf nodes is high compared to cluster heads, network energy efficiency and scalability are significantly improved compared to a flat network.

Since the activity of cluster heads and possible intercluster gateway nodes is much higher than that of leaf nodes, their energy consumption may limit the network lifetime. However, the energy consumption can be balanced by rotating the roles of nodes, as proposed in LEACH (Heinzelman et al. 2002) and PACT (Pei and Chien 2001). Since only a portion of nodes route data, a clustered topology can limit throughput and increase latency. However, the criticality of these drawbacks is low in typical WSN applications.

Due to energy and scalability reasons, we have selected the clustered topology for TUTWSN. For simplicity, intercluster gateway nodes are not used and cluster heads communicate directly with each other. Nodes dynamically alter their roles between a cluster head and a leaf node according to energy resources, loading, and the connectivity of the neighboring cluster heads.

Data forwarding from cluster heads to sinks can be performed by direct communications, as in LEACH (Heinzelman et al. 2002), or as multihop networks using a mesh or a cluster-tree topology. Direct communications severely limit the network coverage area or necessitate the utilization of very high transmission power levels. The reduction of a transmission distance to half decreases the required transmission power to between $\frac{1}{4}$ and $\frac{1}{8}$ depending on the environment. Although, frame receptions consume energy too, a significant energy saving is achieved in large networks by using multihopping.

In a mesh topology, each cluster head can communicate with any other cluster head in its range. This produces a robust and highly connected network, where data can be routed between any two nodes. Yet, this high connectivity is very difficult to achieve in synchronized low duty-cycle protocols, since synchronization with numerous cluster heads is required. This necessitates periodic beacon receptions from all neighboring cluster heads consuming lots of processing and memory resources, and energy.

In a cluster-tree topology, each cluster head maintains connectivity with a single cluster head (parent) one-hop closer to a sink acting as a root of the tree structure. In addition, each cluster head may have several child cluster heads, which form the branches of the tree. The resulting cluster-tree topology is very energy efficient for data routing to one sink node. A disadvantage is that the network robustness against link and node failures is weak due to low connectivity. In addition, data routing between different branches of the tree is inefficient.

For combining the advances of a well-organized and energy-efficient cluster-tree topology with the flexibility of a mesh topology, TUTWSN utilizes a multicluster-tree topology, as presented in Figure 13.1. The multicluster-tree topology consists of several superpositioned cluster-tree structures. In TUTWSN, the cluster heads are called *headnodes*, while end-devices are called *subnodes*. Each headnode associates with a few neighboring headnodes, which may have paths to different sinks with different energy and latency performance metrics. On each hop, a node determines an optimal next-hop node for each routed packet. Thus, packets are forwarded along a mixture of different trees. Sinks are logically similar to headnodes and manage their active data exchange periods. The directions of uplink routing paths (towards a sink) are presented as arrows in the figure. In practice, synchronization with two or three neighbors is feasible depending on the application and routing protocol. The energy consumption of additional beacon receptions is compensated by a more robust network structure, more efficient data routing, and the possibility for load balancing between different routes.

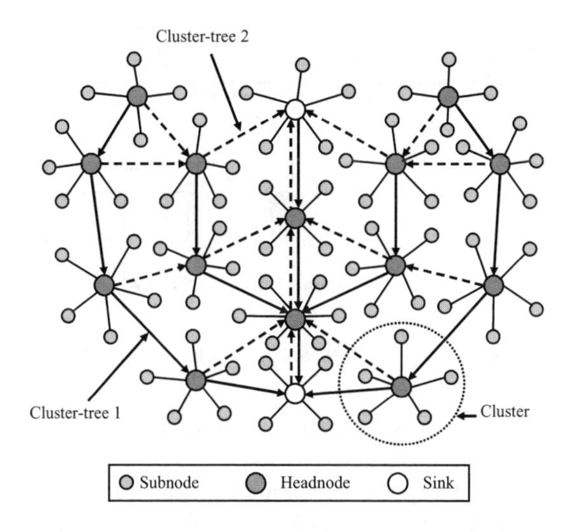

Figure 13.1 TUTWSN multicluster-tree topology.

13.2 Channel Access

Synchronized low duty-cycle MACs can achieve very high energy efficiency in static networks due to low idle listening, as presented in Chapter 7. In dynamic networks, the energy consumption of network scans reduces the achieved energy efficiency. However, with current low-power radio technology, the synchronized approach has the highest potential for fulfilling the energy and performance requirements of WSNs. Next, we will focus on the synchronized low duty-cycle MACs and discuss energy-efficient channel access mechanism in superframes.

Data exchanges in superframes can be performed using contention or contention-free channel access mechanism. The contention mechanism has good support for burst-type traffic and a dynamic network topology, since a source can decide when to transmit data. Energy is consumed in carrier sensing and backoff mechanism prior to transmissions. As the accurate moments for transmissions are not predetermined, a destination node needs to be in reception mode whenever a source node may transmit data. In addition, collisions may be frequent in high congestion conditions causing retransmissions and delays, and increased energy consumption.

The contention-free mechanism reduces collisions, since a destination node assigns non-conflicting time and/or frequency slots for each source node. Energy efficiency increases significantly, since frames are received reliably in a timely manner. One disadvantage is the control packet overheads due to slot reservations. Although slots are typically requested for a continuous bandwidth, the overheads may be severe in a dynamic network, where next-hop nodes change frequently. The slots may be requested during a separate network formation phase, as in SMACS (Sohrabi et al. 2000), or in a short contention access period included in superframes. The latter alternative is more suitable for dynamic networks, where new links should be formed rapidly after link failures. Also, occasional data can be

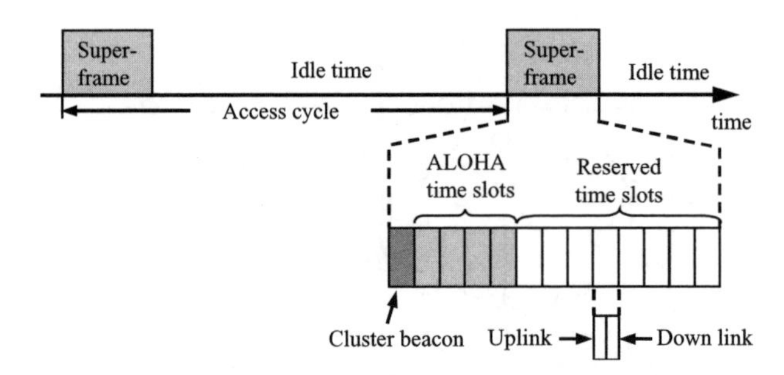

Figure 13.2 TUTWSN access cycle and a superframe.

transmitted in the contention period without a reserved time slot. This results in a hybrid contention/contention-free mechanism.

TUTWSN MAC utilizes the hybrid contention/contention-free mechanism by combining reserved time slots with a slotted ALOHA (Roberts 1975), as presented in Figure 13.2. The channel is accessed in consecutive access cycles, which consist of a superframe and idle time. Each superframe consists of a cluster beacon, a slotted ALOHA contention access period, and reserved time slots for contention-free channel access. At the beginning of each superframe, a headnode transmits a cluster beacon, which contains current assignments for reserved time slots, timing, and cluster status information for network management. ALOHA slots are used mainly for association and slot reservation requests, while reserved time slots are preferred for data frames. Thus, the amount of traffic and collision probability in ALOHA slots remain low.

A headnode allocates the reserved slots in each superframe according to the bandwidth requests of associated nodes. A node may request more bandwidth by setting a control bit in the header of a transmitted data frame. Bandwidth is decreased, if a node does not transmit a frame in a time slot assigned to it. Thus, the reserved bandwidth is rapidly de-allocated if a node moves out of range of its headnode or it dies. For energy efficiency, the numbers of ALOHA and reserved slots are dynamically adjusted in each superframe according to traffic loading.

The slotted ALOHA is selected because of its simplicity and high energy efficiency in low traffic conditions. All data exchanges are accurately synchronized in TUTWSN without carrier sensing, backoff delay, and unnecessary channel reception prior to transmissions, as presented in Figure 13.3. Idle listening is minimized in both source and destination nodes by transmitting all frames accurately on slot boundaries in very short data exchange periods. This is applied to both ALOHA and reserved slots. The relatively long idle time between data exchange periods is well suited for WSNs, since the data processing by low-power MCUs typically causes a significant delay between consequent data transmissions and receptions.

For comparison, the CSMA-CA mechanism of IEEE 802.15.4 is presented in Figure 13.4. Prior to a transmission, a source node waits for a random backoff delay, performs the CCA twice, and transmits a frame. Then, the source node listens to the channel for receiving an ACK frame. The destination node (cluster head) should be active for the entire CAP. Hence, the expected energy efficiency in TUTWSN is much higher than in IEEE 802.15.4 with equal throughput, especially in cluster heads.

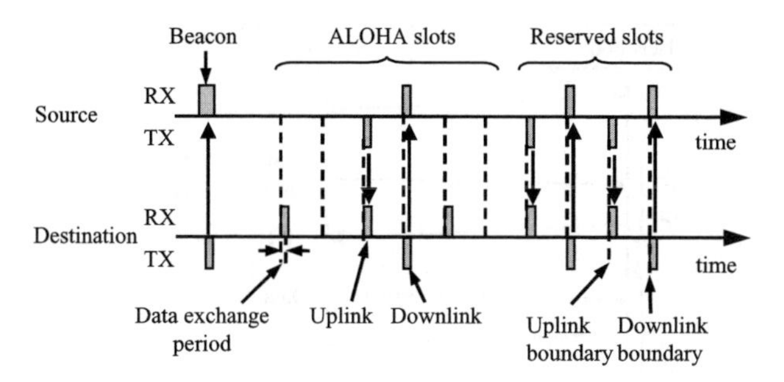

Figure 13.3 TUTWSN channel access with accurate slot timing for minimizing idle listening.

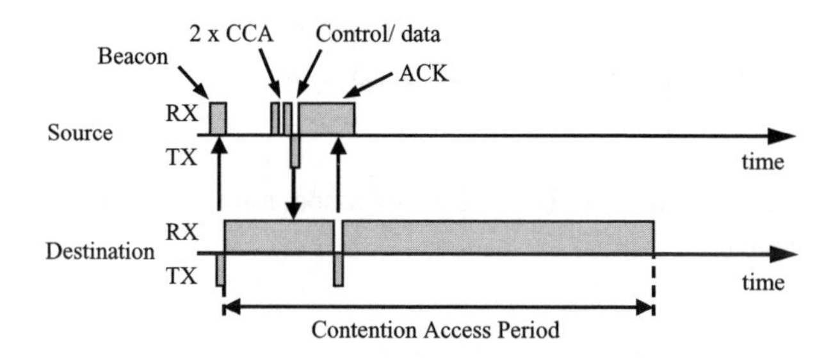

Figure 13.4 The CSMA-CA mechanism of IEEE 802.15.4.

In a TUTWSN MAC superframe, each slot is divided into uplink and downlink slots so that the uplink slot is used for a data or control frame, and the downlink slot is used for an ACK frame from the destination node to the source node of the former uplink slot. Since an ACK message requires only a couple bytes of data, an ACK frame also contains MAC Service Data Unit (MSDU) for a downlink data payload allowing energy-efficient bidirectional communication. A broadcast from a headnode to all cluster members is performed efficiently by reserving one slot for the headnode and transmitting the broadcast frame in the reserved uplink slot (logically to downlink direction). The following downlink slot is not used. Broadcasting is typically utilized for data requests and network-configuration data flooded from the sink nodes to the network.

Performance and energy efficiency are further improved in dynamic networks by piggybacking, as presented in Figure 13.5. When a node (subnode or headnode) is associating with a headnode, it transmits association and slot reservation requests and a data payload in a single frame in ALOHA. The headnode responds with an ACK. If a free slot exists in a current superframe, the headnode may assign it to the node by a control field in ACK. This effectively reduces traffic control overheads and data routing delay after link failures.

13.3 Frequency Division

Although the reserved time slots effectively eliminate collisions inside clusters, interference between neighboring clusters may be high without appropriate superframe interlacing. A

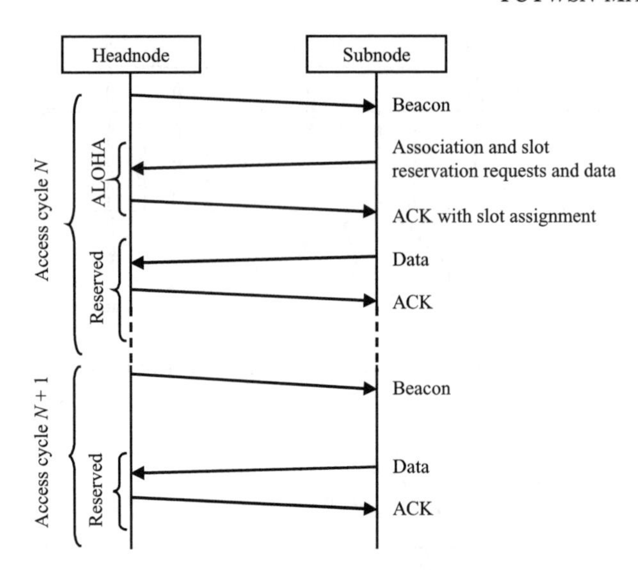

Figure 13.5 Frame exchanges in TUTWSN for association and slot reservation requests, and data transmissions.

time, frequency, or code division can be used. As low-power CDMA transceivers suitable for WSNs are quite rare, we consider only time and frequency division.

In a time division, an entire network operates on a single frequency channel and superframes are positioned consecutively among neighboring nodes, as in ZigBee (Zig 2004) and DMAC (Gang Lu et al. 2004). Although the required network throughput is small, each superframe occupies a relatively long period due to the required time for beacon transmissions, channel access delays, retransmissions, and data processing delays of low-power MCUs. This reduces the maximum number of non-overlapping superframes on a single channel and thus, the allowable number of clusters in the interference range. Still, the avoidance of interference is difficult since the clock frequencies of neighboring headnodes may slightly differ, causing time drifting and potentially overlapped superframes.

Intercluster interference can be reduced by using frequency division between clusters. Although, a crystal inaccuracy may slightly shift the carrier frequency of a radio, the error does not cumulate over the time as in the time division. The maximum number of clusters operating within the interference range equals the number of available non-interfering frequency channels. For narrow band transceivers operating at the 2.4 GHz band and having 1 MHz bandwidth, the number of non-interfering frequency channels is typically around 30.

For maximizing scalability with minimum interferences, TUTWSN utilizes both time and frequency division. As an example, a typical TUTWSN configuration that utilizes a 5 second access cycle resulting space for 15 non-overlapping superframes on each channel is considered. When each cluster has, for example, 12 subnodes, a total of 195 nodes can operate within the interference range in a single channel. Assuming that a transceiver can provide 30 non-interfering frequency channels, a total of 450 clusters and 5850 nodes can operate within the interference range. If nodes are deployed to an area

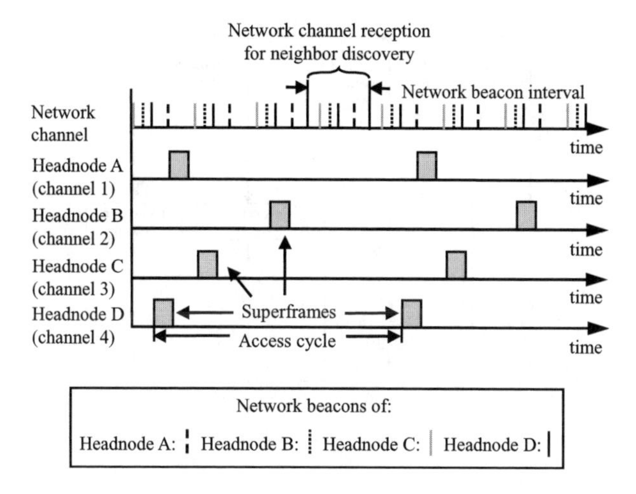

Figure 13.6 Neighbor discovery with network beacons.

that is larger than the interference range, the maximum number of nodes can be even higher.

Although the time and frequency division of superframes improve network scalability, a neighbor discovery may require a long and energy-consuming network scan. For example, when a 5 second beacon interval and 30 frequency channels are used, this results at worst case in 2.5 minutes of continuous channel reception, during which normal node operation is suspended. For avoiding this drawback, TUTWSN utilizes a common network signaling channel (network channel) dedicated for a special type of short network beacons. As presented in Figure 13.6, network beacons are broadcast by all headnodes once per *network beacon interval* during access-cycle idle time. Network beacons contain information for a node to select a suitable headnode and to obtain synchronization with it. Since the network beacons do not interfere with data transmissions, they can be transmitted very frequently compared to the access-cycle length. Hence, all neighboring nodes can be discovered by receiving the network channel for one network beacon interval (Kohvakka et al. 2005b), as presented in Figure 13.6. However, it is often possible to terminate the scan after the required number of neighbors with sufficient signal strengths have been found, which further reduces the energy consumption.

The operation of TUTWSN MAC is illustrated as a simplified state machine in Figure 13.7. Normally, MAC is in sleep state and wakes up upon the interrupt of a scheduler. Possible state transitions from the sleep mode are the node's own superframe, the neighbor's superframe, the node's own network beacon, and a network scan in order of priority. After these operations, execution is always returned to sleep mode. The node's own superframe begins by the transmission of a cluster beacon, after which data exchanges are performed in ALOHA and reserved slots. The neighbor's superframe begins by the reception of the neighbor's cluster beacon. If the beacon reception fails, operation returns to the sleep mode. If the beacon is received, data are exchanged in ALOHA and reserved slots according to the current slot assignment. The node enters a network scan after a link failure or when the number of known neighbors is not sufficient. The scan is terminated

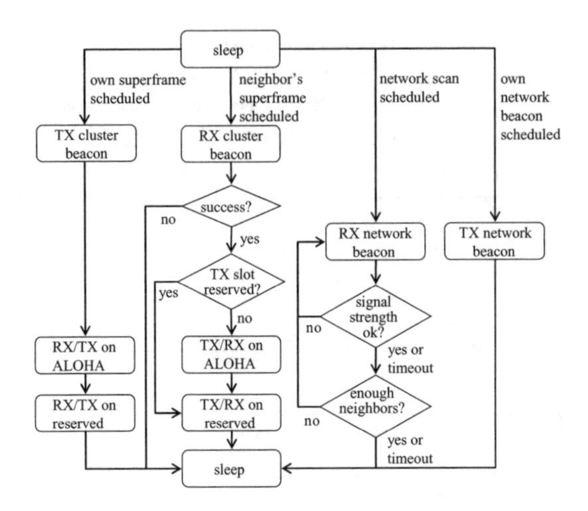

Figure 13.7 State machine representing the operation of TUTWSN MAC.

after the number of new neighbors with sufficient signal strength is adequate (a protocol parameter) or a timeout has occurred.

13.4 Advanced Mobility Support*

WSNs are usually considered to be rather static. In practice, even an immobile network has a dynamic behavior, which is caused by very low transmission power levels and random node failures, together with a dynamic operating environment, such as opened and closed doors, moving objects, and changing weather conditions affecting RF propagation (Polastre et al. 2004b; Suhonen et al. 2006a). Moreover, numerous envisioned WSN applications, such as access control, assets tracking, and interactive games necessitate node mobility causing a very dynamic network topology (Karl and Willig 2005). Thus, WSN nodes should be able to recognize changes in the network topology and to update maintained communication links rapidly, energy efficiently, and with minimum interruptions on data routing. This is a very difficult design problem. In current WSN proposals utilizing synchronized low duty-cycle MAC protocols, new neighbors are discovered by long-term network scans or neighbor discovery periods, which suspend data routing and require constant radio reception. Clearly, this significantly reduces network energy efficiency and data routing performance in a dynamic network.

The operation and energy efficiency in mobile networks can be improved by distributing synchronization information in networks for assisting neighbor discovery. For this, an Energy-efficient Neighbor Discovery Protocol (ENDP) is proposed. The protocol permits mobility for all nodes without significant energy consumption or performance degradation by reducing the need for network scans. The protocol consists of two parts: *proactive distribution of neighbor information*, and *neighbor discovery algorithm* for searching new neighbors according to this information.

*This article was published in Ad Hoc Networks, M. Kohvakka, J. Suhonen, M. Kuorilehto, V. Kaseva, M. Hännikäinen, T.D. Hämäläinen, *Energy-Efficient Neighbor Discovery Protocol for Mobile Wireless Sensor Networks*, Copyright Elsevier (2007).

13.4.1 Proactive Distribution of Neighbor Information

The leading principle in the distribution of neighbor information is the piggybacking of the information in MAC beacons. ENDP operates above a synchronized low duty-cycle MAC and utilizes the payloads of existing MAC beacons for the information distribution. ENDP supports both clustered and non-clustered network topologies. In clustered networks, cluster heads perform the transmission of beacons and the distribution of neighbor information. Yet, all nodes can receive the neighbor information in beacons improving their energy efficiency. In non-clustered networks, all nodes execute ENDP similarly.

The utilization of beacons has several benefits. First, beacon exchanges are synchronized by a MAC protocol. This provides very low-idle listening and energy overheads due to additional control signal exchanges. Second, beacons are quite short, which provides energy-efficient implementation. In addition, there is no need for a new frame type. Most importantly, the beacons are transmitted in any case for maintaining link synchronization.

The distributed neighbor information consists of synchronization information about all parents with which synchronization is currently maintained. Complete synchronization information referring to one node is called a Synchronization Data Unit (SDU). SDU consists of two parts: the channel on which the node is operating, and the time-base difference (offset) between the node and the sender of SDU. The latter is used to avoid the need for a global time and to reduce the required value range, thus fitting the schedule information into fewer bits.

The distribution of information is presented in Figure 13.8. Nodes I and L are parents to node M. Thus, node M receives beacon transmissions from nodes I and L. Similarly, node M is a parent to node K, which receives beacons from M. It is assumed that even

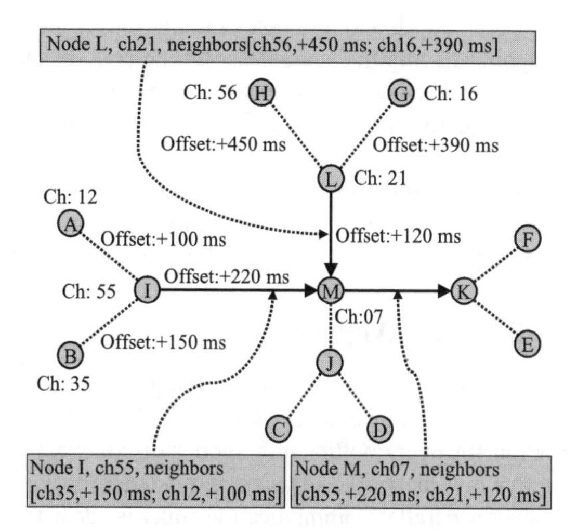

Figure 13.8 Connectivity and distributed neighbor information. Arrows indicate synchronization between nodes ($k = 2$). This article was published in Ad Hoc Networks, M. Kohvakka, J. Suhonen, M. Kuorilehto, V. Kaseva, M. Hännikäinen, T.D. Hämäläinen, *Energy-Efficient Neighbor Discovery Protocol for Mobile Wireless Sensor Networks*, Copyright Elsevier (2007).

if it is physically possible for M to receive beacon transmissions from J, the node should only maintain synchronization with a maximum of k parents, which is in this case 2.

Each node that transmits beacons derives SDUs from its parents and piggybacks them in a beacon payload. In this case, node I transmit the SDUs about nodes A and B. Similarly, L transmits SDUs about nodes H and G. Due to the beacon receptions from nodes I and L, M has accumulated SDUs about nodes A, B, G, and H.

Next, the processing of SDUs that node K receives from node M are considered. Clearly, it would be possible that node M sends all SDUs it knows. Yet, the number of nodes from which SDUs accumulate will rise exponentially on each step towards further nodes. At the same time, the inaccuracy of offset values cumulates degrading the reliability of the protocol. Thus, it is feasible to send SDUs referring to direct parents only, and the beacon transmission of node M contains SDUs referring to nodes I and L. Node K determines the beacon time t_I^{beacon} of node I by employing the SDU sent by neighbor M as

$$t_I^{\text{beacon}} = t_M^{\text{beacon}} + t_{\text{SDU}}^{\text{offset}},$$

where t_M^{beacon} is the known beacon transmission time of node M and $t_{\text{SDU}}^{\text{offset}}$ is the offset stored in the SDU.

Since beacons are broadcast, some of the SDUs node M receives from I and L may be duplicates or refer to already known nodes. However, these situations can be detected since SDU determines the frequency and time moment unambiguously. Upon receiving a new SDU, the timing is compared against other known neighbors. Because beacons are sent periodically, two beacon times (t_1, t_2) are the same, if

$$\exists i, i \in \mathbb{Z} : t_1 - t_2 + i \cdot t_{\text{AC}} < \delta,$$

where t_{AC} is the access cycle length and δ denotes the inaccuracy of the offset value. The SDU is ignored if a neighbor having the same beacon time and channel exists. Otherwise, the neighbor information is updated. As neighbors may move out of communication range, stored information is periodically checked and references to entries that are not recently updated are removed.

Since a SDU consist of only a few bytes of data, several SDUs can be fit into one beacon implying only a minimal energy overhead. By receiving one beacon from each parent, a node gets valid neighborhood information in a two-hop radius, rapidly and energy efficiently.

13.4.2 Neighbor-discovery Algorithm

The neighbor-discovery algorithm operates on three main principles: maintaining redundant communication links, anticipating movement based on the link quality changes, and using the distributed information for energy-efficient neighbor discovery. For assuring a continuous data routing in a dynamic network, adequate redundancy of communication links is essential. The use of several parallel communication links is ideal for a multicluster-tree network topology, like TUTWSN. In the presented design, communication links are maintained with k neighbors. The value of k depends on the frequency of link changes and thus, the physical network topology and the degree of network dynamics.

The neighbor-discovery algorithm is summarized in the following steps:

1. An network scan is performed to detect initial neighbors.

2. Synchronization with selected neighbors is maintained.

3. As the synchronization is maintained, a node receives SDUs regarding its neighborhood.

4. Based on the received SDU information, a node replaces its synchronized neighbors for better connectivity.

5. Steps 2–4 are iterated.

6. If a node loses synchronization to all of its neighbors, Step 1 is resumed.

The neighbor-discovery algorithm is presented more formally as pseudocode in Figures 13.9 and 13.10. The algorithm presentation consists of two functions, NEIGHBOR-DISCOVERY, which maintains connectivity to the neighbors, and SYNCHRONIZE, which performs beacon reception according to the SDU information. The pseudocode presentation considers only Steps 1 and 4, as Step 2 involves simply receiving and handling cluster beacons at the right time, while the SDU maintenance in Step 3 is described in the previous section. The symbols that are used in the pseudocode are summarized in Table 13.1.

In the algorithm, N_* is used to describe a list of neighbors that are discovered with a network scan and the received SDUs. The synchronized neighbors are listed in N_S. Unexpected link breaks are prevented by observing link quality changes and replacing low-quality links with better ones. Because dynamic networks do not allow long-term observation of link quality, RSSI is used for fast link quality evaluation. The algorithm adapts to the movement of nodes by preferring neighbors that are coming closer (N_+) and avoiding nodes that are moving away (N_-). N_+ and N_- lists are updated upon a

```
SYNCHRONIZE(N0, qmin, qend)
 1: N = Ø
 2: for all i ∈ N* do
 3:   if i ∉ NS and source(i) ∩ N0 ≠ Ø then
 4:     receive a beacon from n
 5:     if reception was successful then
 6:       update rssi(i) based on the beacon reception
 7:       if rssi(i) ≥ qend then
 8:         return i
 9:       else if rssi(i) ≥ qmin then
10:         N ← N ∪ {i}
11:       end if
12:     end if
13:   end if
14: end for
15: return i ∈ N that has the highest RSSI
```

Figure 13.9 Beacon reception in neighbor-discovery algorithm. This article was published in Ad Hoc Networks, M. Kohvakka, J. Suhonen, M. Kuorilehto, V. Kaseva, M. Hännikäinen, T.D. Hämäläinen, *Energy-Efficient Neighbor Discovery Protocol for Mobile Wireless Sensor Networks*, Copyright Elsevier (2007).

```
NEIGHBOR-DISCOVERY()
 1: if N_S = ∅ then
 2:   if timer t_s has expired then
 3:     perform network scan until high quality link or k
        nodes is found
 4:     N_S ← nodes found during the scan
 5:     reset timer t_s
 6:   end if
 7: end if
 8: if |N_S| < k then
 9:   N_S ← N_S ∪ SYNCHRONIZE(N_S, q_0, q_+)
10: end if
11: if |N_S| = k then
12:   for all i ∈ N_S do
13:     j = ∅
14:     if i ∈ N_- then
15:       j ← SYNCHRONIZE(N_+, q_0, q_0)
16:     end if
17:     if j = ∅ and i ∉ N_+ and rssi(n) < q_+ then
18:       j ← SYNCHRONIZE(N_S, q_+, q_+)
19:     end if
20:     if j ≠ ∅ then
21:       N_S ← N_S \ {i} ∪ {j}
22:     end if
23:   end for
24: end if
```

Figure 13.10 Neighbor-discovery algorithm. This article was published in Ad Hoc Networks, M. Kohvakka, J. Suhonen, M. Kuorilehto, V. Kaseva, M. Hännikäinen, T.D. Hämäläinen, *Energy-Efficient Neighbor Discovery Protocol for Mobile Wireless Sensor Networks*, Copyright Elsevier (2007).

beacon reception (not presented in the pseudocode). Therefore, a movement is determined according to the changes of RSSI value.

The `SYNCHRONIZE` function tries synchronization based on its parameter values N_0, $q_{\min}$, and q_{end}. The function loops through all known neighbors (rows 2–14 in Figure 13.9) and attempts gaining a synchronization by receiving network beacons from each neighbor i that: a) a node is not already synchronized to ($i \notin N_S$); and b) is advertised by one of the other neighbors listed in parameter N_0. The synchronization is stopped when a neighbor with link quality of q_{end} or higher is found (rows 7–8 in Figure 13.9) or synchronization to all advertised neighbors is attempted. If the `SYNCHRONIZE` function cannot find a neighbor with good link quality, it returns one of the detected neighbors that has the highest link quality. However, when several choices with equal or almost equal link quality are available, the neighbors that advertise different SDUs are selected (not described in the pseudocode). This way, a node gets more comprehensive neighborhood information and thus, has more choices for neighbor selection, which adds robustness.

Table 13.1 Symbols used in the neighbor-discovery algorithm. This article was published in Ad Hoc Networks, M. Kohvakka, J. Suhonen, M. Kuorilehto, V. Kaseva, M. Hännikäinen, T.D. Hämäläinen, *Energy-Efficient Neighbor Discovery Protocol for Mobile Wireless Sensor Networks*, Copyright Elsevier (2007).

Symbol	Description
k	Preferred number of nodes that synchronization is maintained to
q_0	Limit for minimum link quality required for reception
q_+	Limit for good link quality
$rssi(n)$	Link quality to the node n
N_*	List of all known neighbors
N_S	List of synchronized neighbor nodes, $N_S \subseteq N_*$
N_+	List of nodes n whose $rssi(n)$ is increasing, $N_+ \subseteq N_S$
N_-	List of nodes n whose $rssi(n)$ is decreasing, $N_- \subseteq N_S$
$source(n)$	Set of nodes that have advertised node n, $source(n) \subseteq N_S$
t_s	Network scan timer
N_0	List of nodes to which neighbors' synchronization is attempted
q_{min}	Minimum acceptable link quality
q_{end}	Preferred link quality ($q_{end} \geq q_{min}$)
N, i, j	Temporary variables

In the `NEIGHBOR-DISCOVERY` function, a node initially performs a network scan as it does not know any neighbors (rows 1–7 in Figure 13.10). A timer is used to prevent constant scanning if a neighbor within the communication range does not exist. A neighbor having a good link quality has a high probability of advertising other neighbors that are within communication range, whereas a low quality link is unreliable and might break. Therefore, for ensuring that a new energy-consuming network scan is not required, the scan is continued until either a neighbor with a high RSSI or k neighbors are found. After the scan, the node synchronizes to the neighbors with the highest link quality.

The SDU information is used to get the connectivity up to k neighbors (rows 8–10 in Figure 13.10). Based on the given parameters, the `SYNCHRONIZE` function returns immediately after detecting a neighbor that has a good link (q_+). However, if such a neighbor is not found, a neighbor that has sufficient link quality (q_0) is returned.

The latter part of the algorithm monitors neighbor information and attempts to replace these low-quality links with better ones (rows 11–24 in Figure 13.10). First, if a neighbor is moving away, a replacement is searched from the SDUs advertised by neighbors that are coming closer (rows 14–16 in Figure 13.10). The search finishes after detecting a new neighbor with any acceptable link quality (q_0). Even a low-quality link is accepted, since the node is most probably moving towards the advertisers. Thus, it is probable that the link qualities of the accepted neighbors will also increase.

If a node has bad connectivity to a neighbor, the algorithm tries to find a replacement neighbor (rows 17–19 in Figure 13.10). The reason for this replacement is that a high-quality link has a low frame error rate and allows saving energy with transmission power control. However, the neighbors that are coming closer ($i \in N_+$) are ignored, because their RSSI is getting better and thus their link quality will eventually be good.

It should be noted that since the timings are known exactly, a node may turn its radio off when the reception is not expected. Thus, the neighbor detection that utilizes distributed information has an additional benefit over the traditional network scan, as a node can sleep or communicate with its neighbors while waiting for a beacon reception in the synchronization part of the algorithm.

13.4.3 Measured Performance of ENDP Protocol

The performance of the implemented ENDP is measured experimentally with the TUTWSN prototype platforms and the TUTWSN protocol stack. A total of eight nodes are configured to headnodes and deployed in an office environment, as presented in Figure 13.11. One node is mobile in the stationary sensor field as presented in the figure. The radio range varies from 10 m to 120 m depending on the node position. The maximum range is achieved via a long corridor, while walls attenuate RF propagation very significantly.

The average power consumption is determined by powering the mobile node by a 0.2 F (farad) (C), super capacitor and measuring the slope of capacitor terminal voltage. In practice, the elapsed time, (t), when capacitor terminal voltage, (U), decreases from 5.0 V to 4.0 V is measured. Since the prototype platforms are equipped with linear voltage regulators, the relatively high input voltage does not affect the prototype current consumption; except a negligible small increase of a quiescent current. Thus, the average power consumption, (P), is calculated as

$$P = \frac{\partial U}{\partial t} C U',$$

where U' is the nominal battery voltage (3.0 V) of the prototype platform. Thus, the resulting power consumptions are applicable for lithium battery powered nodes.

For the measurements, the mobile node is configured to a subnode operation. This minimizes the data exchange power consumption, which is essential when focusing on the power consumption of network maintenance operations. The mobile node transmits a 32 B diagnostics data frame on a contention-free time slot at 8 s intervals. In addition, control frames for association and slot reservation requests are transmitted after communication link

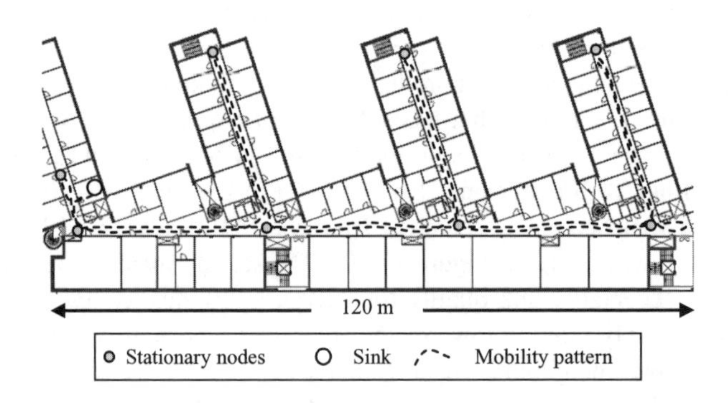

Figure 13.11 Network topology for ENDP performance measurements. This article was published in Ad Hoc Networks, M. Kohvakka, J. Suhonen, M. Kuorilehto, V. Kaseva, M. Hännikäinen, T.D. Hämäläinen, *Energy-Efficient Neighbor Discovery Protocol for Mobile Wireless Sensor Networks*, Copyright Elsevier (2007).

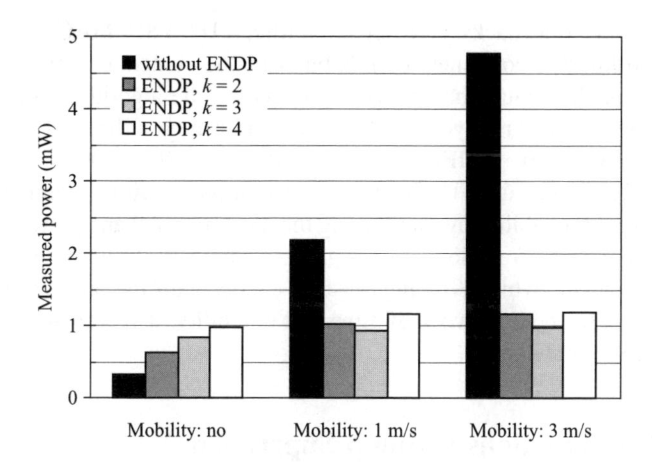

Figure 13.12 Measured power consumption using ENDP at three levels of node mobility, and without ENDP. This article was published in Ad Hoc Networks, M. Kohvakka, J. Suhonen, M. Kuorilehto, V. Kaseva, M. Hännikäinen, T.D. Hämäläinen, *Energy-Efficient Neighbor Discovery Protocol for Mobile Wireless Sensor Networks*, Copyright Elsevier (2007).

changes. The transmission interval of network beacons is fixed to 500 ms, thus defining the maximum duration of a single network scan.

The power consumption of the mobile node is measured while executing ENDP and varying k between 2 and 4. For comparison, power consumption is also measured without ENDP when new neighbors are searched by network scans. Measurements are performed with node velocities of 1 m/s (meters per second) and 3 m/s, and without mobility.

The results are presented in Figure 13.12. The results prove that ENDP effectively reduces the power consumption of a mobile node. In this case, ENDP operates most effectively when k is three. Then, the resulting energy saving is 57%–80% compared to the situation without ENDP. When all nodes are stationary, ENDP increases node power consumption 300 μW–640 μW.

13.5 Advanced Support for Bursty Traffic

The usage of reserved slots is often desirable over contention-based channel access because collisions within a cluster are avoided and energy-consuming idle listening is minimized. However, the traditional problem with the reservations is their inflexibility with the dynamic traffic. Fixed reservations work well in Constant Bit Rate (CBR) traffic, in which reservations match accurately with the actual traffic. However, a WSN often contains extremely bursty traffic, which is triggered by environmental events. When the traffic varies, unused reservations are wasted. Further traffic variation is caused by node mobility and network dynamics due to unreliable wireless communications. Still, reservations should be supported since certain applications, such as surveillance data, require bandwidth guarantees. As there may be spatial and temporal variances in traffic, adaptation and load-balancing mechanisms are required.

To overcome the drawbacks in reserved traffic, TUTWSN MAC uses a novel slot reservation algorithm that combines the flexibility of the contention-based channel access with the benefits of the contention-free channel access. The algorithm itself is targeted at locally synchronized, low duty-cycle WSN MACs and can be utilized in any MAC having both a CAP and a CFP, like in IEEE 802.15.4 LR-WPAN (IEE 2003b). Unlike the other proposed MAC algorithms, the TUTWSN slot reservation algorithm significantly increases energy-efficiency and reliability by minimizing the need for CAP and preferring contention-free slots. The contention-free slots are assigned dynamically to a on-demand basis, thus supporting traffic bursts, while avoiding wasting energy with the unused reservations. As the algorithm increases both network lifetime and reliability, it is well suited, for example, to monitoring WSN applications.

13.5.1 Slot Reservations within a Superframe*

The dynamic slot reservation algorithm consists of two parts that control CFP usage on the MAC layer, an on-demand and a traffic-adaptive CFP slot reservation. These slot reservation methods complement each other by using different approaches to minimize the need for energy-inefficient CAP, but can also be used independently (Suhonen et al. 2007).

The on-demand slot reservation supports unpredictable and bursty traffic by allowing a node to make a request for several dedicated CFP slots within an access cycle. In traffic-adaptive slot reservation, CFP slots are assigned to member nodes based on their long-term reservations (guaranteed traffic) and traffic patterns (adapted traffic) as shown in Figure 13.13 (Suhonen et al. 2007).

This method is used with stable links and is targeted at continuous traffic. The traffic guarantees are obtained with a separate handshake, in which a node asks for certain capacity. Although the guaranteed traffic is not analyzed here, it is important as it allows a resource reservation algorithm to provide end-to-end throughput guarantees by reserving capacity on each hop. The adapted traffic avoids using the CAP by granting CFP slots without explicit request.

To reduce contention channel access, a node may transmit only one frame during each CAP. Also, a node does not use the CAP if it has a CFP slot. A cluster head signals the

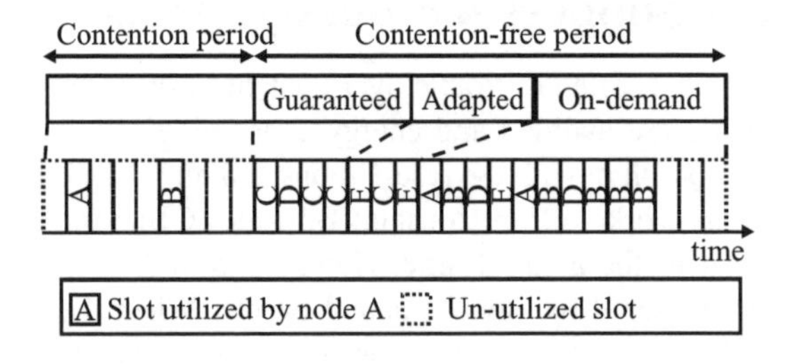

Figure 13.13 Superframe and slot reservation between nodes A–E. © 2007 IEEE. Reprinted, with permission, from *Proceedings of the Design, Automation and Test in Europe*.

*© 2007 IEEE. Reprinted, with permission, from *Proceedings of the Design, Automation and Test in Europe*.

guaranteed and adapted slot usage in a cluster beacon at the beginning of each access cycle. The contention-free slot assignment may be different on every access cycle.

13.5.2 On-demand Slot Reservation*

In on-demand slot reservation, a node makes a requests for a CFP slot from the cluster head by setting a reservation flag in its transmitted frame. This allows a node to use collision-free transmissions. The cluster head replies with an acknowledgment frame that contains the slot index of the granted slot. As the required information is carried in headers, the reservation is performed within a normal data transmission and no additional signaling is required (Suhonen et al. 2007).

A member can make the slot request in any frame sent during the CAP or in its last reserved slot, which allows the efficient use of any excess capacity. For example, in Figure 13.13 node B first asks for a contention-free slot first during the CAP. Then, it asks for an additional reservation four times in the contention-free slots. If the cluster head cannot assign a contention-free slot during the same access cycle, it will grant it on the next access cycle. This way, a member does not need contention access on the next access cycle.

On-demand reservation can also be used with variable-length frames, allowing compatibility with several WSN MACs, such as LR-WPAN. Variable-length frames allow efficient bandwidth usage, because the frame length can be optimally adjusted to the required data length. In an on-demand reservation request, a member uses an estimate of its transmission time instead of setting the reservation flag. The cluster head replies with an exact time stamp of the transmission.

13.5.3 Traffic-adaptive Slot Reservation*

In traffic-adaptive slot reservation, a cluster head keeps a record of the average throughput sent by its member nodes. The CFP slots are assigned to each member according to the recorded throughput. Because a node might not get a CFP slot during every access cycle, the sending of a frame is postponed until the slot is granted or D_P access cycles elapse. This way, a node can benefit from the adapted traffic and avoid sending in CAP. The use of the delay D_P prevents too long waiting delays. The value of D_P can be configured per frame basis, thus allowing service differentiation. If a node has delay-critical data or the cluster head does not grant a reservation in time, CAP is used to send the frame. As other frames might be buffered during the postponing, the on-demand slot reservation is used to request for additional slots. While the postponing introduces an additional delay, it reduces the usage of CAP and thus prefers the more reliable CFP transmissions (Suhonen et al. 2007).

The reservations during a CFP are managed as reservation periods that consist of R_A access cycles. The total number of slots r granted to a node during a reservation period is expressed as $r = r_a + r_c$, where r_a is the number of adapted slots and r_c is the number of requested guaranteed slots. These slots are divided equally between access cycles. This

*© 2007 IEEE. Reprinted, with permission, from *Proceedings of the Design, Automation and Test in Europe.*

keeps average delay low, since a node does not have to wait long for the next reserved slot. The average delay caused by the postponing of a frame transmission to the next granted CFP slot is

$$delay = min\left(\frac{R_A}{r}, D\right) \cdot t_{AC},$$

where t_{AC} is the access cycle length. If $r > R_A$, an additional delay penalty does not occur.

If a node does not have data to send on its contention-free slot, it transmits an empty frame. The cluster head does not reply but frees all the remaining reserved slots during the access cycle. By sending the empty frame, the cluster head can distinguish between a failed transmission and the case of when a node has nothing to send. Because the slots are alternated, other nodes have a better chance of benefiting from the freed slots. For example, if the node C in Figure 13.13 sends an empty frame on its first transmission, the node D could utilize the guaranteed slots originally reserved for the node C.

13.5.4 Performance Analysis*

The performance of the TUTWSN MAC slot reservation algorithm is compared against ZigBee (IEEE 802.15.4), based on the models presented in Chapter 18. The comparison does not use constant reservations, thus emphasizing the dynamic behaviour of the slot reservation algorithm. The analysis does not consider hidden nodes, network maintenance (network scans), or data aggregation. In general, hidden nodes have the greatest impact on contention channel access, especially on CSMA-based algorithms, where the failure of a clear channel assessment leads to collision. Therefore, it is expected that TUTWSN MAC will perform better in respect to IEEE 802.15.4 (LR-WPAN) on the presence of hidden nodes.

The parameters for the models are shown in Table 13.2. The analysis uses a fixed-data frame length of 105 B (L_L) and acknowledgment frame length of 33 B (L_S), both values include a 25 B header overhead. Slots are large enough to allow frame reception, processing, and sending of an acknowledgment. As the purpose of this analysis is to emphasize bandwidth allocation, link failures are not modeled. Furthermore, downlink transmissions are not used.

IEEE 802.15.4 LR-WPAN uses SO set to 2 and BO parameter set to 8, which results in a 61.44 ms superframe and 3.932 s access cycle. TUTWSN MAC uses 250 ms superframe and 4 s access cycle. These values are selected to achieve comparable results in respect of throughput, energy, and delay. For comparison, results with traditional slotted ALOHA are provided. This equals to the case, in which the active period consists solely of CAP (Suhonen et al. 2007).

The analysis is performed with a cluster-tree multihop network, in which network depth (d) is four, each cluster head has two child cluster heads and 5 subnodes (n_s), resulting in a total of 186 nodes. Each node is synchronized exactly to one next-hop cluster and transmits data frames to the root of the cluster-tree topology (sink) (Suhonen et al. 2007).

Table 13.2 Parameters used in the dynamic slot reservation analysis.

Symbol	Parameter	Value
I_{U}	Uplink data transmission interval	1..100 s
I_{D}	Downlink data transmission interval	∞
L_{F}	Beacon/ACK frame length	33 B
L_{P}	Data frame length	105 B
R	Radio data rate	250 kbps
T_{AC}	Access cycle length	4 s
T_{AP}	Active period length	0.25 s
T_{CAP}	CAP length	$0.3 \cdot T_{\mathrm{AP}}$
T_{CFP}	CFP length	$0.7 \cdot T_{\mathrm{AP}}$
t_{I}	Synchronization inaccuracy	100 μs
t_{IA}	Transceiver idle-to-active transient time	192 μs
t_{SI}	Transceiver sleep-to-idle transient time	970 μs
t_{S}	Slot length	10 ms
ϵ	Crystal tolerance	20 ppm
p_{F}	Link failure probability	0

The TUTWSN MAC slot utilization equations in Chapter 18 are slightly modified for the slot reservation algorithm. The ALOHA slot utilization (U_A) is modeled as

$$U_A = min\left(\frac{1}{1 + D_{\mathrm{P}}}\frac{T_{\mathrm{REQ}}}{L_P}, 1\right),$$

where D_{P} is the number of access cycles that a transmission is postponed, T_{REQ} is the generated traffic, and L_P is data frame length. The ALOHA slot utilization is used to derive the reserved slot utilization as

$$U_{\mathrm{R}} = \frac{T_{\mathrm{REQ}}}{L_P} - U_{\mathrm{A}}.$$

The energy consumption analysis is based on the values measured on a real hardware consisting of Microchip PIC18F8722 MCU (Microchip Technology Inc. 2004) and IEEE 802.15.4 compliant Texas Instruments CC2420 transceiver (Texas Instruments Inc. 2007). The static power consumptions on the used hardware are presented in Table 18.3. The -5 dBm value for the transmission power is used. For fair comparison, both TUTWSN MAC and IEEE 802.15.4 analyses use the same equipment.

Two variants of the slot reservation algorithm, denoted as $R(D)$, are considered. Variant R(0) tries to transmit all buffered frames immediately, while R(2) waits two access cycles for a reserved slot. As waiting increases buffering, a node sends only one frame during the CAP, while the other buffered frames are sent in on-demand contention-free slots. Thus, the rationale behind waiting is to reduce the proportion of less reliable CAP transmissions in order to favor CFP transmissions (Suhonen et al. 2007).

The achieved goodput with 3 retransmissions at sink is presented in Figure 13.14. R(2) variant performs slightly better than R(0), because frames are buffered by delaying

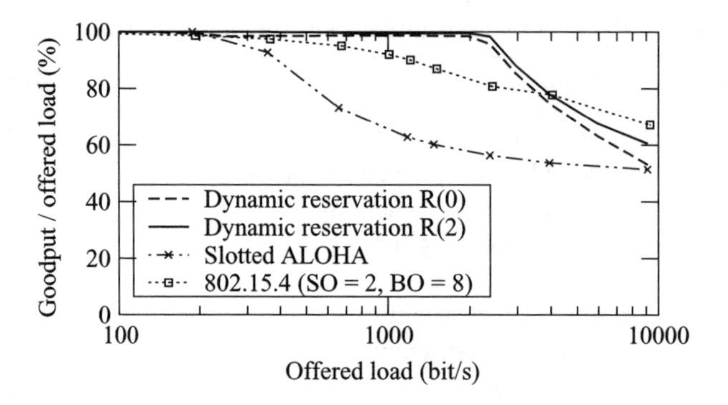

Figure 13.14 Goodput at a sink as the function of total offered load. © 2007 IEEE. Reprinted, with permission, from *Proceedings of the Design, Automation and Test in Europe*.

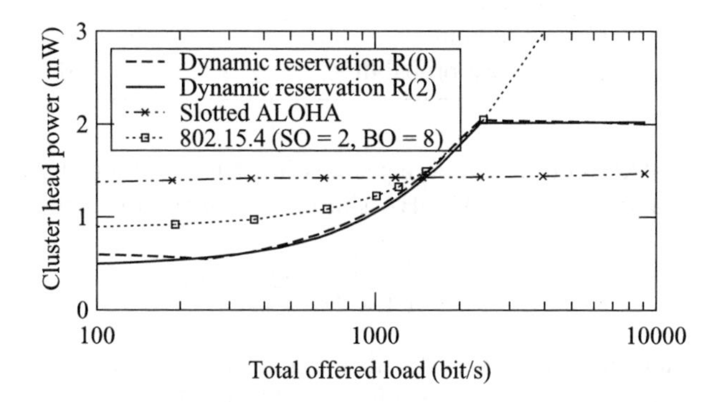

Figure 13.15 Cluster head power consumption with slot reservation algorithm. © 2007 IEEE. Reprinted, with permission, from *Proceedings of 2006 IEEE 17th International Symposium on Personal, Indoor and Mobile Radio Communications*.

sending. When the T_A is exceeded, a node sends one frame in CAP and the rest of the buffered frames in CFP, which decreases the traffic and collision probability in CAP. The goodput of 802.15.4 decreases as the channel becomes congested and backoff times increase. Traditional slotted ALOHA performs poorly, because collisions are common and a node can send only one frame per access cycle (Suhonen et al. 2007).

Cluster head power consumption at 3 V operating voltage is presented in Figure 13.15. On a lightly loaded network, both variants have 50% smaller power consumption than a LR-WPAN coordinator. After the network gets congested, the power consumption of the dynamic slot reservation algorithm approaches LR-WPAN. The nearly constant power consumption of traditional slotted ALOHA is not comparable at high traffic loads as its goodput is low.

As most of the delay in low duty-cycle WSNs is caused by periodic communications, it is expected that LR-WPAN and the dynamic slot reservation algorithm with R(0) have similar delay characteristics. However, R(2) variant will have up to $2 \times t_{AC} = 4$ s higher delay. Thus, as R(2) variant has best goodput characteristics, it can be concluded that adjusting the value of D_P allows making a trade-off between delay and energy efficiency.

13.6 TUTWSN MAC Optimization

13.6.1 Reducing Radio Requirements*

An energy efficient and robust WSN operation requires that nodes are capable of measuring radio wave attenuation (path loss) to neighboring nodes. This information is crucial for efficient network self-configuration and routing; nodes can form and maintain robust routing paths with sufficient radio link quality and energy efficient hop lengths (Sikora et al. 2004). A weakened radio link can be detected and the route changed before an actual link failure. In addition, transmission power levels can be adjusted dynamically for minimizing energy consumption and interferences with other nodes. As radio wave attenuates with the transmission distance (Rappaport 1996), measured path loss can be used for estimating distances between nodes (ranging), and further for node localization (Sichitiu et al. 2003). Since the location of each sensor can be determined automatically, the applicability of gathered data is improved in a mobile network, and even random node deployments are allowed (Kohvakka et al. 2006c).

In practice, path loss can be estimated according to the measured signal strengths of received frames and their known transmission power levels (Sichitiu et al. 2003). Conventionally, signal strength is determined by a RSSI mechanism of a receiver. The RSSI mechanism samples the signal strength after a channel filtering, and stores the value in a register. Yet, the simplest and lowest energy commercial off-the-shelf radio transceivers lack the RSSI mechanism. To be able to utilize the lowest energy transceivers, the signal strength metering should be otherwise implemented (Kohvakka et al. 2006c).

Next, a novel signal strength metering mechanism for low-power WSNs is presented. The method utilizes receiver sensitivity as a simple signal strength meter, and estimates path loss according to successful and failed receptions of beacon frames transmitted at different transmission power levels. This is possible since almost every transceiver supports adjusting the transmission power.

The path loss A is defined as the signal attenuation between the transmitter output and receiver input ports including antenna gain, and the losses in impedance matching network. Thus, A is determined as the ratio of transmission power level P_T to the received signal strength P_R that is measured according to successful and failed frame receptions and receiver sensitivity S_{min}. Receiver sensitivity is defined as the minimum mean power received at the input port at which the Bit Error Rate (BER) does not exceed a specified value (typically 0.1%). This yields that a frame of l bits can be received successfully exactly at the probability of $(1 - BER)^l$, when $P_R = S_{\text{min}}$. For simplification, it is approximated that $P_R \geq S_{\text{min}}$, when a frame is received successfully. The upper and lower limits for the path loss are estimated by retransmitting a frame with different transmission power levels and detecting the lowest transmission power resulting a successful reception $P_{T(\text{successful})}$

*

Table 13.3 Resulted path loss according to successful and failed beacon receptions. © 2006 IEEE. Reprinted, with permission, from *Proceedings of 2006 IEEE 17th International Symposium on Personal, Indoor and Mobile Radio Communications.*

TX		power	level	
−20 dBm	−10 dBm	−6 dBm	0 dBm	Resulted path loss
1	1	1	1	≤65 dB
0	1	1	1	65 − 75 dB
0	0	1	1	75 − 80 dB
0	0	0	1	80 − 85 dB

1 = successful reception, 0 = failed reception.

and the highest power resulting a failed reception $P_{\mathrm{T(failed)}}$ as

$$\frac{max\left(P_{\mathrm{T(failed)}}\right)}{P_{\mathrm{R}}} < A \leq \frac{min\left(P_{\mathrm{T(successful)}}\right)}{P_{\mathrm{R}}}.$$

As an example, calculated path-loss values for a transceiver having $S_{\mathrm{min}} = -85$ dBm and beacon transmission powers from −20 dBm to 0 dBm are presented in Table 13.3 (Kohvakka et al. 2006c). Since path-loss is typically fairly equal in both directions (Reijers et al. 2004), the same path-loss is assumed to apply to transmissions to neighbors.

The path-loss metering is implemented on MAC protocol beacons due to the following reasons. First, the interval of beacon transmissions of low duty-cycle MAC protocols is highly suitable for path-loss measurements. In addition, beacon transmissions are synchronized, which allows the minimization of reception time. Beacons are also fairly short, which provides energy-efficient implementation. Moreover, path-loss information is typically needed together with the node status information transmitted in beacons for network management.

The operational principle of the path-loss metering method is presented in Figure 13.16 (Kohvakka et al. 2006c), where node 3 determines the path losses between itself and two other nodes (node 1 and node 2) by receiving their beacons. The ranges of beacon transmissions with the four power levels are indicated with circles. As node 1 is close to node 3, a beacon transmitted at the second lowest transmission power level can be received, resulting in low path loss. As the distance from node 2 is much longer, only the highest power beacon is successfully received. This indicates a high path loss.

For reducing energy consumption, beacons are transmitted in the increased order of transmission power. This enables that beacon frames are received until one reception succeeds, after which the receiver may be switched to the power down mode. As the contents of the beacons are similar, the reception of higher power beacons does not bring any additional information, and they can be ignored.

Accuracy measurements

The accuracy of the path-loss metering method is measured experimentally by two TUTWSN temperature sensing platforms. The platform is presented in Section 4.3.1. Since

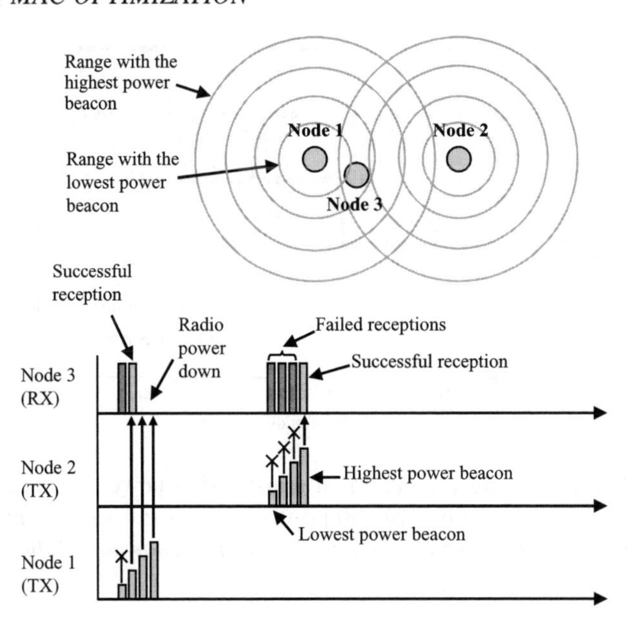

Figure 13.16 The operational principle of path-loss metering. © 2006 IEEE. Reprinted, with permission, from *Proceedings of 2006 IEEE 17th International Symposium on Personal, Indoor and Mobile Radio Communications.*

the transceiver lacks the RSSI mechanism, the platform is suitable for testing the performance of the path loss metering.

The transmission ranges of beacons with four different transmission power levels (−20 dBm, −10 dBm, −5 dBm and 0 dBm) were measured outdoors in a park in a fairly open space, and on a roadside mostly line-of-sight conditions. One platform was placed 1.5 m above snowy ground and configured to transmit beacons periodically. Another node was receiving beacons, while it was moved away from the transmitter around 2 m above the ground. The beams of loop antennas were directed towards each other. The minimum transmission power levels of successful beacon receptions ($min(P_{\mathrm{T(successful)}})$) were recorded as the function of the distance between the transmitting and receiving node, as presented in Figure 13.17 (Kohvakka et al. 2006c). In the open space, a path loss increases quite proportionally to the distance. Beacons transmitted at −20 dBm, −10 dBm, −5 dBm, and 0 dBm power levels are received until around 25 m, 190 m, 350 m, and 370 m distances, respectively. On the roadside, significant radio wave fading and gaining is caused by reflection from the ground and the surrounding snowy trees and rocks. Beacons transmitted at −20 dBm, −10 dBm, −5 dBm, and 0 dBm power levels are received until around 60 m, 130 m, 340 m, and 440 m distances, respectively. The results are comparable to RSSI measurements, such as presented in Reijers et al. (2004).

According to the measurements, antenna directivity and radio wave reflections have a higher effect on the ranging accuracy than the path-loss measurements mechanism itself. Localization accuracy would be highly improved by a more omnidirectional antenna, such as a monopole.

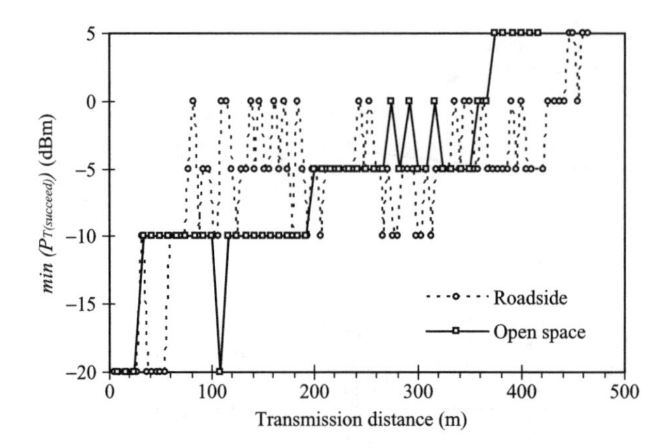

Figure 13.17 Test points and the minimum transmission powers of received beacons in an outdoor environment. © 2006 IEEE. Reprinted, with permission, from *Proceedings of 2006 IEEE 17th International Symposium on Personal, Indoor and Mobile Radio Communications.*

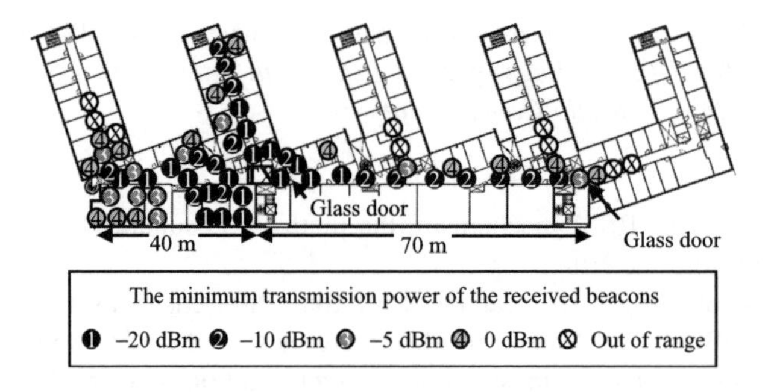

Figure 13.18 Test points and the minimum transmission powers of received beacons in an office environment. © 2006 IEEE. Reprinted, with permission, from *Proceedings of 2006 IEEE 17th International Symposium on Personal, Indoor and Mobile Radio Communications.*

Indoor measurements were conducted in an office environment, presented in Figure 13.18 (Kohvakka et al. 2006c). A beacon-transmitting node is marked with TX and placed on a corridor around 2 m above the floor such that the antenna beam was directed along the long corridor. The measuring points are marked with the numbers 1 to 4, according to the minimum power level of the transmissions that were the successfully received by the beacons. According to the results, around 40 m distance was achieved with the smallest −20 dBm transmission power. A glass door and a wall attenuated the radio wave from 5 to 10 dB.

The accuracy of the path-loss metering is sufficient for managing a robust network operation. Since walls and doors cause significant signal attenuation and reflection, high node density, omnidirectional antennas, and very low radiation power provide the highest localization accuracy.

Power consumption analysis

First, the average power consumed by the beacon transmissions and receptions (beacon exchange) caused by the path-loss metering mechanism is analyzed. Only the transceiver power consumption is analyzed, ignoring all other hardware components. For the analysis, a Nordic Semiconductor nRF24L01 transceiver is selected. The transceiver operates at the 2.4 GHz frequency band and has 2 Mbps data rate. This is the most energy efficient commercial off-the-shelf transceiver. The transceiver has a carrier detection functionality, but lacks RSSI. In the analysis, the beacon transmission power level is adjusted between −18 dBm and 0 dBm. The beacon length is set to 24 B, which is similar to IEEE 802.15.4 standard (IEE 2003b). Power consumption analysis utilizes the models presented in Chapter 18 (Kohvakka et al. 2006c).

During a beacon interval, one set of 1 – 4 beacons were transmitted and received. Transceiver start-up transient t_{ST}, crystal inaccuracy (20 ppm), and sleep-mode power consumption are included in the analysis. For comparison, the same analysis is performed for an IEEE 802.15.4 compliant Texas Instruments CC2420 transceiver. Since the transceiver has an RSSI mechanism, only one beacon transmission and one reception during a beacon interval are required for the basic MAC operation.

The power consumption results are presented in Figure 13.19 (Kohvakka et al. 2006c). The beacon-exchange power consumption is typically far below 10 μW. With a 4 s beacon interval (I_B), a basic MAC operation with a single beacon transmission and reception consumes 3.51 μW with the nRF24L01 transceiver. If two beacons are sufficient for the path-loss metering method, the increase of power consumption is 2.78 μW resulting in 6.29 μW beacon-exchange power. With four beacons, the beacon-exchange power is 13.6 μW. For comparison, the CC2420 transceiver beacon-exchange power consumption with single beacons and RSSI measurements is 20.1 μW. Clearly, the increase of transmitted beacons increases measurement accuracy but raises beacon-exchange energy consumption. Energy can be saved by transmitting two beacons but randomizing the transmission power of the

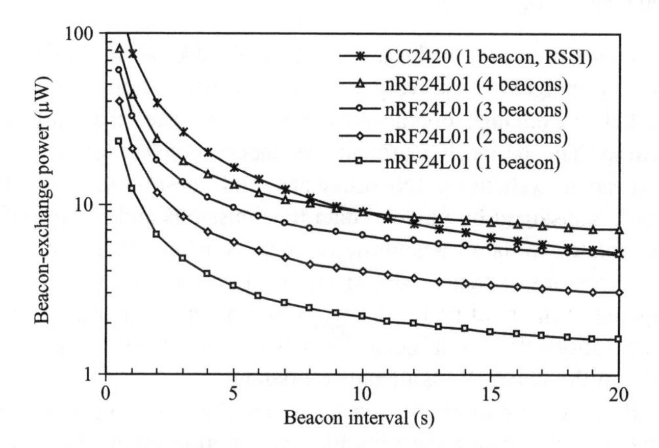

Figure 13.19 Beacon exchange power consumption as the function of a beacon interval with one to four beacons. © 2006 IEEE. Reprinted, with permission, from *Proceedings of 2006 IEEE 17th International Symposium on Personal, Indoor and Mobile Radio Communications.*

lower power beacon. For maintaining network connectivity, the range of beacon transmissions must not vary and the higher power beacon is always transmitted with the highest power. Thus, a node can accurately measure path loss using less beacons. Nevertheless, the method can provide the functionality of RSSI for the lowest complexity and power transceivers with very low power consumption.

13.6.2 Network Beacon Rate Optimization*

In synchronized low duty-cycle MAC protocols beacons are usually utilized at the beginning of an active period for scheduling transmissions, obtaining time synchronization, and informing node status. Beacons are periodically received for maintaining synchronization. When searching for new neighbors, a node performs a long-term channel reception (network scan) for receiving the beacons of neighboring nodes. To be able to detect all neighbors, the network scanning time should be longer than beacon transmission interval (Kohvakka et al. 2005b).

In very low data-rate WSNs, the duty cycle of nodes is reduced by extending the sleep times between data exchange periods. This usually reduces the beacon transmission rate, but also proportionally increases the required network scanning time and energy. The network scanning may consume energy equal to the transmission of thousands of data packets. Hence, the energy optimization of the beacon transmission rate is important (Kohvakka et al. 2005b).

The beacon transmission rate is energy optimized . For making the beacon rate optimization independent of data exchanges, the energy consumed in a wireless sensor node is divided into three classes:

- Node start-up energy
- Network maintenance energy
- Data-exchange energy

These energies are presented in Figure 13.20 (Kohvakka et al. 2005b). The node start-up energy consists of neighbor discovery and network association operations. The network maintenance and data-exchange operations are executed after the start-up period during the node's lifetime. The network maintenance energy consists of beacon transmissions and receptions (beacon exchange), network scans, and possible re-associations. The data-exchange energy is consumed by payload data transmissions and receptions, and the MAC signaling frames related to data transmissions, such as acknowledgments.

As node lifetimes are expected to last from months to years, the start-up energy is negligible compared to the total node energy consumption. Furthermore, it is assumed that data-exchange operations are not affected by network maintenance operations. Hence, we can focus purely on the network maintenance operations.

Next, we will energy optimize the beacon transmission rate in two cases. Firstly, in a simple case without ENDP, where the network scan duration equals the beacon transmission interval. Secondly, in a more energy-efficient case, where ENDP is used and the network scan duration is minimized.

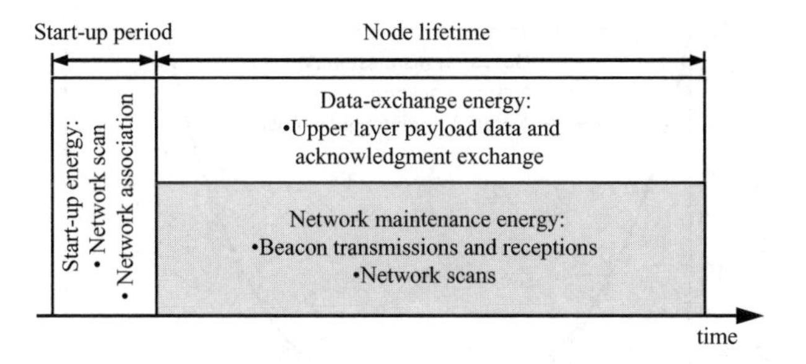

Figure 13.20 The classification of a WSN node energy consumption. © 2005 IEEE. Reprinted, with permission, from *Proceedings of 2005 IEEE 16th International Symposium on Personal, Indoor and Mobile Radio Communications.*

Optimal beacon rate without ENDP*

The average network maintenance power P_M is defined as the sum of the network scanning power P_{NS} and the beacon-exchange power P_B. The network scanning power P_{NS} depends on the energy of a single scanning procedure E_{NS}, and the network scanning interval I_{NS}. I_{NS} depends on the lifetime of maintained communication links. In dynamic networks, where nodes are mobile or RF propagation conditions are altering extensively, I_{NS} is shorter increasing the average network scanning power consumption. A long-time average network scanning power is obtained by (Kohvakka et al. 2005b)

$$P_{NS} = \frac{E_{NS}}{I_F}.$$

Beacons are transmitted at rate f_{BTX} and received at f_{BRX}. Beacons may be received from multiple neighbors, which increases f_{BRX} proportionally. The average power consumed by the beacon exchange is

$$P_B = E_{TX} f_{BTX} + E_{RX} f_{BRX}.$$

The network scanning energy E_{NS} consists of a radio start-up transient t_{ST} and RF channel reception time. To be able to receive beacons from all neighboring nodes, the reception time equals the whole beacon interval, which is an inverse of the beacon transmission rate f_{BTX}. Each received beacon is loaded from the radio to MCU, and an association request may be transmitted. These energies are difficult to estimate, but they are negligibly small compared to the channel reception energy and therefore ignored in the model. Thus, the network scanning energy E_{NS} can be modeled as

$$E_{NS} = \left(t_{ST} + \frac{1}{f_{BTX}}\right) P_{RX}.s$$

When f_{BTX} is set to 1 Hz, the resulting network scanning energy $E_{NS} = 44.99$ mJ, when determined for the TUTWSN temperature-sensing platform (Section 4.3.1). This equals to the energy of 2793 beacon transmissions at the highest transmission power. Clearly, the

*© 2005 IEEE. Reprinted, with permission, from *Proceedings of 2005 IEEE 16th International Symposium on Personal, Indoor and Mobile Radio Communications.*

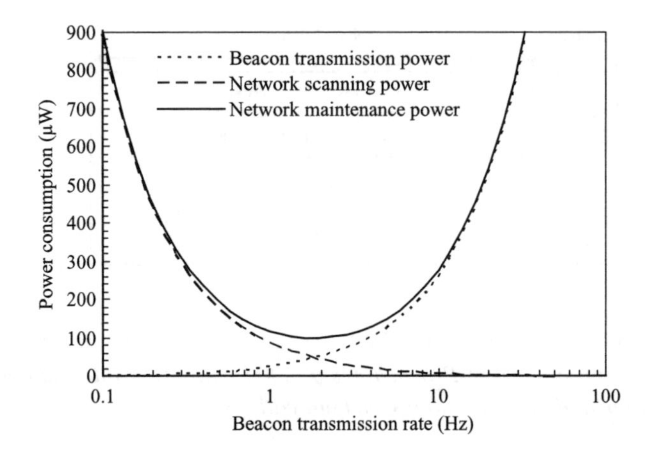

Figure 13.21 Beacon exchange, network scanning, and network maintenance power consumptions as the function of beacon transmission rate ($I_{NS} = 500$ s, $f_{\mathrm{BRX}} = 0.25$ Hz). © 2005 IEEE. Reprinted, with permission, from *Proceedings of 2005 IEEE 16th International Symposium on Personal, Indoor and Mobile Radio Communications.*

effect of scanning on the whole energy consumption of a node is very significant, and becomes more critical at slower beacon transmission rates.

The network scanning, beacon transmission, and network maintenance powers as the function of the beacon transmission rate f_{BTX} are presented in Figure 13.21 (Kohvakka et al. 2005b). I_{NS} and f_{BRX} are fixed to 500 s and 0.25 Hz. The figure clearly shows how the network scanning power P_{NS} decreases rapidly as f_{BTX} increases. A decrease from 900 μW to 90 μW is obtained as the beacon transmission rate increases from 0.1 Hz to 1.0 Hz.

In contrast to P_{NS}, the beacon transmission power P_B can be reduced by minimizing f_{BTX}. P_{B} decreases from 269 μW to 26.9 μW as f_{BTX} decreases from 10 Hz to 1 Hz. As shown in the figure, the network maintenance power P_{M} has the minimum (98 μW) at the 1.9 Hz beacon transmission rate. At low beacon transmission rates below 1 Hz, P_{M} typically doubles as the beacon transmission rate halves. The effect is reversed at high beacon transmission rates above 10 Hz.

Next, we consider the effect of the network scanning interval on network maintenance power consumption. Clearly, the increase of network scanning interval reduces the power for network scanning and shifts the P_{M} minimum to lower beacon transmission rates. This is plotted in Figure 13.22 (Kohvakka et al. 2005b), where P_{M} varies with f_{B}, while I_{NS} equals to 20 s, 100 s, and 500 s. The network scanning interval affects the maintenance power more significantly at lower beacon transmission rates. When $f_{\mathrm{B}} = 1$ Hz, P_{M} increases from 137 μW to 2.3 mW as I_{NS} decreases from 500 s to 20 s. Next, we will determine the optimal beacon transmission rate in respect of the network scanning interval.

An optimal beacon transmission rate (f^*_{BTX}) is determined by minimizing the network maintenance power with respect to the beacon transmission rate. The whole network maintenance power can be summarized as

$$P_{\mathrm{M}} = \frac{P_{\mathrm{RX}}}{I_{\mathrm{NS}}}\left(t_{\mathrm{ST}} + \frac{1}{f_{\mathrm{BTX}}}\right) + E_{\mathrm{TX}} f_{\mathrm{BTX}} + E_{\mathrm{RX}} f_{\mathrm{BRX}}.$$

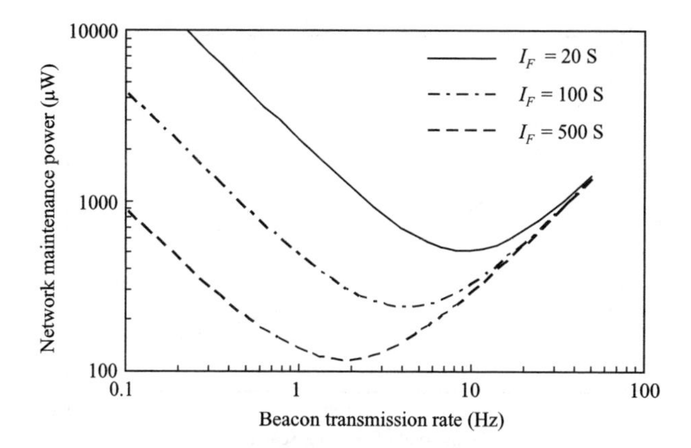

Figure 13.22 Network maintenance power as the function of beacon transmission rate with different network scanning intervals ($f_{\text{BRX}} = 0.25$ Hz). © 2005 IEEE. Reprinted, with permission, from *Proceedings of 2005 IEEE 16th International Symposium on Personal, Indoor and Mobile Radio Communications*.

It can be shown that a unique minimum at (f^*_{BTX}) exists, which is obtained by setting $\partial P_{\text{M}}/\partial f_{\text{BTX}} = 0$ in the previous equation. This yields

$$f^*_{\text{BTX}} = \sqrt{\frac{P_{\text{RX}}}{E_{\text{TX}} I_{\text{NS}}}}.$$

An optimal beacon transmission rate is determined by a network parameter (scanning interval) I_{NS}, and the radio parameters E_{TX} and P_{RX}. Figure 13.23 presents the variation of (f^*_{BTX}) with I_{NS}. For the TUTWSN temperature-sensing platform, the optimal network beacon transmission rate increases from 1.3 Hz to 12.9 Hz, as I_{NS} decreases from 1000 s to 10 s.

Optimal network beacon transmission rate with ENDP*

Next, we determine an optimal network beacon transmission rate for TUTWSN utilizing ENDP. In addition, network scan is terminated after a beacon is received from a new neighbor with a sufficient signal strength. These algorithms increase the complexity of energy optimization.

First, we model the probability of successful neighbor discovery by SDU information. Let's consider the situation presented in Figure 13.24, where a node A maintains synchronization and receives SDUs from nodes B and D. The distance between nodes A and B is b, and their radio ranges form circles with a radius r. Moreover, node B maintains synchronization with nodes C and E located in its range. Since the node E is in the intersection area of the ranges of nodes A and B ($S_{A\cap B}$), node A can receive its (beacon) transmissions and the signaled SDU is valid. Since node C is outside the area $S_{A\cap B}$, node A cannot detect its transmissions resulting in invalid SDU. The size of the intersection area $S_{A\cap B}$ is

*This article was published in Ad Hoc Networks, M. Kohvakka, J. Suhonen, M. Kuorilehto, V. Kaseva, M. Hännikäinen, T.D. Hämäläinen, *Energy-Efficient Neighbor Discovery Protocol for Mobile Wireless Sensor Networks*, Copyright Elsevier (2007).

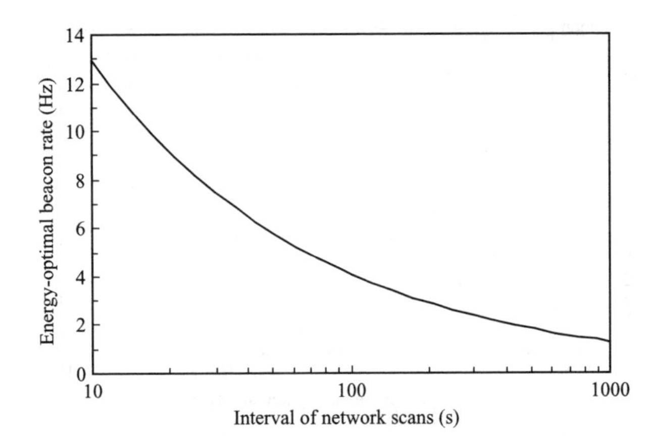

Figure 13.23 Optimal network beacon transmission rate as the function of the network scanning interval. © 2005 IEEE. Reprinted, with permission, from *Proceedings of 2005 IEEE 16th International Symposium on Personal, Indoor and Mobile Radio Communications.*

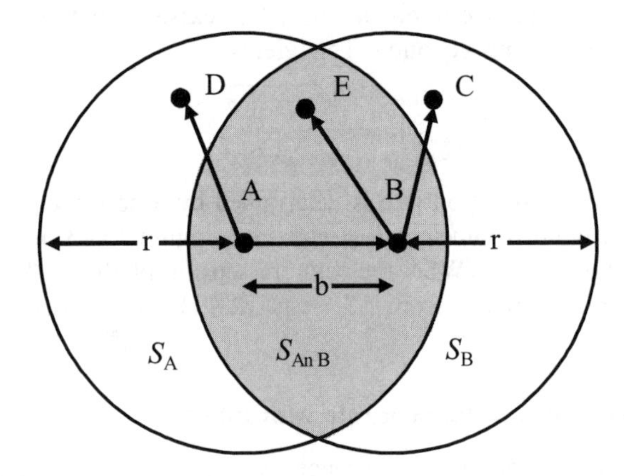

Figure 13.24 Intersections of the communication ranges of nodes A and B. Arrows show the direction of synchronization. This article was published in Ad Hoc Networks, M. Kohvakka, J. Suhonen, M. Kuorilehto, V. Kaseva, M. Hännikäinen, T.D. Hämäläinen, *Energy-Efficient Neighbor Discovery Protocol for Mobile Wireless Sensor Networks*, Copyright Elsevier (2007).

calculated by the radius r and the distance b as (Tseng et al. 2003)

$$A_{A\cap B}(b) = 4\int_{b/2}^{r} \sqrt{r^2 - x^2}dx.$$

Thus, the probability p_I that a node located in S_B is also in the intersection area $S_{A\cap B}$ equals to $A_{A\cap B}(b)/\pi r^2$. Moreover, let nodes A and B be randomly located neighbors, where b gets a value in the range $[0, r]$, and node A receives SDUs from node B. The probability that a received SDU is valid is determined by integrating the probability $A_{A\cap B}(b)/\pi r^2$ over the circle of radius b centered at A for b in $[0, r]$ so we obtain (Tseng et al. 2003)

$$p_I = \int_0^r \frac{2bA_{A\cap B}(b)}{\pi r^4} db \simeq 59$$

As each node maintains synchronization with k other nodes, which all generate k SDUs, the probability (q_S) that none of the received k^2 SDUs is valid and a network scan is required is

$$q_S = \prod_{a=1}^{k} \left(1 - \frac{np_I}{n-(a-1)}\right)^k.$$

Finally, the required network scan interval (I_{NS}) is determined according to q_S and the average rate f_F of link failures

$$I_{NS} = \frac{1}{f_F q_S}.$$

In case of a link failure, a node attempts to receive beacons according to SDUs until a new neighbor with sufficient signal strength is detected. We define that the range of sufficient signal strength in proportion to maximum radio range (r) is ρ. The expected number (n_{BS}) of beacon receptions by SDU information, until one beacon within the distance of ρr is successfully received, is modeled by weighted average as

$$n_{BS} = p_I\rho^2 + \sum_{a=2}^{k^2-1} a \left(\prod_{b=1}^{a-1} 1 - \frac{np_I\rho^2}{n-(b-1)}\right) \frac{np_I\rho^2}{n-(a-1)}$$
$$+ k^2 \left(\prod_{a=1}^{k^2-1} 1 - \frac{np_I\rho^2}{n-(a-1)}\right).$$

Because duplicated SDUs are improbable in dense WSNs, SDUs referring to a same node or a current neighbor are not considered in the model. If no new neighbors are detected with SDUs, a network scan is performed. Similarly, the network scan is performed until a node within a range of ρr is detected. The number of beacon receptions (n_B) by a network scan, until one with sufficient signal strength is received, is

$$n_B = \rho^2 + \sum_{a=2}^{k^2-1} a \left(\prod_{b=1}^{a-1} 1 - \frac{n\rho^2}{n-(b-1)}\right) \frac{n\rho^2}{n-(a-1)}$$
$$+ k^2 \left(\prod_{a=1}^{k^2-1} 1 - \frac{n\rho^2}{n-(a-1)}\right).$$

Moreover, the required network scan duration (t_{NS}) for detecting a new node with sufficient signal strength is

$$t_{NS} = \frac{n_B}{f_N n}.$$

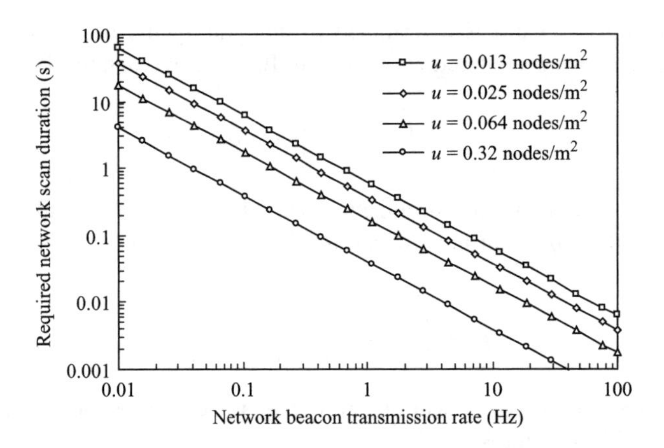

Figure 13.25 The effect of node density and network beacon transmission rate to the required network scan time ($r = 10$ m, $\rho = 0.5$). This article was published in Ad Hoc Networks, M. Kohvakka, J. Suhonen, M. Kuorilehto, V. Kaseva, M. Hännikäinen, T.D. Hämäläinen, *Energy-Efficient Neighbor Discovery Protocol for Mobile Wireless Sensor Networks*, Copyright Elsevier (2007).

The network scan duration as the function of the network beacon transmission rate f_{N} is plotted in Figure 13.25. If the number of nodes within radio range (n) is high, t_{NS} is reduced even one order of magnitude compared to the entire network beacon interval.

A network maintenance power consumption P_{M} is defined as a sum of network scan power P_{NS} and a beacon-exchange power P_{B}. The network scan power consumption P_{NS} is defined as the energy of a single scan procedure E_{NS} divided by the average network scan interval I_{NS}. Long-time average network scan power consumption is obtained by

$$P_{\mathrm{NS}} = f_{\mathrm{F}} q_{\mathrm{S}} \left(t_{\mathrm{ST}} + \frac{n_{\mathrm{B}}}{f_{\mathrm{N}} n} \right) P_{\mathrm{RX}}.$$

The transmission rates of cluster and network beacon are f_{C} and f_{N}, respectively. In addition, both network and cluster beacons are received at the beginning of superframes at rate f_{C} from the k neighbors with which synchronization is maintained. Moreover, link failures occur at rate f_{F} after which averagely n_{BS} cluster beacons are received. The average power consumed by the beacon exchange is

$$P_{\mathrm{B}} = (f_{\mathrm{N}} + f_{\mathrm{C}})\, E_{\mathrm{TX}} + (2 f_{\mathrm{C}} k + f_{\mathrm{F}} n_{\mathrm{BS}})\, E_{\mathrm{RX}}.$$

Network maintenance power consumption is plotted as the function of the network beacon transmission rate and with 0.1 m/s, 1 m/s, and 10 m/s node velocities in Figure 13.26, Figure 13.27, and Figure 13.28 As seen in the figures, node mobility significantly increases the network maintenance power consumption when ENDP is not used. Power consumption can be reduced to some extent by adjusting the network beacon transmission rate according to mobility, but the lowest power consumption is achieved by ENDP. Typically, the highest energy efficiency is achieved by selecting $k = 3$.

An optimal network beacon transmission rate f_N^* is determined by minimizing the network maintenance power with respect to the beacon transmission rate. It can be shown

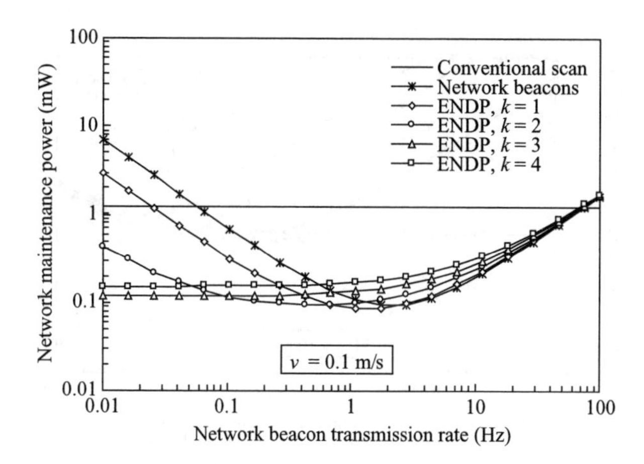

Figure 13.26 Network maintenance power at 0.1 m/s node velocity, as k varies ($r = 10$ m, $\rho = 0.5, n = 0.1$ nodes/m^2). This article was published in Ad Hoc Networks, M. Kohvakka, J. Suhonen, M. Kuorilehto, V. Kaseva, M. Hännikäinen, T.D. Hämäläinen, *Energy-Efficient Neighbor Discovery Protocol for Mobile Wireless Sensor Networks*, Copyright Elsevier (2007).

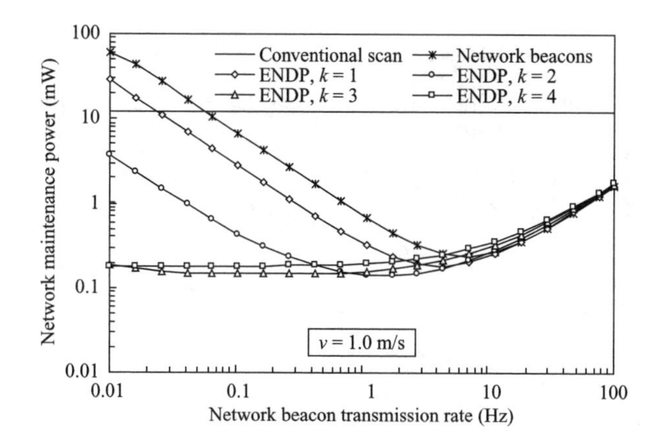

Figure 13.27 Network maintenance power at 1 m/s node velocity, as k varies ($r = 10$ m, $\rho = 0.5, n = 0.1$ nodes/m^2). This article was published in Ad Hoc Networks, M. Kohvakka, J. Suhonen, M. Kuorilehto, V. Kaseva, M. Hännikäinen, T.D. Hämäläinen, *Energy-Efficient Neighbor Discovery Protocol for Mobile Wireless Sensor Networks*, Copyright Elsevier (2007).

that a unique minimum for f_N^* exists, which is obtained by writing $P_{\mathrm{M}} = P_{\mathrm{NS}} + P_{\mathrm{B}}$ and setting $\partial P_{\mathrm{M}}/\partial f_{\mathrm{N}} = 0$. This yields

$$f_{\mathrm{N}}^* = \sqrt{\frac{P_{\mathrm{RX}} n_{\mathrm{B}}}{E_{\mathrm{TX}} I_{\mathrm{NS}} n}}.$$

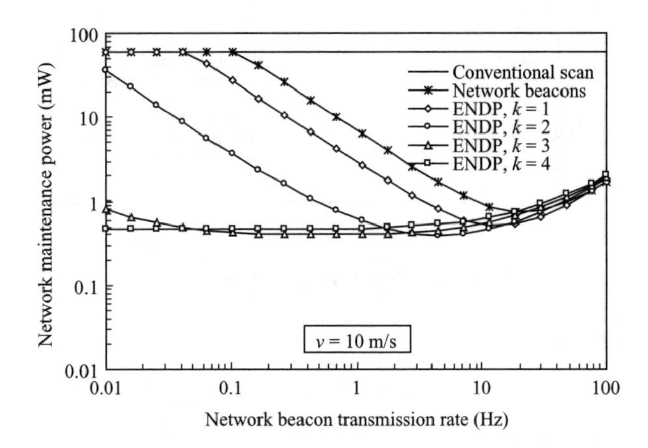

Figure 13.28 Network maintenance power at 10 m/s node velocity, as k varies (r = 10 m, $\rho = 0.5, n = 0.1$ nodes/m^2). This article was published in Ad Hoc Networks, M. Kohvakka, J. Suhonen, M. Kuorilehto, V. Kaseva, M. Hännikäinen, T.D. Hämäläinen, *Energy-Efficient Neighbor Discovery Protocol for Mobile Wireless Sensor Networks*, Copyright Elsevier (2007).

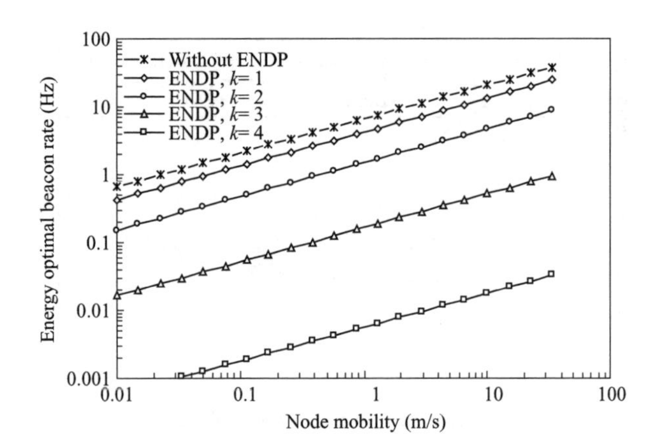

Figure 13.29 Optimal network beacon transmission rate as the function of node mobility (r = 10 m, $\rho = 0.5, u = 0.1$ nodes/m^2). This article was published in Ad Hoc Networks, M. Kohvakka, J. Suhonen, M. Kuorilehto, V. Kaseva, M. Hännikäinen, T.D. Hämäläinen, *Energy-Efficient Neighbor Discovery Protocol for Mobile Wireless Sensor Networks*, Copyright Elsevier (2007).

An optimal network beacon transmission rate is determined by a network scan interval (I_{NS}), beacon frame transmission energy (E_{TX}), radio power consumption in reception mode (P_{RX}), number of nodes in range (n), and the range of sufficient signal strength (ρ). Furthermore, I_{NS} is the function of node speed (v), radio range (r), and the number of nodes with which synchronization is maintained (k). The optimal network beacon transmission rate is

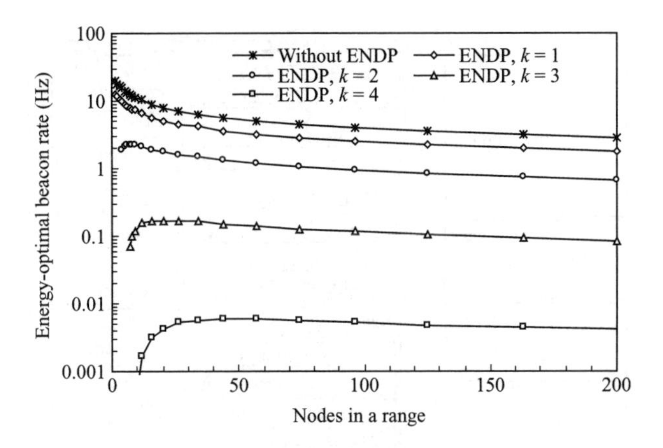

Figure 13.30 Optimal network beacon transmission rate as the function of neighboring nodes in a range ($v = 1$ m/s). This article was published in Ad Hoc Networks, M. Kohvakka, J. Suhonen, M. Kuorilehto, V. Kaseva, M. Hännikäinen, T.D. Hämäläinen, *Energy-Efficient Neighbor Discovery Protocol for Mobile Wireless Sensor Networks*, Copyright Elsevier (2007).

plotted as the function of node speed in Figure 13.29. The optimal beacon transmission rate increases proportionally to the node speed. At 1 m/s node speed, an optimal beacon rate is around 6 Hz without ENDP. By using ENDP, an optimal network beacon transmission rate is even three orders of magnitude less. Figure 13.30 presents the optimal network beacon transmission rate as the function of nodes in a range. At higher node densities the optimal network beacon transmission rate decreases, since the required network scan time (t_{NS}) reduces. In very sparse networks the optimal beacon rate also decreases, especially when k is 3 or 4. This is caused by fewer beacon receptions (n_{BS}) until one beacon with sufficient signal strength is received. With ENDP the optimal beacon rate is up to four orders of magnitude less than without the protocol. These results indicate that ENDP can effectively reduce the need for network scans.

The results are lower than the measured power consumption presented earlier in this chapter. This is partly caused by the fact that the energy consumption of data processing and sensing are not included in the analysis. Moreover, the office environment caused significant fading, gaining, and shadowing on RF propagation making link changes more frequent and protocol operation more difficult than in an open space.

13.7 TUTWSN MAC Implementation

The architecture of the TUTWSN protocol stack and the TUTWSN MAC layer are presented in Figure 13.31. Service Access Points (SAPs) present the functional interfaces for accessing MAC protocol services and the physical layer. Management procedures are executed on demand for forming and maintaining network topology and managing data transfer. Vertical management signaling with routing and physical layers is handled via a Management Service Access Point (MSAP) and a Physical Management Service Access Point (PMSAP). Functions associated to user data processing are grouped to frame control, queue control, error control, and channel access. The higher layer (routing) is accessed

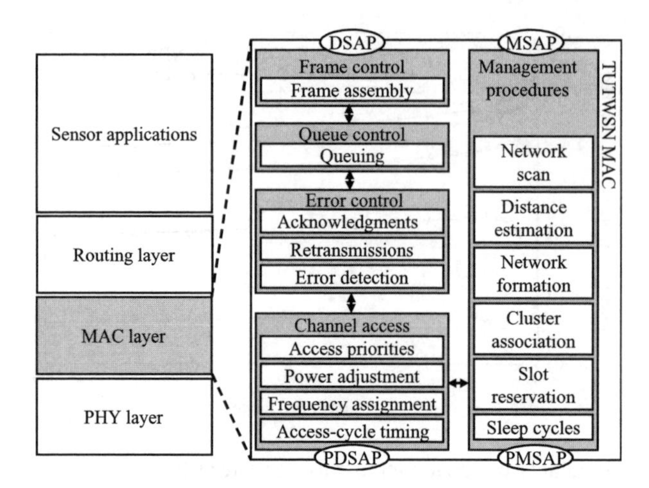

Figure 13.31 TUTWSN protocol stack and MAC layer architecture.

by Data Service Access Point (DSAP) and the transceiver of the physical layer through Physical Data Service Access Point (PDSAP).

MAC frames are built by a frame assembly function, which also manages addressing. User data payloads are arranged in queues according to their priorities. The error control performs error detection, acknowledgments, and retransmissions. The channel accesses control frame transmissions and receptions on RF channels using access-cycle timing and frequency assignment functions. A power adjustment function selects a suitable transmission power level according to transmission distance. Frame priorities are controlled by the access priorities function.

The MAC implementation on a PIC18 microcontroller (Microchip Technology Inc. 2004) takes 25 kB of program memory. On very resource-constrained nodes, only the subnode functionality is compiled. Because the routines required to maintain an access cycle are not included, the MAC implementation on a subnode requires only 14 kB of program memory. The data memory that is used on a subnode consists mainly of structures storing the synchronized cluster information. This information contains the cluster channel and timing, address, link-quality measurements, and the state information supplied by the received beacons. Each cluster structure requires 28 B of data memory. Additionally, a fully functional MAC maintains information about its associated member nodes, requiring 8 B of data memory per member.

13.8 Measured Performance of TUTWSN MAC*

The performance of TUTWSN MAC protocol was measured by employing TUTWSN Linear Position Sensor prototype, presented in Section 21.1 (Kohvakka et al. 2005a). The platform consists of a Nordic Semiconductor nRF2401A transceiver, Semtech XE88LC02 MCU, and a linear position sensor. The employed MCU has very high energy efficiency,

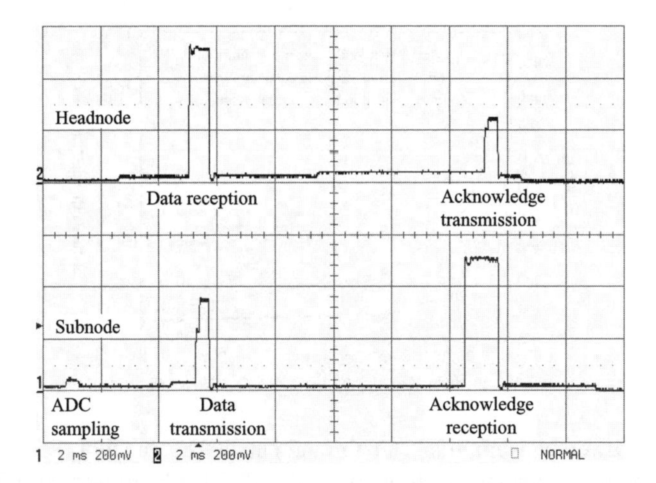

Figure 13.32 Oscilloscope screenshot of the headnode and subnode current waveforms during ADC sampling, data transmission, and reception. © 2005 IEEE. Reprinted, with permission, from *Proceedings of the 8th Euromicro Conference on Digital System Design.*

but low memory resources severely limit the implementation of protocols and applications. The TUTWSN implementation that was measured contained a TUTWSN MAC protocol without ENDP, and very simple upper protocol layers. Thus, the data processing in upper protocol layers was minimal and the measured power consumption mostly caused by MAC and PHY layers. Measurements were performed in a static network without network scans. The network beacon interval was fixed to 500 ms. All nodes sampled their linear position sensors at a 1 Hz sample rate and routed the data to a sink node. The utilized frame size was 32 B containing a 16 B protocol overhead and 16 B MAC payload.

An oscilloscope screenshot of the current waveforms of a headnode and a subnode is presented in Figure 13.32 (Kohvakka et al. 2005a). The waveforms present a 20 ms time period, which contains ADC sampling, and TUTWSN data transmission and acknowledgment reception. A peak current during reception is about 20 mA.

The power consumption of a subnode and a headnode were measured using four different access cycle lengths: 1 s, 2 s, 4 s, and 10 s. The measured headnode routed data from itself and one child node to a parent. Power consumptions were measured at 3.0 V supply voltage. The measured power consumption results with a maximum throughput in one route are presented in Figure 13.33 (Kohvakka et al. 2005a). The minimum power was obtained with 10 s access cycle, when subnode and headnode power consumptions were 59.6 μW and 302 μW, respectively. The available throughput is 102 bits/s. The throughput was inversely proportional to the access cycle length. With 1 s access cycle, throughput was increased to 1024 bits/s, but the power consumptions of a subnode and a headnode were raised to 249 μW and 968μW, respectively.

Then, the average headnode power consumption was measured, while the number of subnodes per each headnode was increased from 1 to 5. The TUTWSN MAC access cycle length was fixed to 2 s. By increasing the number of subnodes per headnode, the power consumption of the headnode increased. This increase was caused by the higher utilization

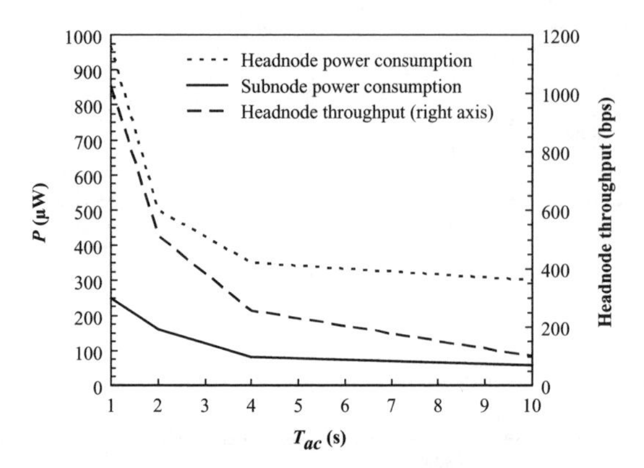

Figure 13.33 Power consumption and throughput as the function of TUTWSN MAC access cycle length (1 subnode per headnode). © 2005 IEEE. Reprinted, with permission, from *Proceedings of the 8th Euromicro Conference on Digital System Design.*

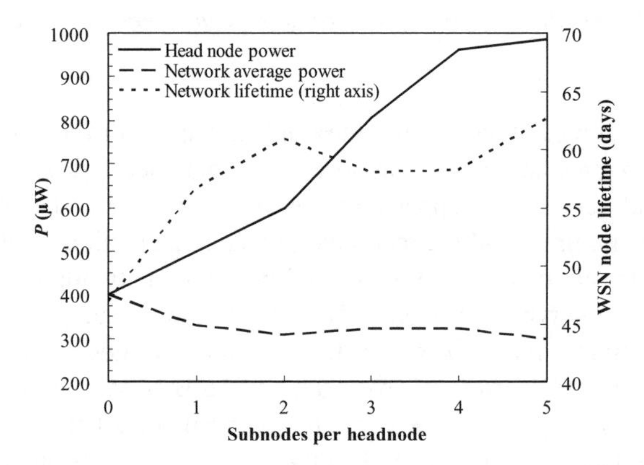

Figure 13.34 Headnode, subnode, and network average power consumptions as the function of the number of subnodes per headnode (TUTWSN access cycle length = 2 s). © 2005 IEEE. Reprinted, with permission, from *Proceedings of the 8th Euromicro Conference on Digital System Design.*

of reserved slots. Yet, the power consumption of subnodes remained in 162 µW. The resulting headnode and network average power consumptions are presented in Figure 13.34 (Kohvakka et al. 2005a). The increase of subnodes per headnode from 0 to 5 reduced the average node power consumption from 400 µW to 299 µW. Using two AA-type alkaline batteries (3000 mAh), in a completely static network the network lifetime of this implementation is estimated to be 2.5–3.4 years.

The results indicate that the energy efficiency of TUTWSN MAC protocol is low enough for long-lived operation by small batteries or even scavenging the supply energy purely from ambient energy, for example using low-level vibrations, ambient light, and temperature gradients (Roundy et al. 2003).

14

TUTWSN Routing Protocol

14.1 Design and Implementation

The configuration of the TUTWSN Routing Protocol (TUTWSNR) is guided by three design goals. First, the network lifetime should be maximized. Second, as different applications need different services, a certain level of QoS must be supported. Third, the network must be able to operate autonomously without any configuration when adding new nodes.

To meet the design requirements, TUTWSNR uses cost field-based routing and table-driven routing. The design benefits from the asymmetric characteristics of a WSN, which makes maintaining routing tables even in large networks feasible. The cost field-based routing minimizes route maintenance messaging and allows selecting the lowest energy route by using energy as a cost parameter. Furthermore, load balancing between nodes can be achieved by using cost penalty on highly loaded nodes.

TUTWSNR uses a unique method to calculate the cost field from a set of QoS parameters that form a *traffic class*. Energy is used as one parameter, thus allowing trade-off between other QoS parameters and the energy. A node may choose a suitable traffic class to send a packet.

A cross-layer design between TUTWSN MAC and routing is used to reduce message exchange and improve performance. TUTWSNR is implemented and tested on a TUTWSN prototype platform. However, although the protocol is implemented on TUTWSN MAC, its design is MAC independent.

14.2 Related Work

The related work is presented in Chapter 8. Unlike other presented WSN routing proposals, the cost-field calculation in TUTWSNR is not based on a single decision method, but includes several cost metrics. This allows traffic differentiation by choosing between delay, throughput, reliability, and energy metrics.

At first, routes are established with a method similar to the DD. However, TUTWSNR is more dynamic than DD as the nodes can also discover and advertise their routes locally.

Ultra-Low Energy Wireless Sensor Networks in Practice: Theory, Realization and Deployment

In addition, the sink does not need to refresh its interest and the required protocol message exchange is small.

14.3 Cost-Aware Routing*

A cumulative cost to the sink is used to select the next-hop in a route. A node directs its forwarding gradient towards nodes that advertise the lowest costs. TUTWSNR uses two methods for route discovery: *sink initiated* (setup phase) and *node initiated* (maintenance phase). The sink-initiated method allows fast network build-up and is used when the sink defines interests. In the node-initiated method, a node queries available routes from its neighbors, which allows mobility and adds robustness as the nodes do not need to wait for a sink to refresh route information (Suhonen et al. 2006b).

The protocol message exchange used in routing is presented in Figure 14.1. A node requests for a list of known sinks with route request (RREQ) packet, which is replied with route advertisement (RADV). RADV contains a list of sinks, cost information to reach each sink, and a per sink list of the sequence number of active interests. If the route advertisement contains a sink/interest pair that is previously unknown, a node makes a request for detailed information about the interest with interest request (IREQ) containing the address of the sink. IREQ is replied with interest advertisement (IADV). IADV contains a sequence number and an application-specific description of interests, such as the type of the data and collection interval. The interest description itself is transparent to the routing and is passed to the application layer. Thus, the routing protocol can be used with different sensor applications. In addition, having separate packets for the route and interest advertisements makes the protocol more flexible. Usually, routes change more often than configured interests, so separating routing and interest information packets substantially reduces the required routing overheads (Suhonen et al. 2006b).

A separate routing table entry is maintained for each sink and traffic class combination. The next-hop that minimizes the calculated cost is selected. As costs are calculated separately for each traffic class, different traffic classes may yield different routing paths between same source and destination nodes.

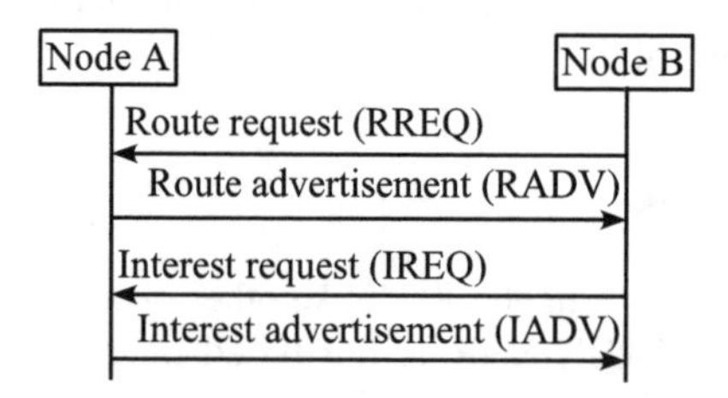

Figure 14.1 Routing information and interest exchange between sensor nodes. © 2006 IEEE. Reprinted, with permission, from *Proceedings of 2006 IEEE 17th International Symposium on Personal, Indoor and Mobile Radio Communications*.

*© 2006 IEEE. Reprinted, with permission, from *Proceedings of 2006 IEEE 17th International Symposium on Personal, Indoor and Mobile Radio Communications*.

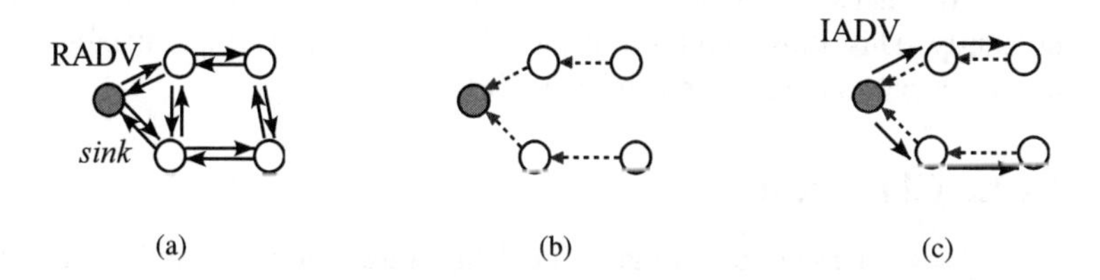

Figure 14.2 Sink-initiated route establishment: a) Sink advertises its presence with RADV; b) nodes establish gradients towards the sink; and c) sink broadcasts interests in the reverse direction of the gradients. © 2006 IEEE. Reprinted, with permission, from *Proceedings of 2006 IEEE 17th International Symposium on Personal, Indoor and Mobile Radio Communications*.

14.3.1 Sink-initiated Route Establishment

A sink advertises its presence by flooding RADV packets into the network as shown in Figure 14.2. A node that receives a RADV compares included cost against old cost. If the cost decreases, the node redirects its gradient and transmits a RADV with updated costs to its neighbors. The sink requests data from the network by sending an IADV packet in the reverse direction of the established gradients. A node forwards an IADV if it was previously unknown. A node always stores the sequence numbers of active interests, but can disregard storing the interest if the available memory is low (Suhonen et al. 2006b).

14.3.2 Node-initiated Route Discovery

A node initiates route discovery when it starts up, loses its next-hop links, or detects that its neighbors have changed (due to mobility or link errors). At first, a node broadcasts RREQ as shown in Figure 14.3. Next, the node queries interests from the node that has the lowest cost. If the neighbor does not have the interest cached, it forwards the query to the next-hop along its gradient. The forwarding is proceeded until a node with an interest is found or the query proceeds to the sink (Suhonen et al. 2006b).

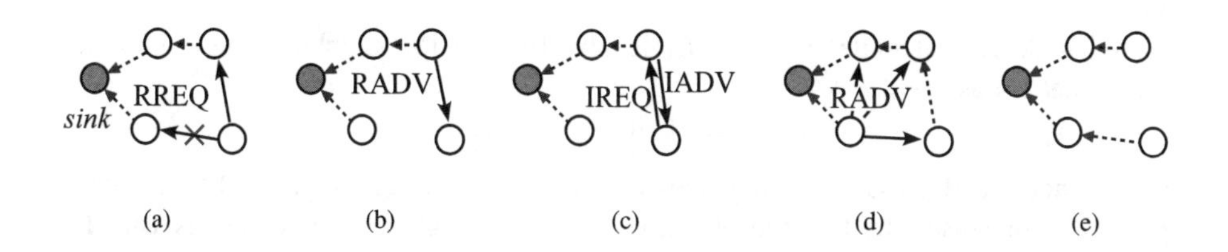

Figure 14.3 Node-initiated route discovery: a) routes are requested with RREQ; b) neighbors reply with RADV; c) a node requests for interests; d) periodic RADV broadcasts; and e) RADV reveals a better path. © 2006 IEEE. Reprinted, with permission, from *Proceedings of 2006 IEEE 17th International Symposium on Personal, Indoor and Mobile Radio Communications*.

Each node in the network broadcasts its RADV periodically or when a route cost changes significantly. This allows nodes to detect changes in interests and makes the node-initiated route discovery tolerant to transmission failures.

14.3.3 Traffic Classification

A traffic class defines separate cost weights for delay, throughput, reliability, and energy QoS metrics. The preferred route has properties fulfilling the metrics that have the highest weights. For example, by maximizing one of the weights while setting others to zero the user can:

- maximize route *reliability*,
- emphasize *energy usage* by either selecting a route that uses least energy or a route that balances overall network energy usage,
- select a route with most available *bandwidth*, or
- minimize *end-to-end delay* by choosing shortest path and next hop clusters that can be reached fastest.

By defining non-zero weights on several metrics, a route that makes a compromise between these extremes is selected. Several cost functions can be defined, and a individual route is maintained for each cost function.

The next-hop cost is calculated in two phases. First, a node N_i selects best next-hop by minimizing the cost C_{ij} that is required to reach the sink through node N_j. C_{ij} is calculated as

$$C_{ij} = C_j + \alpha(e_{ij}) + \beta(r_{ij}) + \gamma(a_{ij}), \tag{14.1}$$

where e_{ij} is the required transmission power to reach the next-hop, r_{ij} is measured packet error rate, a_{ij} is access delay, and α, β, γ are scaling functions. The transmission power metric is used to minimize energy consumption along the route. The packet error rate affects energy consumption and delays, because a packet sent via an unreliable link is likely to be retransmitted. The access delay is the required forwarding time between nodes. In TDMA-based MACs, the access delay is caused by the time difference between time slots. In CSMA-based MACs, the access delay is caused by possible backoff times (Suhonen et al. 2006b).

Lastly, node N_i calculates the cost C_i it advertises to its neighbors from (14.1) and local cost metrics as

$$C_i = C_{ij} + \delta(L_i) + \epsilon(E_i),$$

where L_i is node load, E_i is remaining energy, and δ, ϵ are scaling functions. The remaining energy metric increases the lifetime of a node by increasing the cost for nodes with low energy. Node load is derived from average load (throughput/buffer states) and equalizes energy consumption in cluster heads. Furthermore, the node load is used to select the route with the most available bandwidth (Suhonen et al. 2006b).

The scaling functions α, β, γ, and δ are linear and defined as $f(x) = a \cdot x$, where a determines the weight of the cost metric. The scaling function for the remaining energy is exponential and defined as

$$\epsilon(x) = \begin{cases} a * (x \quad b)^c & x > b \\ 0 & x \leq b \end{cases},$$

where a is a weight, b determines threshold energy for calculation, and c determines the steepness of the cost. The purpose of the exponential energy cost is to ensure that data is routed through almost depleted nodes only if there are no other possibilities.

14.4 Implementation

The TUTWSN routing implementation on a prototype utilizing PIC18F4620 MCU uses 9500 B code memory and 265 B data memory. The data memory is used by routing table (20 B/routing entry) and neighbor table (33 B/entry). The implementation limits the maximum number of routes and neighbors to 5. The neighbor table contains the routes that each neighbor has advertised allowing fast error recovery, as the node can select a new next-hop immediately on link break.

14.4.1 Protocol Architecture

The architecture of the TUTWSN protocol stack and the routing layer are presented in Figure 14.4. Management signaling with the application and MAC layers is handled via a MSAP and a PMSAP. The actual management procedures are executed as required based

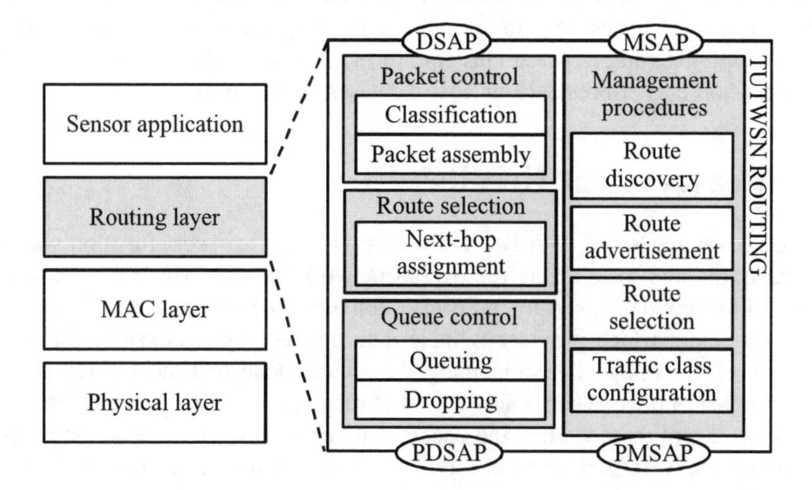

Figure 14.4 TUTWSN protocol stack and routing layer architecture.

on neighbor and link information. Route advertisement and route discovery management procedures are responsible for spreading the route information between neighbors. The route advertisement procedure generates the periodic advertisement beacons, whereas the route discovery procedure detects new routes that are advertised in the neighborhood. The route selection procedure updates the state of the routing table based on the received route advertisements and traffic class configuration.

SAPs present the functional interfaces for accessing routing protocol services and the MAC layer. An application layer payload is classified according to the parameters given in DSAP and assembled to a routable packet. Next, the packet is assigned with a next hop address according to the routing table and enqueued for MAC.

14.4.2 Implementation on TUTWSN MAC

A node performs network scans periodically in order to detect new clusters. The scan interval depends on the node connectivity. If a cluster is not known, the scan interval is smaller. In addition, a node increases the interval when its energy level decreases. When a node detects a new cluster during the scan, it sends a RADV packet to the cluster in question (Suhonen et al. 2006b).

A node that receives a new RADV during the sink-initiated route discovery performs a network scan. Next, the node unicasts the RADV packet to the detected neighbor clusters. The sending of IADV packets is much more efficient as network scanning is not involved. Each cluster head broadcasts IADV in its cluster channel. Members that act as headnodes retransmit the IADV to their members, while subnodes only process the IADV. Thus, the required number of transmissions to spread the interest to the whole network is less than or equal to the network size.

A node synchronizes to the next-hop clusters that have the lowest costs and joins the clusters with a brief handshake. In addition, a node synchronizes to a few clusters that provide sub-optimal routes. These clusters are used as alternative routes if a link to the primary next-hop cluster breaks or its cost increases significantly.

14.5 Measurement Results*

The protocol was measured with two cost functions. The first cost function was targeted for normal traffic and maximizing the overall network lifetime. The second cost function was targeted for low-delay, end-to-end packet delivery. The results were compared against traditional flooding, which provides reliable data delivery by redundancy. In flooding, a node maintains a routing table containing the source node and the last-received packet sequence number entries. When a new packet is received, the node transmits the packet to its neighbors, except the node that the packet was received from (Suhonen et al. 2006b).

Measurements were performed in a typical office environment. The test scenario contained a sink, five subnodes generating traffic, and 32 cluster heads forwarding the traffic. The layout of nodes and typical routes from subnodes to sink is shown in Figure 14.5 (Suhonen et al. 2006b).

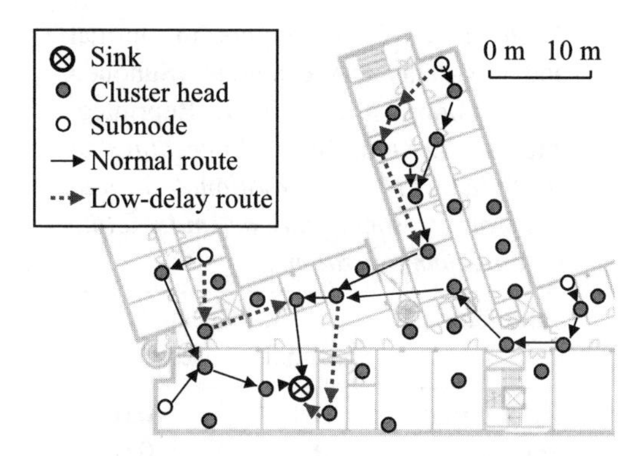

Figure 14.5 Node layout in the measurements and typical routes from subnodes to the sink. © 2006 IEEE. Reprinted, with permission, from *Proceedings of 2006 IEEE 17th International Symposium on Personal, Indoor and Mobile Radio Communications.*

Nodes were practically deployed in rooms instead of a grid formation. The actual routes varied during the measurement depending on the used cost function and network conditions such as link errors. The results were obtained during a one day measurement period. The environment contained WLAN access points operating on the same frequency band as the sensor nodes. Although WLAN activity was light, it introduced interference to the measurement and resulted in a few packet losses. However, the activity was fairly constant throughout the measurements.

14.5.1 Network Parameter Configuration

The measurements used a 2 s access cycle length that causes on average a 1 s per hop delay when the access delay is ignored in the next-hop selection. A network beacon interval determines the minimum time required for network scanning and was set to 0.5 s. The MAC layer tried to retransmit a failed frame up to three times. A node was synchronized to at least two clusters (Suhonen et al. 2006b).

The parameter weights used with the two cost functions are presented in Table 14.1. The remaining energy parameter was not used in the cost functions, because the lifetime of the prototype node is long (years) compared to testing period (days). Thus, it would not have had a significant effect on results.

14.5.2 Network Build-up Time

The efficiency of the cross-layer design and the route-discovery methods was tested by measuring the network build-up time. Figure 14.6 shows the delay between sink start-up and the reception of the first packet from a node. Also, the average hop count at the time of the first transmitted packet is presented. For example, nodes 16–24 had on average four hops to the sink. A node sent a packet immediately after it joined a cluster (Suhonen et al. 2006b).

The nodes form routes to the sink gradually, as the RADV packets are flooded through the network. The delays in packet reception from the 13th, 16th, and 24th nodes are caused

Table 14.1 Cost weights in normal and low-delay traffic classes (Suhonen et al. 2006b). © 2006 IEEE. Reprinted, with permission, from *Proceedings of 2006 IEEE 17th International Symposium on Personal, Indoor and Mobile Radio Communications*.

Parameter	Traffic class	
	Normal	Low delay
Throughput	0.4	0.0
Energy	0.6	0.0
Reliability	0.0	0.5
Delay	0.0	0.5

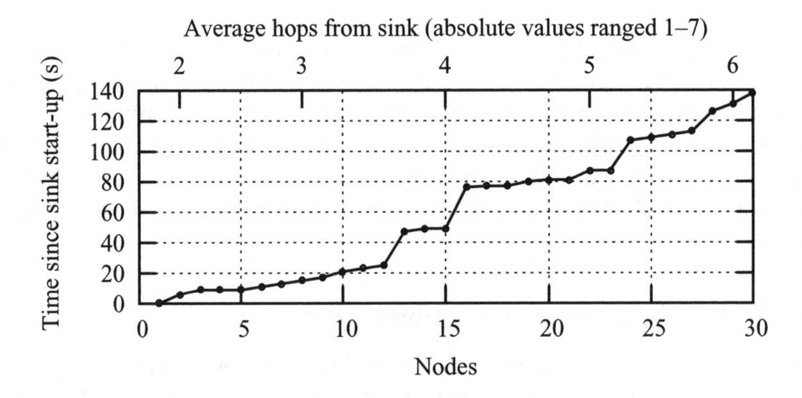

Figure 14.6 The reception time of the first packet after sink start-up. © 2006 IEEE. Reprinted, with permission, from *Proceedings of 2006 IEEE 17th International Symposium on Personal, Indoor and Mobile Radio Communications*.

by sending failure of RADV or IADV packets. However, as the nodes are synchronized to only a few neighbors, they obtain route information from cluster beacons and recover from failures with node-initiated route discovery. It should be noted that the presented times include the node-to-sink packet delivery delay. Therefore, the time when a node has a valid route and interests is smaller than presented.

14.5.3 Distribution of Traffic

The distribution of traffic between nodes is shown in Figure 14.7. In flooding, centrally located nodes significantly forward more traffic because they have more neighbors. Also, the traffic is much higher than in TUTWSNR because each packet is flooded to every node

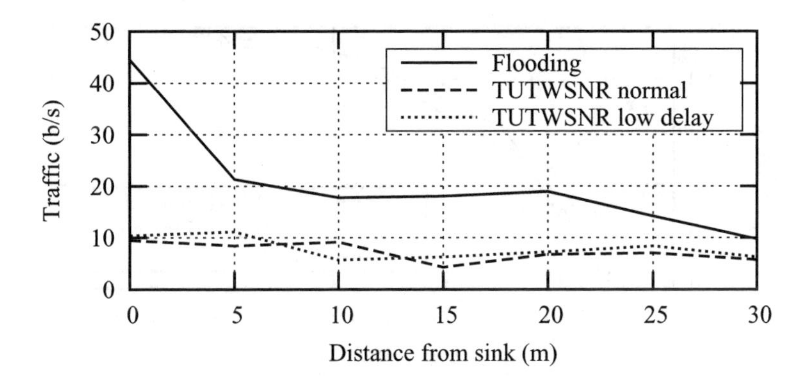

Figure 14.7 End-to-end delays of packets from a selected subnode using on average 7 hops to the sink. © 2006 IEEE. Reprinted, with permission, from *Proceedings of 2006 IEEE 17th International Symposium on Personal, Indoor and Mobile Radio Communications.*

in the network and duplicates are received often. On average, 75% of the received data in flooding consisted of duplicates. TUTWSNR had an average of 7.8% duplicates that were caused by packet errors as a failed reception of an acknowledge packet forces a sender to retransmit (Suhonen et al. 2006b).

The routing activity measured as received and the forwarded packets per minute is presented in Figure 14.8. The data received by the sink was equal in all cases. As both reception and transmission cause energy, the nodes with the highest activity are likely to die first. This is especially true in the prototype platform where the reception of a packet consumes as much energy as transmission, because the transmission powers are relatively low. TUTWSNR's normal routing is energy efficient and balances the load between nodes.

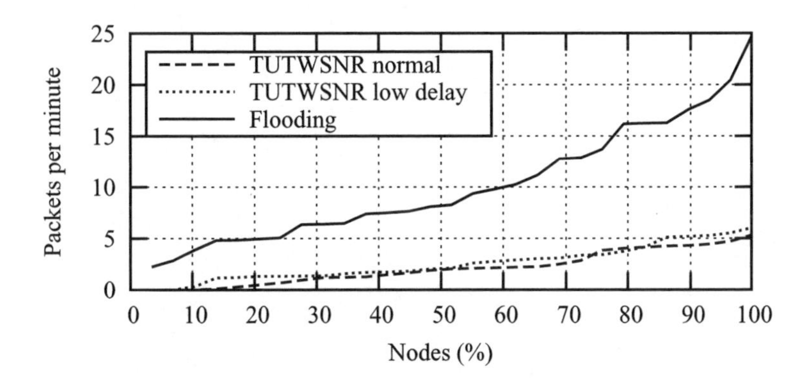

Figure 14.8 Average forwarded traffic per node. © 2006 IEEE. Reprinted, with permission, from *Proceedings of 2006 IEEE 17th International Symposium on Personal, Indoor and Mobile Radio Communications.*

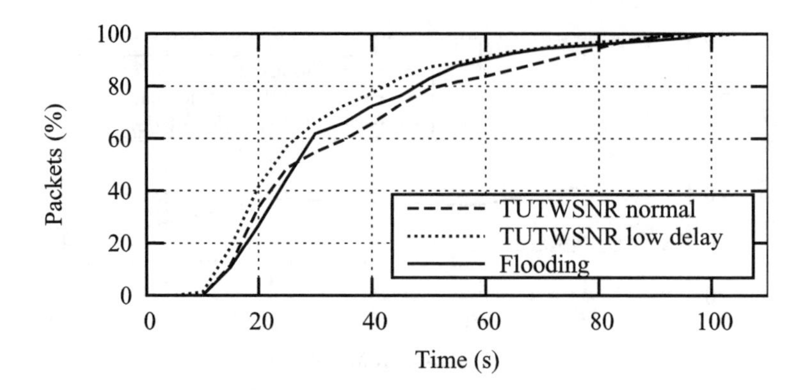

Figure 14.9 Routing activity measured as time (s) transmitted and received packets per minute.

14.5.4 End-to-end Delays

The distribution of end-to-end packet delay on a selected subnode is shown in Figure 14.9. The end-to-end delay varies due to link errors that force MAC layer retransmissions and introduce extra delays. A packet from the selected node was received in all routing cases on average after 7 hops (Suhonen et al. 2006b).

15

TUTWSN API*

Potential application scenarios for TUTWSN vary from small monitoring networks to full-feature industrial control systems. These set different requirements to the WSN itself but also to the client infrastructure utilizing the gathered data. The term *client* in this context means a SW component or a human user that commands, gathers, visualizes, or utilizes WSN data. A client can be *internal* if it is an integrated part of the WSN, or *external*, in which case the client is connected to the WSN through a *gateway*.

In general, TUTWSN application scenarios can be divided into three classes according to their dependencies to external clients:

1. A closed system, in which gathered data are processed within the network by internal clients. For example, node applications can react to detected events and activate actuators for affecting the surrounding conditions.

2. A full-featured TUTWSN, the data of which is gathered by external TUTWSN client tools. In this case TUTWSN is complemented with customized PC tools and support infrastructure.

3. A WSN with an access and data gathering interface to external clients that are not part of TUTWSN framework. This scenario enables third-party clients to connect TUTWSN and query its services and data.

In the first alternative, no external access is needed while the data are used by internal clients. Instead, in the latter two cases WSN data are exploited by external clients. In the second case, the access is performed by SW components that are tightly integrated to the TUTWSN. In the last one, assumptions about the client behavior cannot be made, but a generalized API is needed.

TUTWSN offers a unified API to WSN data. The same API is used by both TUTWSN client SW and external clients. In order to facilitate integrated application development

*This chapter was first published entitled "WSN API: Application Programming Interface for Wireless Sensor Networks" (Juntunen et al. 2006). © 2006 IEEE. Reprinted with permission, from *Proceedings of 2006 IEEE 17th International Symposium on Personal, Indoor and Mobile Radio Communications*.

for both sides – the measuring application on WSN nodes and the gathering or visualization application on the external client – a corresponding API is present also on the node side.

The main design objectives for TUTWSN API are to abstract the internals of the network and to offer a simple-to-use interface for WSN service discovery and data access. The abstraction conceals the complexities of WSN communication protocols and architectures, and makes TUTWSN easily extendable for new applications. TUTWSN API provides a well-defined and easy-to-use way to collect data from WSN nodes, and to give commands to them. In addition, service discovery mechanisms and attribute-based queries are supported (Juntunen et al. 2006).

15.1 Design of TUTWSN API

The basic architecture of TUTWSN API consists of two interfaces, as depicted in Figure 15.1 (Juntunen et al. 2006). Since external clients communicate with WSN through gateway nodes, these nodes are a logical architectural location for a *Gateway API*. The Gateway API provides interfaces for service discovery, data access, and network configuration. The interface between the sensor applications running on individual nodes and the WSN communication protocols is referred to as *Node API*. It abstracts the WSN communication and offers a congruent API to node applications with that of Gateway API.

15.1.1 Gateway API

The Gateway API conceals the underlying WSN communication protocols. From an external client point of view, the Gateway API presents the TUTWSN as a set of sensor applications. Each sensor application is viewed as a distinct service to the external clients.

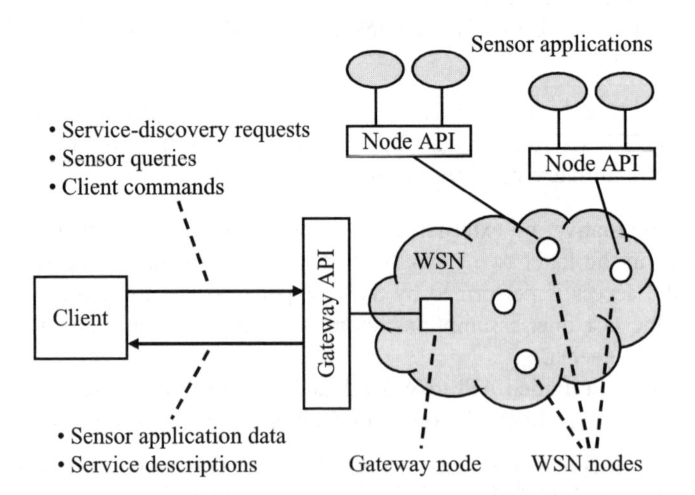

Figure 15.1 Architectural overview of TUTWSN API. (Juntunen et al. 2006. © 2006 IEEE. Reprinted with permission from *Proceedings of 2006 IEEE 17th International Symposium on Personal Indoor and Mobile Radio Communications.*)

Thus, a service is for example a temperature measurement service, humidity measurement service, or data processing service. All services are accessed through the Gateway API.

A service is identified by a tuple ⟨*ApplicationId, Description*⟩, in which the *ApplicationId* is a unique identifier for the application implementing the service. The *Description* defines the data content of the service. An *ApplicationId* is dynamically assigned by the Gateway API when a new WSN node with a new sensor application is discovered. It is assumed that there are no global, static values for *ApplicationId*. In order to list the services of TUTWSN, a client performs a service discovery process.

An external client accesses WSN data using a sensor query ⟨*ApplicationId, Location*⟩. The *Location* field defines the area from which the data related to the sensor application specified by the *ApplicationId* is requested. *Location* can define either a single node, a group of nodes, or coordinates specifying a geographical location.

The responses to data queries and periodical data updates are received in a format ⟨*NodeId, ApplicationId, ApplicationData*⟩. The *NodeId* is the unique identifier of a WSN node, which can be considered as the MAC address of the node. The *ApplicationData* is the output of the sensor application specified by the *ApplicationId*. The external client decodes the received data according to the service description obtained during the service discovery process. The service discovery either floods a discovery request to the network or, if available, queries a local cache maintained by Gateway API.

An example TUTWSN data access through the Gateway API is presented in Figure 15.2. First, an external client sends a service discovery request to the Gateway API (1), which disseminates the discovery to the network (A). WSN nodes reply with service descriptions (B), which are forwarded to the client by the Gateway API. Next, the client queries data from the network by injecting a query specifying the application and the location (3). The query is sent to the WSN (C) and once the related nodes reply (D), the Gateway

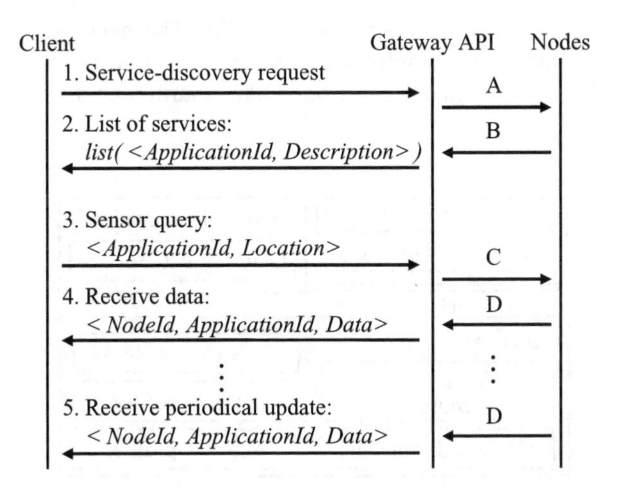

Figure 15.2 An example usage sequence showing a service discovery process and sensor data query performed by an external client. (Juntunen et al. 2006. © 2006 IEEE. Reprinted with permission from *Proceedings of 2006 IEEE 17th International Symposium on Personal Indoor and Mobile Radio Communications*.)

API passes the results to the client (4). A query can either specify a one-time operation or initiate periodical data updates (5).

15.1.2 Node API

The structure of Node API is presented in Figure 15.3 (Juntunen et al. 2006). The Node API implements a flexible architecture that allows a transparent addition and removal of new sensor applications. In addition to the application information, Node API includes information about the physical characteristics of the node and about the role of the node in TUTWSN.

The Node API operates on top of SensorOS (see Chapter 16) and the TUTWSN protocol stack. The communication between Node API, sensor applications, and underlying components is implemented with SensorOS IPC routines. The Node API gathers operational parameters from SensorOS and WSN protocols and implements an interface to the applications for querying these parameters. These parameters include statistical information about the node performance and actions in the network, these include, for example, remaining energy level, buffer (memory) usage, node role, and neighbor information. The utilization of the information is left to the sensor applications.

In addition to the OS interface, Node API also implements an interface to the WSN communication protocols. The Node API maintains a cache of active data interests of the WSN. These define global data requests that are served periodically. When the Node API receives a data packet from the WSN protocol stack, it redirects it to the correct sensor application. Similarly, a sensor application uses the Node API interface to send a packet to the network.

The services of a WSN node are described by profiles. The Node API contains a *node profile*, a set of *sensor and actuator profiles*, and a distinct *application profile* for each sensor application. The contents of profiles are presented in Table 15.1 (Juntunen et al. 2006). These profiles are used for producing the service description passed to an external client through the Gateway API during the service discovery process.

A node profile is associated to each node in WSN. The node profile consists of basic node identification and location information together with the purpose of the node. For example, the Role field may contain a description like "fourth floor, room 406" or "building automation node".

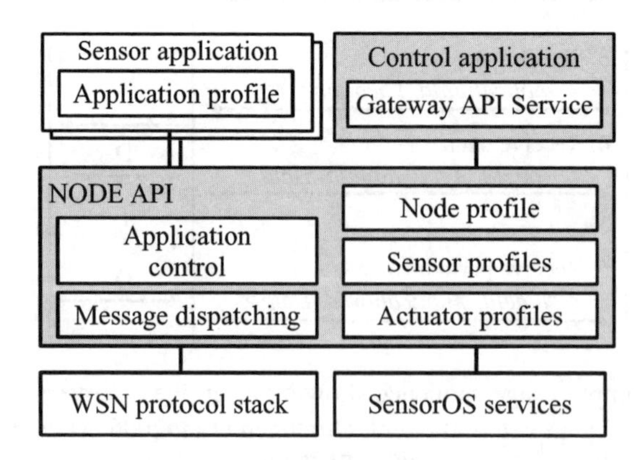

Figure 15.3 An overview of Node API architecture.

Table 15.1 Node, sensor, and application profile fields. (Juntunen et al. 2006. © 2006 IEEE. Reprinted with permission from *Proceedings of 2006 IEEE 17th International Symposium on Personal Indoor and Mobile Radio Communications.)*

Node profile field	Description
NodeId	Unique ID of the node
Location	Coordinates of the node
Role	A descriptive text defining the role of the node
Area code	Nodes with the same code form a group
Applications	List of sensor applications executed on the node

Sensor profile field	Description
Type	Sensor type (e.g. temperature)
Manufacturer	Sensor manufacturer
Model	Sensor model name
Sample size	The size of the generated sample
Sample type	The type of sample (e.g. integer)
ADC resolution	ADC resolution if needed
Sample rate	The maximum sample rate

Application profile field	Description
Name	Application name
Role	A descriptive text defining the purpose of the application
Sensors/Actuators	List of sensors and actuators used by the application
Remote UI	Remote user interface

A sensor profile is associated to each physical sensor present in a node. The sensor profile defines the characteristics of the sensor (Krco et al. 2005). The fields of the sensor profile are set by the component manufacturer. An actuator profile with similar contents is associated to each actuating component present in the node.

Each application present in a node is described by an application profile. It describe the purpose and capabilities of the sensor application. The profile fields define what sensors and actuators the application uses (if any). In addition, the application may specify a remote UI for external clients in its profile. The remote UI consists of tuples ⟨*CommandId*, *CmdDescription*⟩, in which *CommandId* is a unique identifier for a command and *CmdDescription* a textual description defining the purpose of the command.

15.2 TUTWSN API Implementation

The TUTWSN API is implemented according to the design. The implementation consists of two parts; a Gateway API implementation on the gateway node and a Node API implementation for the sensor applications on the TUTWSN nodes. The constraints for the Node API and Gateway API implementations are set by the limited memory resources of WSN nodes.

15.2.1 Gateway API

The communication between the Gateway API located on the TUTWSN gateway node and an external client SW is implemented by a set of messages. These messages carry the service discovery requests, sensor queries, service descriptions, application data, and remote UI commands between the client and the gateway.

The message interface to the Gateway API uses a predefined encoding. The Gateway API in the node contains a message interpreter that handles escape coding, parses messages, and converts them to network packets. The physical transmission medium between the gateway node and the external client is not limited. It can be, for example, a wired Ethernet or RS-232 link, or a Bluetooth or General Packet Radio Service (GPRS) allowing wireless communication between client and gateway node.

The Gateway API maps the requests and queries made by a client to the TUTWSN protocol stack. The service discovery is embedded in the Gateway and Node API layers. When the client issues a service discovery request it is flooded to all nodes of the network. When the Node API of a TUTWSN node receives the request, it replies with a service description consisting of the profile information. Depending on the location specifier, a sensor query maps either to a single target node, to a group of nodes, or to a network wide data interest.

15.2.2 Node API

The implementation of the Node API on a TUTWSN protocol stack is shown in Figure 15.4. A sensor application communicates with the Node API through a MSAP interface. The management procedures implement node and sensor profiles and maintain the interest cache. For data transfers, a sensor application accesses the TUTWSN protocol stack through the DSAP interface. Node API abstracts the network access by selecting the traffic class for the outgoing packet based on the application profile and assigning the destination address from the interest that the application is serving. Thus, each application can operate in a data-centric nature by generating data related to its interest.

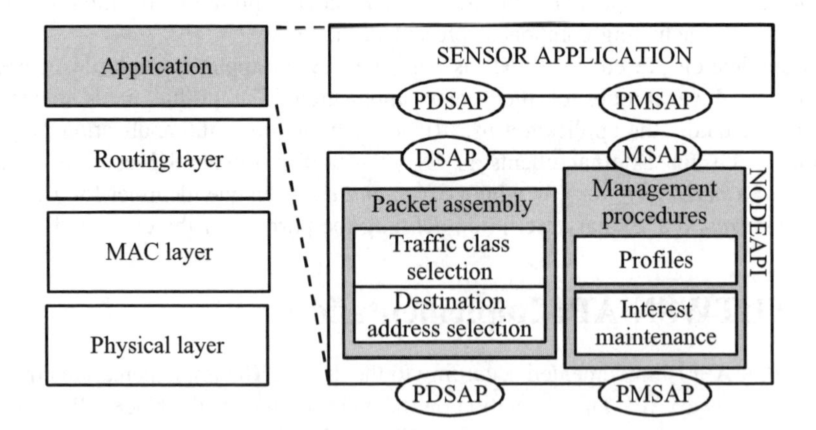

Figure 15.4 TUTWSN protocol stack and application layer architecture.

The DSAP of the Node API is presented in Table 15.2. The DSAP is implemented with SensorOS IPC messages. An application constructs a Protocol Data Unit (PDU) and sends it to the TUTWSN protocol stack for processing with IPC message **MSG_SEND**. Received data is relayed to an application with **MSG_RECEIVE**. Data aggregation in TUTWSN is controlled by message **MSG_FORWARD**, which indicates that received data should be forwarded as is, without further processing. An application prevents data forwarding, if it considers data potential for aggregation.

The control interface between sensor applications and Node API is by functions that realize the MSAP; MSAP is defined in Table 15.3. During the initialization, each application registers itself with the **api_register_application** function. The registration makes the Node API layer aware of the application. The function **api_get_area_code** allows the application to query the group code field set to the node profile. With this information, the application decides whether it performs actions targeted to a certain group of nodes. An application can also join the node to a group with the function **api_set_area_code**.

Also, the state changes in the underlying WSN protocol stack are indicated to the Node API and sensor applications through the PMSAP. The messages realizing the interface are presented in Table 15.4. When a new sink node is detected or routes to a known sink are lost, messages **MSG_API_SINK_FOUND** or **MSG_API_SINK_DROPPED** are used, respectively. When a route to a sink changes, or a new route is found, the protocol stack informs Node API with message **MSG_API_ROUTE_CHANGED**. New network-wide data interests are indicated with **MSG_API_INTEREST_CHANGED**. Finally, a new member node and its parameters are passed to the Node API with message **MSG_API_MEMBER_JOINED**.

Node and sensor profiles are stored in the EEPROM, while application profiles are part of the program code and therefore located in the program memory. The profiles are compact in order to diminish their transmission cost over WSN. An application profile consists of a

Table 15.2 IPC messages implementing Node API DSAP. (Juntunen et al. 2006. © 2006 IEEE. Reprinted with permission from *Proceeding of 2006 IEEE 17th International Symposium on Personal Indoor and Mobile Radio Communications.*)

Message	Parameters
MSG_SEND	`pdu_t* sendPdu`
MSG_RECEIVE	`pdu_t* recvPdu`
MSG_FORWARD	`pdu_t* forwardPdu`

Table 15.3 Node API MSAP interface targeted at sensor applications. (Juntunen et al. 2006. © 2006 IEEE. Reprinted with permission from *Proceedings of 2006 IEEE 17th International Symposium on Personal Indoor and Mobile Radio Communications*).

`bool`	**api_register_application**	`( application_profile_t* prof )`
`uint8_t`	**api_get_area_code**	`( void )`
`void`	**api_set_area_code**	`( uint8_t code )`

Table 15.4 The event-driven PMSAP interface between Node API and WSN protocol stack.

Message	Parameters
MSG_API_SINK_FOUND	`addr_t* address`
MSG_API_SINK_DROPPED	`addr_t* address`
MSG_API_ROUTE_CHANGED	`sink_t* sink`
MSG_API_INTEREST_CHANGED	`addr_t* source, uint8_t interestId, interest_t interest`
MSG_API_MEMBER_JOINED	`addr_t address, assoc_data_t* assocData`

descriptive name and a list of used sensors. A sensor profile comprises of sensor component type, name of the manufacturer, model name, size of the sample, and type of the sample. The node profile consists of a node ID, a short description, GPS coordinates, and a group code.

15.3 TUTWSN API Evaluation

The main point of evaluation in APIs is the ease of use. However, this a subjective qualifier, which makes its evaluation difficult. Instead, the evaluation focuses mainly on the node and network resources consumed by the main characteristics of the Node API.

15.3.1 Ease of Use

The main design objective for the TUTWSN API has been the extensible and easily comprehended interface. The Gateway API provides a well-defined and easy-to-use interface to TUTWSN, which facilitates the development of external applications. An addition of a new application to the WSN requires only the definition of application data to the messaging interface. All the other information can be gathered from profile fields deployed on the nodes.

The design choices allow a modular implementation of Node API through which new applications and sensors can be introduced to TUTWSN without interfering with the existing ones. In nodes, an addition of a new application or sensor requires only the definition of the corresponding profile and the compilation and linking of the code with the existing applications.

15.3.2 Resource Consumption

The memory consumptions of the Gateway API, Node API, and profile implementations are shown in Table 15.5. The program memory used by Gateway and Node API are moderate. For example, for a TUTWSN PIC nodes with 64 kB of program memory, Node API uses only 8% of the available memory.

The memory used by profiles depends on their contents. The descriptive text fields are limited to 32 B, but only the necessary amount of characters are stored. Still, the values given in Table 15.5 with one sensor component and two sensor applications give good

Table 15.5 Memory usage of TUTWSN API and profile implementations on a TUTWSN PIC prototype platform. (Juntunen et al. 2006 © 2006 IEEE. Reprinted with permission from *Proceedings of 2006 IEEE 17th International Symposium on Personal Indoor and Mobile Radio Communications*.)

API	Memory	Type of memory
Gateway API	7074 B	Program
Node API	5298 B	Program

Profile	Memory	Type of memory
Node	37 B	EEPROM
Sensor	19 B	EEPROM
Application	40 B	Program

approximations. Basically, the objective is to minimize the sizes of profiles in order to minimize the communication cost caused by their transmission and needed storage space.

15.3.3 Operational Performance

Compared to overall networking delay, the time consumed by internal Node API operations is minimal. Therefore, the timing instrumentation of these operations is omitted. Similarly, the bandwidth of the medium between a gateway node and an external client is always more than adequate compared to the throughput of the WSN. Therefore, the operational times of Gateway API operations are not measured.

The network-wide operations that relate to the TUTWSN API, such as service discovery, data queries, or commands, are highly dependent on the size of the network. It is evident that a service discovery process on a 20-node network is significantly faster than the same operation on a 1000-node network. The performance of these operations depends solely on the communication protocols and their parameters, as the networking delay dominates the operation time. Since the performance of underlying protocols does not depend on the TUTWSN API operation, further performance evaluation is omitted.

16

TUTWSN SensorOS

A full-featured TUTWSN consisting of protocols and several applications constitutes an extremely complex system. In a typical TUTWSN deployment there are several sensing, diagnostics, and control applications on top of the TUTWSN protocol stack and API. The managing of timing, IPC, and resources in such a system requires OS control and coordination. Therefore, SensorOS, a preemptive multithreading kernel has been implemented for TUTWSN nodes.

The main design points for SensorOS have been the extremely accurate time concept and efficient resource usage. The timing accuracy required by the TUTWSN MAC channel access is guaranteed in SensorOS by a priority-based realtime capable scheduler. The small-memory footprint SensorOS kernel offers a rich API that simplifies application development. Further, energy efficiency is obtained through the careful use of power-saving modes (Kuorilehto et al. 2007a).

However, in some deployments a single application with a very simple configuration of the TUTWSN protocol stack can be adequate. For such configuration, a preemptive SensorOS kernel is an overkill but a more lightweight kernel is sufficient. Therefore, a lightweight version of SensorOS kernel offers similar OS services but replaces the preemptive kernel by a simple run-to-completion scheduler.

Both versions of SensorOS can be supplemented by a separate bootloader service. The bootloader allows a runtime reprogramming of both SensorOS kernel and applications. Runtime reprogramming is a desired option in WSNs, since the manual reprogramming of large deployments is tedious, in some situations even impossible.

16.1 SensorOS Design*

SensorOS design objective is a realtime kernel that supports features required by WSN protocols and applications. Different WSN protocol and application tasks are executed as separate threads that communicate with SensorOS IPC methods. The composition of threads implementing protocols and applications is not restricted (Kuorilehto et al. 2007a).

*With kind permission of Springer Science and Business Media.

In the following, a *task* is a functional entity, whereas a *thread* is the OS context, in which a task is executed. A task can be divided into multiple threads, but on the other hand, several tasks can be executed within a single thread.

16.1.1 SensorOS Architecture

The architecture of SensorOS is divided into components as depicted in Figure 16.1. Tasks access OS services through API. The main components in the kernel are *scheduler*, *message passing IPC*, *timer*, *synchronization*, and *memory and power management*. Interrupt-driven device drivers are also integrated into the kernel, whereas context-related drivers are executed in the context of a calling thread without a relation to the OS kernel. In general, devices that are accessed only by a single thread are implemented as context-related, while common devices are included in the kernel. Hardware resources are accessed through a Hardware Abstraction Layer (HAL), which enables SensorOS portability (Kuorilehto et al. 2007a).

Each thread in SensorOS has a Thread Control Block (TCB) for storing per thread state information. A thread can be in three different states. When a thread is executed on MCU it is *running*. The state of a thread is *ready* when it is ready for execution but another thread is running, and *wait* when it needs an event to occur before execution.

A thread can be waiting for multiple different types of events in SensorOS. The relation between a running thread, a ready queue, and different wait queues is depicted in Figure 16.2. When a thread is created it is put to the ready queue, and it can explicitly exit when running. Threads waiting for a timeout are in timer queues and those waiting for IPC are in message set. Synchronization is waited in per mutex queues and a completion of peripheral operation in a peripheral specific item (Kuorilehto et al. 2007a).

16.1.2 OS Components

The SensorOS components maintain TCBs of threads accessing their services. The interrelations between components are kept to a minimum, but clearly the scheduler is dependent on other components (Kuorilehto et al. 2007a).

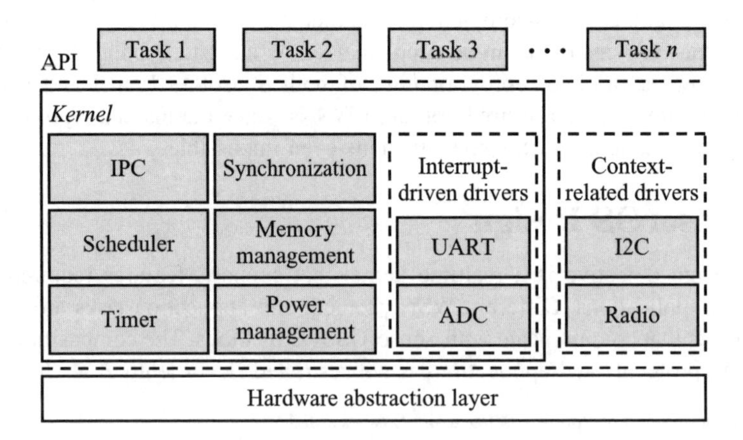

Figure 16.1 Overview of SensorOS architecture. With kind permission of Springer Science and Business Media.

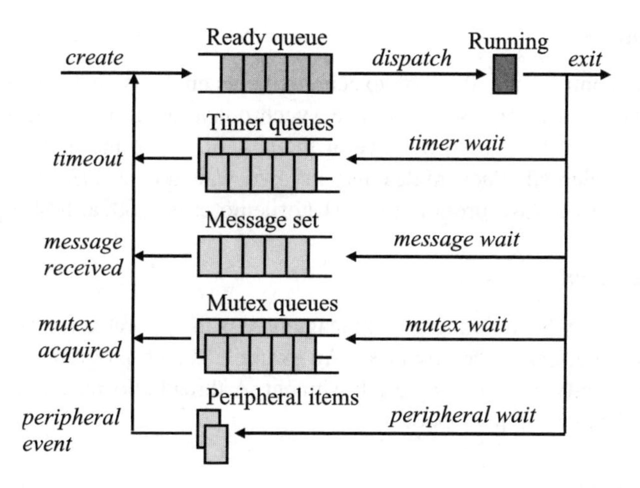

Figure 16.2 Thread queues and events moving a thread from one queue to another. With kind permission of Springer Science and Business Media.

Scheduler

SensorOS incorporates a priority-based preemptive scheduling algorithm. Thus, the highest priority thread ready for execution is always running. A round-robin algorithm is used for scheduling threads at the same priority level. The time-slicing of thread execution at the same priority level is not supported but each thread is run to completion (Kuorilehto et al. 2007a).

When an event changes a thread to the ready state, the scheduler checks whether the thread should be dispatched. If its priority is higher than that of a running thread, contexts are switched. When the running thread enters a wait state, the highest priority thread from the ready queue is dispatched. If the ready queue is empty, power management procedures are activated.

Timer

Timer component implements timeout functionality. The local time in SensorOS is microseconds since system reset. Timing is divided into two separate approaches that have their own timer queues. A fine granularity timing provides microsecond accuracy for applications and communication protocols that need exact timestamps. The coarse timing is for tasks that tolerate timeout variations in order of millisecond.

Inter-Process Communication (IPC)

The method for communication between tasks in SensorOS is message-passing IPC. A thread allocates a message envelope and fills it, after which it is sent to the recipient. In contrast, a recipient can either poll for available messages or block its execution until a message is received. The message must always be assigned to a certain thread. Broadcast messages can be implemented using unicast messages.

Synchronization

Synchronization controls the flow of execution between tasks and access to peripheral devices and other hardware resources. The synchronization is implemented with binary mutexes. A mutex can be waited by several threads, of which the highest priority thread acquires it when released. Each mutex has its own wait queue. Priority inversion is not considered but is left to task programmers (Kuorilehto et al. 2007a; Stallings 2005).

Memory management

In SensorOS, a simple dynamic memory management is incorporated for message envelopes and for temporary buffers needed by tasks. An example is a cryptographical algorithm executed only infrequently during a key establishment. A thread allocates and frees previously reserved blocks from a memory pool.

Power management

The power management of peripherals is implemented in the device drivers. The power saving activation of context-related devices, such as radio, is left to the task that controls the device because the task is aware of the device activation patterns. Instead, the power modes of MCU and integrated peripherals, such as ADC, are managed by OS.

When there are no threads to schedule for execution, MCU is set to power-saving mode that depends on the platform and the length of the sleep period. The MCU is woken up by an external event. Integrated peripherals are powered off when not used.

Peripherals

The interrupt-driven device drivers integrate peripherals, such as UART and ADC, tightly to SensorOS kernel. They have separate functions for open, close, control, read, and write operations. The data transmissions for read and write operations are controlled by interrupts. A thread can block its execution on such peripherals until it has read or written the specified number of bytes. The context-related device drivers are either non-blocking or can use an external interrupt source for controlling read and write operations.

16.2 SensorOS Implementation*

SensorOS has been implemented to TUTWSN PIC nodes. The implementation follows the architecture presented above. Common functionality is implemented separately, whereas hardware dependent parts are included in HAL. The common functionalities and most of HAL are implemented in American National Standards Institute (ANSI) C. Only a small portion of the lowest level HAL, such as context switch, is implemented in assembly (Kuorilehto et al. 2007a).

16.2.1 HAL Implementation

Lowest level context switching, power saving, timer, and peripheral access are detached from the SensorOS kernel to the HAL implementation. Internal registers, which need to be

*With kind permission of Springer Science and Business Media.

saved when thread contexts are switched, are MCU dependent. Also, power-saving modes need low-level register access. Each peripheral has a HAL component that implements interface to dedicated I/O ports and interrupt handlers.

Each MCU has its own set of hardware timers together with control registers. HAL timer implementation consists of time concept, interrupt handlers, and time management routines. The time concept is implemented in HAL so that hardware timer-dependent data types can be used efficiently. SensorOS utilizes two different time concepts implemented by HAL; a microsecond resolution timer for accurate timing, and a millisecond resolution timer for timeouts. The interrupt handlers update the internal time and when a time limit expires it indicates this to the OS timer through a callback function. The time management routines are for getting and manipulating internal time, setting timeout triggers, and atomic spinwait for meeting an exact timeline.

16.2.2 Component Implementation

SensorOS API consists of system calls as listed in Table 16.1. Peripheral system calls are for character devices (e.g. UART) while context-related devices have dedicated interfaces. SensorOS is initialized in `main`-function, which issues `user_main`-function after OS components have been initialized. In `user_main`, threads for application tasks and required mutexes are initialized. After the `user_main` returns, scheduling is started.

Scheduler

A thread is created with **`os_thread_create`** that takes the memory for stack and Process Identifier (PID) as parameters. This simplifies the implementation but prevents runtime creation and deletion of threads. Yet, the modification of the kernel for such a support is straightforward. When a thread is created it is inserted into the ready queue but the scheduler is not invoked until the running thread releases the processor.

Instead of a completely modular approach, the scheduling decisions are distributed to kernel components. This complicates the changing of scheduling algorithm but improves context switching performance. When an event moves thread(s) to the ready queue, the OS component checks whether one of the threads has a higher priority than the running one. If true, an OS service for swapping threads' contexts is invoked. The context of a thread is stored in its stack. A running thread can release processor with **`os_yield`** or it can permanently exit. When there are no threads to schedule, an *idle thread* is scheduled for activating MCU sleep mode through HAL.

Event waiting in SensorOS is implemented by a single interface that allows a thread to wait simultaneously for multiple events. The events include timeout, message received, character device read and write, peripheral device, and user-generated events. Function **`os_poll_event`** loops actively while **`os_wait_event`** blocks the thread until any of the events occur. When an event for a thread is raised, the scheduler checks whether the thread is waiting for the event, and if so, performs scheduling.

Timer

Timer operation is mainly implemented in HAL but API and scheduling on timeouts are provided by the OS component. The system time is obtained with the function **`os_get_time`**.

Table 16.1 SensorOS system call interface, categorized by components. With kind permission of Springer Science and Business Media.

Thread and scheduler management system calls

```
void os_thread_create( os_proc_t *p, os_pid_t pid,
  os_priority_t pri, char *stack, size_t stackSize,
  prog_counter_t entry )
void os_yield( void )
os_eventmask_t os_wait_event( os_eventmask_t events )
os_eventmask_t os_poll_event( os_eventmask_t events )
```

Timer system calls

```
uint32_t os_get_time( void )
os_uperiod_t os_get_entryperiod( void )
int8_t os_wait_until( os_uperiod_t event )
void os_set_alarm( uint16_t timeout )
```

IPC system calls

```
os_status_t os_msg_send( os_pid_t receiver, os_ipc_msg_t *msg )
os_ipc_msg_t* os_msg_recv( void )
int8_t os_msg_check( void )
```

Synchronization system calls

```
void os_mutex_init( os_mutex_t *m )
void os_mutex_acquire( os_mutex_t *m )
void os_mutex_release( os_mutex_t *m )
```

Memory management system calls

```
void* os_mem_alloc( size_t nbytes )
void os_mem_free( void *ptr )
```

Character device system calls

```
os_status_t os_open( os_cdev_t dev )
void os_close( os_cdev_t dev )
int8_t os_write( os_cdev_t dev, const char *buf, uint8_t count )
int8_t os_read( os_cdev_t dev, char *buf, uint8_t count )
void os_close( os_cdev_t dev )
```

Accurate timestamps for events are set with **os_get_entryperiod**, which returns the internal time at the moment of the function call. Both utilize a microsecond resolution timer.

The accurate microsecond resolution wait is implemented by **os_wait_until**. The thread is blocked until a threshold before the deadline. The atomic spinwait in HAL is used to suspend the operation until the timestamp. In the current implementation, only one thread can issue **os_wait_until** at a time to guarantee accurate timing.

To initialize a millisecond resolution wait, a thread issues **os_set_alarm**. The thread is put to the timer queue that is sorted according to the timeouts. The first item in the queue is passed to HAL in order to trigger a callback function when the timeout expires. The callback function sets the timer event for the first thread in the queue. A zero timeout period can be used with **os_wait_event** to check a status of other events.

Message passing IPC

The memory allocation for message envelopes and the contents of messages are left to the application. A message is sent to a specified recipient with **os_msg_send**. The function inserts the message to the recipient's message queue and sets the message-received event. Each thread has an own message queue in its TCB. A thread can check whether its queue is empty with function **os_msg_check**. A message is removed from the queue by calling function **os_msg_recv**.

Synchronization

When a mutex is created with **os_mutex_init**, its wait queue and owner are cleared. If the mutex is blocked by another thread when **os_mutex_acquire** is issued, the calling thread is inserted to the wait queue of the mutex; otherwise, the caller becomes the owner of the mutex. When the owner calls **os_mutex_release** and the wait queue is not empty, the highest priority thread is moved to the ready queue, or scheduled immediately if its priority is higher than that of the running thread.

Memory management

Currently, there are two alternatives for memory management. A *binary buddy* algorithm allows the allocation of different-sized blocks, while a more lightweight alternative uses static-sized blocks and is mainly targeted to message envelopes. Memory is allocated with **os_mem_alloc** and released with **os_mem_free**.

Peripherals

The interrupt-driven character device drivers are opened and closed by **os_open** and **os_close**, respectively. The device handle contains the owner, type, and event information and defines the HAL routines and data pipe for communication between HAL and OS. Data to the device is sent with **os_write** and received with **os_read**, which return the number of bytes handled. The completion of a pending operation can be waited either by **os_flush** or **os_wait_event**.

16.3 SensorOS Performance Evaluation*

The performance of SensorOS is evaluated with a test application that benchmarks OS operation. The objectives of evaluation are the verification of correct functionality and the measuring of OS resource consumption and performance (Kuorilehto et al. 2007a).

16.3.1 Resource Usage

The portable implementation in ANSI C results in the slightly more inefficient use of resources than an assembly optimized one. The code and data memory consumptions of each OS component are depicted in Table 16.2. Help routines include implementations for internal OS lists and a small set of library functions.

The code memory usage of SensorOS with static block memory management is 6964 B and with binary buddy 7724 B, which are 10.6 % and 11.8 % of the total memory, respectively. These do not include an optional I/O library, which implements `printf` type routines. Static data memory used by SensorOS is 115 B or 118 B, depending on the used memory management. These do not include thread stacks and TCBs. A thread context takes on average 36 B but during interrupts an additional 35 B is stored. Since the context is kept in the thread's stack, a typical stack size is 128 B. The size of TCB is 17 B. Thus, over 20 SensorOS threads can be active simultaneously in TUTWSN PIC platform.

16.3.2 Context Switch Performance

The performance of SensorOS is evaluated by measuring the context switch overheads and the execution times of the main kernel operations. These are given in Table 16.3 with timing accuracy results. The MCU is run with 2 MIPS and loaded by five threads that have a 2 ms activation interval on average. The results are gathered over 50000 iterations.

Table 16.2 Code and data memory usage of different SensorOS components. With kind permission of Springer Science and Business Media.

OS component	Code memory (B)	Data memory (B)
Scheduler	728	38
Thread	184	0
Event handling	384	1
Timer	646	6
IPC	248	0
Mutex	428	0
Binary buddy memory management	1048	5
Static block memory management	288	2
Character device	414	0
HAL	2266	68
Help routines	1378	0
I/O library	862	16

*With kind permission of Springer Science and Business Media.

Table 16.3 SensorOS kernel operation times and timing accuracy. With kind permission of Springer Science and Business Media.

Operation	Time (μs)		
	Mean	Min	Max
HAL context swap	92	92	92
os_wait_until timeout initialization	125	125	125
os_wait_until spinwait time after thread wake-up	1110	680	1310
os_set_alarm timeout initialization	270	222	324
os_wait_event context switch from timer interrupt	532	486	558
IPC from lower priority thread to higher one	346	346	346
os_wait_until timing accuracy error (absolute)	1.8	0.0	4.2

Context swap is the time needed to store old and restore new threads to the MCU. The initialization of `timer_wait_until` includes a trigger setting to HAL and scheduling, while the `spinwait` is required after the trigger expires in order to get the accurate timestamp. IPC context switch is the time elapsed from a message sent from a lower priority thread to its processing at a higher priority thread. The last time is the delay from a timer interrupt to the scheduling of a thread to execution when it has issued `timer_wait_for`. Thus, the last two also include the context swap time.

Context swap time includes the storing of an old and restoring of a new thread to MCU. The initialization of **os_wait_until** sets a trigger to HAL. The thread is woken up 2 ms before the deadline and after a scheduling delay the rest of the time is spent in spinwait. The time in **os_set_alarm** is consumed in timer queue handling and a trigger setting. The **os_wait_event** time is the delay from a timer interrupt to the scheduling of the thread. The IPC delay is measured from the sending of a message from a lower priority thread to its processing in a higher priority one.

The accuracy of **os_wait_until** is measured as the absolute error between the resulted timing and the real-world time. The maximum inaccuracy is below 5 μs and typically the error is less than 2 μs. The variance is caused by thread atomicity consideration when returning from the spinwait.

16.4 Lightweight Kernel Configuration

The main motivation for the design and implementation of a lightweight SensorOS kernel is the data memory consumption of thread stacks. The lightweight kernel uses a run-to-completion scheduler, which needs only one stack for the whole system. This makes it possible to utilize the data memory required for per thread stacks in preemptive kernel, e.g. for data queuing or neighbor information.

16.4.1 Lightweight OS Architecture and Implementation

The architecture of a lightweight SensorOS kernel is depicted in Figure 16.3. Compared to the full-feature SensorOS, the main differences are in scheduling and in timer concept.

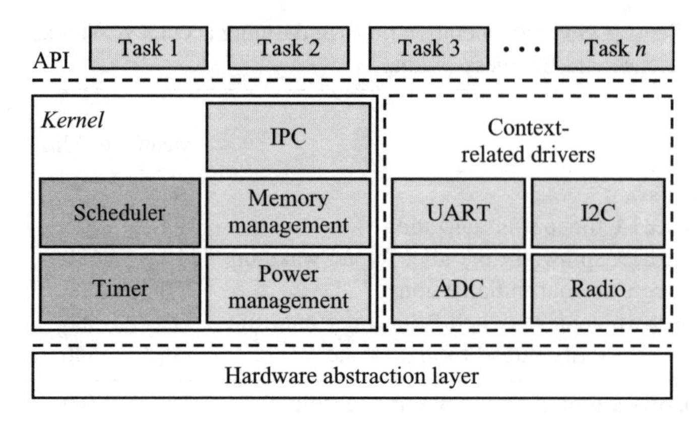

Figure 16.3 Architecture overview of the lightweight version of SensorOS kernel.

Lightweight kernel incorporates a simple run-to-completion scheduler. The scheduler does not support any kind of preemption. Instead, it offers a service for lower priority threads to check whether a higher priority thread should be scheduled, and a routine for yielding the processor in such cases.

The timer concept in lightweight kernel is also a simplified version. A microsecond resolution timer similar to the full-feature SensorOS kernel is provided for high priority threads but other threads are served by a second resolution timer. In TUTWSN, the high priority thread is MAC protocol, while other tasks are run in lower priorities.

The OS routines for IPC and memory and power management are similar to those in full-feature SensorOS kernel. Interrupt-driven device drivers, such as UART, are not included in the lightweight kernel but they are implemented as context-related drivers. Yet, the interface to for example UART is kept similar, which also allows the utilization of I/O library on top of the lightweight kernel.

Most parts of HAL are identical in both versions. The main differences are in hardware timer interfacing routines that differ due to the divergent resolutions. Furthermore, the context switching service is not required.

Due to the differences in programming models, API of the lightweight kernel differs considerably from that of full-feature SensorOS. Therefore, application tasks need to be reprogrammed separately for both kernels. While the core functionality of the task can be left untouched, the interfacing of resources and OS routines must be adapted.

16.4.2 Performance Evaluation

It is evident that the lightweight kernel cannot guarantee realtime operation. Yet, through the use of sophisticated operational parameters, decent timing accuracy can be offered for a high priority thread. Yet, this requires large execution margins that cause drawbacks in other aspects, e.g. in power saving.

The resource consumption statistics given in Table 16.2 also apply fairly accurately for the lightweight kernel. Scheduler, thread, event handling, and mutex components are absent but the differences in other components are minimal. The main benefit is obtained in data memory consumption. While in the full-feature kernel each thread requires typically a 128 B stack, a lightweight kernel can operate with a single 256 B stack.

16.5 SensorOS Bootloader Service

Traditionally, a bootloader is a SW component that loads the actual OS from a permanent storage to the memory, which transfers the execution control to OS after the loading is properly completed. Thus, it is a small-memory footprint software executed after system reset.

In resource-constrained MCUs, instructions are typically executed directly from a flash type memory that is also a permanent storage. Thus, a bootloader is not needed in a traditional sense. However, in WSNs it is beneficial to load the code image from other nodes over a wireless link. This allows the migration of a new code image to the whole WSN without user intervention.

16.5.1 SensorOS Bootloader Design Principles

The main function of the bootloader service is the cloning of a code image from another node to the node itself. The code image includes both OS routines, networking protocols, and application tasks. Therefore, the bootloader allows a complete changing of network operation and functions.

The main design objective for the bootloader is the small code size of the bootloader itself. As the bootloader code cannot be utilized during the actual WSN operation, the code memory it uses is wasted from the application point of view. On the other hand, data memory consumption of the bootloader is not an issue because it can be reused.

As the SensorOS bootloader loads a new code image over a wireless link using the same radio transceiver as TUTWSN protocols, integrating the bootloader into the normal TUTWSN operation could be easily justified. However, the bootloader operation is kept completely distinct from the normal networking. This allows the utilization of the bootloader even if the networking principles or algorithms change between the code image versions updated through the bootloader.

A crucial aspect that arises when a code image is loaded over a wireless link is security. A corrupted code image results in an operational failure that may corrupt the whole WSN. Further, a malicious party can attempt to attack the network by loading a hostile code image to the network. Therefore, checksums and MICs need to be integrated into the bootloader operation to prevent such failures.

16.5.2 Bootloader Implementation

The operation principle of SensorOS bootloader is illustrated in Figure 16.4 for a TUTWSN PIC node with a 64 KB code memory. The bootloader overwrites complete code memory excluding reset vectors and the bootloader code itself. The image header contains identification information that defines applicable platforms, an image version, and MIC for the image.

The basic operation of the bootloader is a simple request/response type protocol that transfers the code image in small chunks from a server to a client. Each node may operate as a server when it has a valid code image. A server can simultaneously serve more than

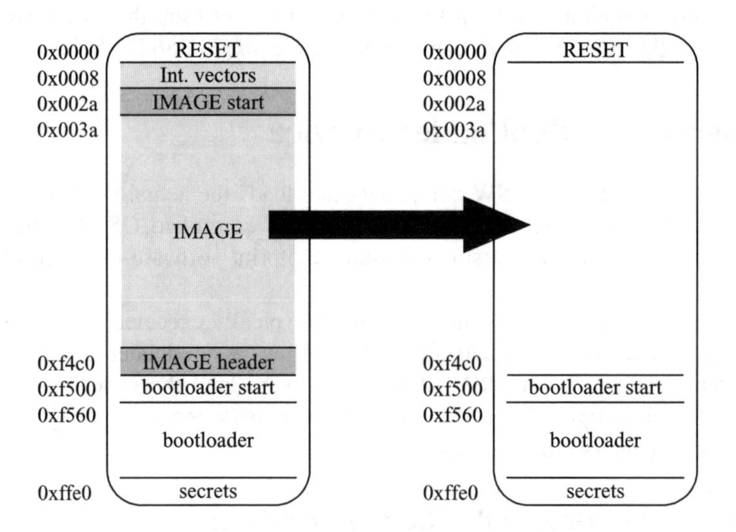

Figure 16.4 SensorOS bootloader operation principle.

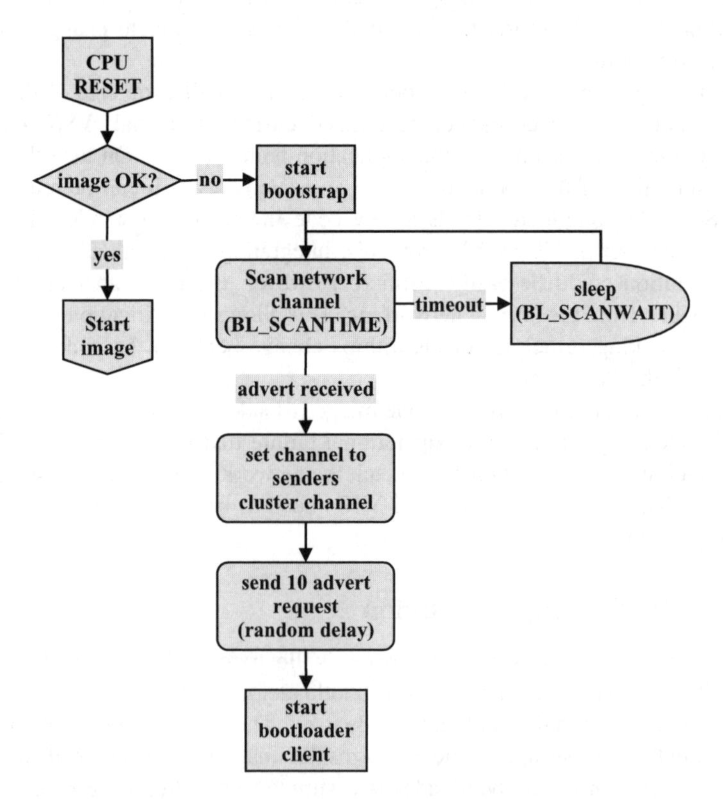

Figure 16.5 Bootloader operation initiation.

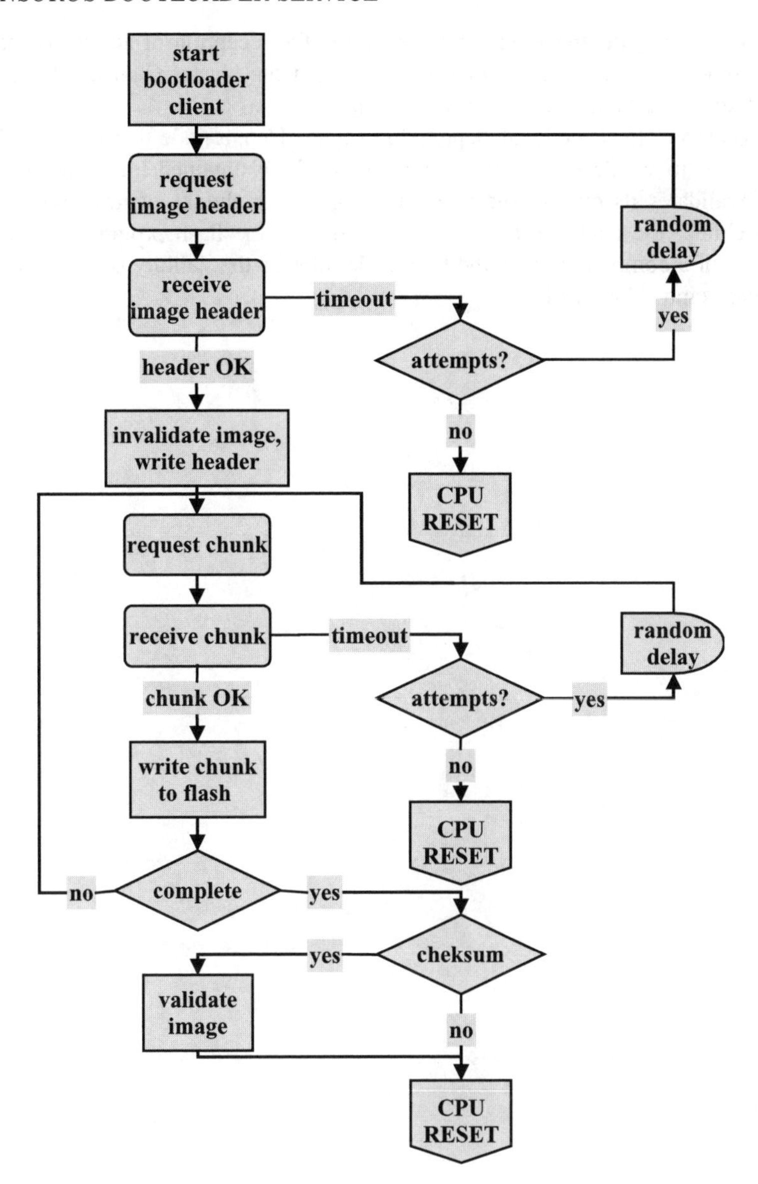

Figure 16.6 Bootloader client operation.

one client, but a client may update its image only from a single server. All operations in the bootloader are initiated by a client request. A client may request either bootloader initiation, image header, or a chunk of image starting from a specific address. A server sends a response to the client request immediately after reception. In case of errors (e.g. a collision), a client retransmits the request after a random delay.

A flow graph presenting the phases in the initiation of a bootloader operation is depicted in Figure 16.5. If the bootloader detects an invalid image during CPU reset, it starts scanning a predefined channel for image advertisements. On the reception of an advertisement, the bootloader sends ten subsequent initiation requests at the cluster channel of the potential server, and starts the bootloader client operation.

In a bootloader client mode, as depicted in Figure 16.6, a node first requests the image header in order to validate the server parameters. If the obtained image header is valid, the node invalidates its own image and starts requesting chunks of new image. When it receives a chunk, the chunk is immediately written to the flash program memory. After the reception of a complete image, the bootloader checks the validity of the image using a simple stream cipher based MIC.

17

Cross-layer Issues in TUTWSN

In the cross-layer design, the protocol layers cooperate to achieve a common goal, such as a certain QoS. In resource-constrained sensor nodes, cross-layer design also targets at increasing efficiency by sharing functionality between hardware components and communication protocols.

TUTWSN utilizes cross-layed design in several ways. First, the node configuration that meets the deployment specific QoS requirements is determined by selecting hardware components and TUTWSN network protocol settings as show in Figure 17.1. This way, overlapping or contradicting functionality is avoided. For example, if CRC calculation is supported on a transceiver module, the MAC protocol does not implement that functionality. Second, the energy efficiency is increased by piggybacking data to the packet belonging to another layer. Thus, the amount of messaging between nodes can be reduced. Third, a node makes self-configuration decisions based on the cross-layer information. The purpose of the information combining is to achieve a better QoS, as several decisions cannot be made solely based on information belonging to one layer.

17.1 Cross-layer Node Configuration

The effect of configuration parameters on QoS is presented in Table 17.1. The runtime field expresses that the category setting can be modified during the normal operation. Some of the categories, such as network address, are determined on compile time as changing them on runtime would require complex negotiations between nodes. Obviously, the physical components are chosen on design time and thus their properties cannot be changed later. The other categories are compiled with default values, but modified on runtime to achieve the desired behavior. The runtime configuration of traffic-class weights and application categories is performed from a sink node with the Gateway API. In addition, the node protocols adjust the access-cycle length and the amount of slots dynamically according to the network load. For these parameters, the initial values and allowed value range is defined on compile time.

Ultra Low Energy Wireless Sensor Networks in Practice: Theory, Realization and Deployment

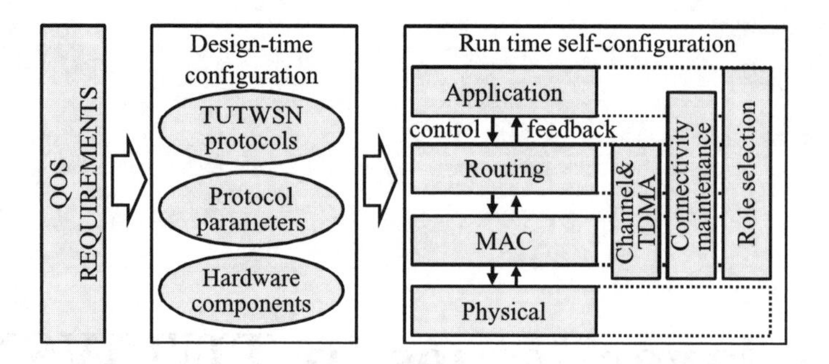

Figure 17.1 Cross-layer design in TUTWSN.

Table 17.1 The effect of configuration parameters to the QoS in TUTWSN.

Configuration parameter / change	Runtime	Throughput	Reliability	Latency	Security	Mobility	Data accuracy	Lifetime
Application								
Measurement rate / increase	Yes						+	−
Transmission rate / increase	Yes	−	+					−
Aggregation / increase	Yes	+	−	+			−	+
Routing								
Alternative routes / increase	Yes		+			+		−
Traffic-class weights	Yes	+/−	+/−	+/−				+/−
Multipath routing / increase	Yes	−	+			+		−
MAC								
Access-cycle length / increase	Yes	−		+		−		+
ALOHA slots / increase	Yes	+	+			+		−
Reserved slots / increase	Yes	+						
Network beacon rate / increase	No					+		−
Network address / increase	No	−	+					
Acknowledgments	Yes		+					−
Retransmissions / increase	Yes		+	−				
Reception margins / increase	Yes		+			+		−
Encryption	No	−			+			−
Authentication	No				+			−
Cluster density / increase	Yes	+	+			+		−
Physical								
Communication range / increase	No	−				+		−
Data rate / increase	No		−					+

+ = positive effect on the QoS parameter; − = negative effect on the QoS parameter.

17.1.1 Application Layer

The measurement interval defines how often an application senses data. Thus, the interval allows a trade-off between the accuracy and the required sensing energy. The transmission interval describes how often the sensed values are sent to a sink. If the measurement interval is shorter than the transmission interval, a node performs aggregation by either combining several values (e.g. by calculating an average) or inserting several measurement values into a single packet. In the former case, the aggregation reduces the data accuracy, but at the same time reduces the throughput requirements. The aggregation has the drawback of increasing latency as a node caches several measurements before sending the aggregated value.

17.1.2 Routing Layer

The routing layer allows a trade-off between reliability and network lifetime with the usage of alternative routes and multipath routing. The use of alternative routes requires maintaining synchronization with many neighbors, thus increasing the energy usage. However, reliability and mobility are increased, because a replacement route is discovered faster when a link breaks. In the multipath routing, a packet is transmitted via multiple paths, therefore increasing reliability and mobility, but requiring more bandwidth and energy.

In TUTWSN routing, the traffic-class weights have a significant effect on how a route is selected. The traffic weights define the preference between throughput, reliability, latency, and energy usage. A more detailed discussion about the weights is presented in Chapter 14.

17.1.3 MAC Layer

The access-cycle length and the superframe structure have the most significant effect on the performance of the MAC layer. Because a cluster beacon is sent every access cycle, a long access cycle reduces energy usage. However, as the number of ALOHA and reserved slots per access cycle is fixed, increasing access cycle length reduces the throughput. In addition, the latency is significantly increased because a member node must wait longer to send its data.

Increasing the ALOHA slot count decreases collision probability, therefore increasing reliability. However, as each ALOHA slot causes energy-consuming idle listening, the lifetime of a cluster head is reduced. The use of the reserved slots is preferred because collisions within a cluster are avoided and unused reserved slots do not consume energy. Still, the ALOHA is required for joining the cluster and making initial reservations. Moreover, as a mobile node can only communicate with a cluster for a short time until moving outside the communication range, long-term reservations cannot be made and ALOHA slots are required. A mobile node also requires a high network beacon rate for detecting new neighbor clusters rapidly. Thus, the network beacon rate has a trade-off between the energy consumption of cluster advertisement and cluster discovery as discussed in Chapter 13.

Network address width and reception margins control the reliability of a frame transmission. In TUTWSN, the network address is used as a preamble on a frame transmission. A longer address increases reliability by reducing spurious receptions, but it also increases overheads. The reception margins compensate for synchronization inaccuracy by determining time values before and after the expected reception that a frame is listening to.

The MAC layer recovers from failed transmissions with acknowledgments and retransmissions. The drawback is that an acknowledgment frame significantly increases the energy usage. As data frames are usually small, the acknowledgment frame may be as large as the actual data frame. Thus, the use of acknowledgments essentially doubles the energy consumption. In TUTWSN MAC, the acknowledgments do not affect overall throughput because of the TDMA slot structure, in which a slot is divided into uplink and downlink phases. Thus, the acknowledgment can be sent in the downlink phase that would otherwise remain unused. Furthermore, when data is sent to both directions, the acknowledgment bits are piggybacked into the downlink data frame following the uplink transmission. A failed acknowledgment triggers a retransmission. The retransmissions have a drawback of increasing latency because new frames have to be buffered, while an old frame is retransmitted.

The security is controlled with encryption and authentication, which usually increases throughput requirements. The frame encoding often enlarges payload, whereas the authentication adds an additional header that signs the packet contents. However, TUTWSN replaces the usual per frame CRC with the authentication header and thus the throughput is unaffected. Encryption and authentication have small effect on lifetime due to increased processing. However, it should be noted that the processing energy is small when compared to the transceiver energies and is not therefore a significant drawback.

17.1.4 Physical Layer

As the selection of a transceiver has the most significant effect on the trade-off between QoS parameters, the physical layer presented in Table 17.1 contains only transceiver-related parameters. Although the other physical components, such as sensors or MCU, also affect to the energy usage, they are selected mainly according to the required capabilities.

A transceiver has two important properties, communication range and data rate. While a long communication range allows covering a large area with a few hops, the required transmission energies increases because the power requirement is proportional to the square of the distance. An increase in the communications range also reduces throughput as more robust modulation or error correction are needed.

The data rate parameter has a trade-off between reliability and lifetime. Although a higher data rate slightly increases the required transmission power per time unit, the overall energy requirement is lowered as more data can be sent at the same time interval. However, increasing data rate often reduces reliability, as the sensitivity at the receiver decreases due to practical limitations in hardware.

17.1.5 Configuration Examples

In this section, examples of selecting configuration parameters in outdoor environmental monitoring and indoor localization are presented. The QoS requirements for these examples have been determined as shown in Figure 17.2. The configuration parameters are selected based on these QoS requirements.

The resulting configuration parameters are summarized in Table 17.2. In both cases, the amount of reserved slots is configured to be enough for the expected traffic. Also, both cases use a network address length of 3 B, which minimizes spurious receptions.

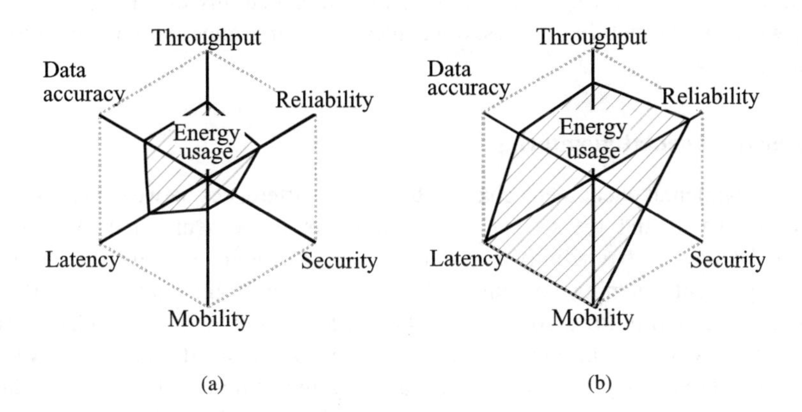

Figure 17.2 QoS parameters in two configuration examples: (a) environmental monitoring; (b) localization.

Table 17.2 The configuration parameters in two deployment examples.

Configuration parameter	Environmental monitoring	Localization
Application		
Measurement rate	Low	High
Transmission rate	Low	High
Aggregation	Yes	No
Routing		
Alternative routes	No	Yes
Traffic-class weights	Low energy	Low delay
Multipath routing	No	No
MAC		
Access-cycle length	Long	Short
ALOHA slots	Few	Many
Network beacon rate	Low	High
Acknowledgments	Yes	Yes
Retransmissions	Few	Few
Encryption	No	Yes
Authentication	No	Yes
Cluster density	Low	High
Physical		
Communication range	Long	Short
Data rate	Low	High

The environmental monitoring requires long reception margins due to temperature differences, whereas the mobility causes slight inaccuracy that must be compensated in the localization.

Outdoor environmental monitoring

Because environmental phenomena change slowly, the differences between successive measurements are small and the measurements need not to be sent often. Therefore, the throughput, latency, and reliability requirements are moderate. However, if the network is monitoring life-threatening conditions, like in fire detection, it must support low latency and high reliability for alarm packets. In this example, we only consider monitoring of non-critical environmental data. As the environmental values do not change often, the measurement interval may be low, e.g. once every few minutes. To further reduce the throughput requirements, aggregation is performed. Due to relaxed latency requirements, the access cycle length can be long (8 s) to save energy.

The network dynamics are mainly caused by slowly changing environmental conditions and mobility support is not essential. Therefore, only a few ALOHA slots (changed dynamically between 2–4) and a low beacon rate (once per second) are required. Furthermore, as the links are expected to be long term, only one alternative route is maintained and the multipath routing is not used. Due to the nature of the environmental monitoring, the measurement information is rarely considered as secret. Thus, the security requirement is low, and encryption and authentication are not used.

The network lifetime is identified as the most important factor in environmental monitoring. This can be seen in the energy usage parameter shown in Figure 17.2(a). To achieve long lifetime and load balancing, the traffic-class weights prefer long lifetime. The energy usage weight is set 0.8, while the remaining weights are divided between throughput, reliability, and latency. The energy usage is further diminished by having a very low cluster density, which is enabled by using long-range transceivers. Although a long communication range increases transmission energy, it reduces total energy consumption in this example, as more nodes can act as energy-efficient subnodes.

Indoor localization

The indoor localization is a much more demanding application than environmental monitoring. For enabling the realtime tracking of objects, the latency, and the mobility requirements are critical. As a missing localization packet can cause losing track of the target, reliability is also important. Due to the more demanding requirements, the energy usage shown in Figure 17.2(b) is quite high. The security requirement depends on the nature of the tracked objects. Encryption and authentication are used to prevent unauthorized personnel from obtaining sensitive localization information.

The data accuracy requirement ranges from moderate to high depending on the desired level of tracking. Still, the localization information packets are simple, indicating only current position and a timestamp, and thus only moderate throughput is required. For realtime tracking, the localization application uses short measurement and transmission intervals. The application layer does not perform aggregation, as it would result in loss of accuracy and increased latency.

On the routing layer, 2–3 alternative routes are maintained to allow fast recovery from a broken link. The traffic-class weights are configured to prefer low latency, but also reliability. The used weights are 0.7 for latency and 0.3 for reliability.

The latency is reduced on the MAC layer by using a short, 2 s access cycle length, which minimizes the forwarding delays. The mobility is maximized by using several ALOHA slots (4–8) and high network beacon rate (4 times per second). The MAC layer utilizes acknowledgments and retransmission to increase reliability. However, only one retransmission is allowed to prevent an increase in latency due to the buffering. Another reason for the retransmission limit is the use of multipath routing. As several routes are used, a packet is probably received via an alternative route even if the transmission on one route fails.

The localization is performed by calculating the position based on the neighbor information rather than positioning chips, such as GPS, which might not work indoors. To achieve more accurate positioning, low-communication range and high-cluster density is used. The high-cluster density has an additional benefit as it allows several next-hop alternatives for a mobile node. This way, a node can select the least-loaded cluster, and thus obtain higher throughput and experience lower latency. Although maintaining many clusters increases energy usage, the energy consumption is offset in the mobile nodes, as the reliability increases, link breaks are less common, and fewer network scans are required.

17.2 Piggybacking Data

The purpose of piggybacking is to make the design more efficient. Each packet transmission causes additional overheads due to packet headers and energy inefficiency that are due to the required reception margins. Therefore, piggybacking reduces the overheads per packet. In addition, as the MAC protocol uses fixed size frames due to radio limitations, the piggybacking allows utilizing unused space.

The TUTWSN routing piggybacks the routing information to the periodic cluster beacons that are received by a synchronized node on the MAC layer. As nodes form a synchronization tree rooting at the sink node, the routing information is slowly propagated towards the leaf nodes. This adds robustness, as a missing explicit route advertisement (RADV) packet on the routing layer does not cause a failure in the routing process. The routing information contains a sink address, the sequence number of the route to the sink, and route costs. The sink address allows detecting a new route, whereas an increase in the sequence number indicates changed interests. By advertising the cost on the cluster beacon, a node can rapidly detect changes in the routing conditions, and thus select another neighbor for its next-hop. For faster route acquisition, explicit route request (RREQ) and route advertisement (RADV) packets are still used on the routing layer when a node loses its neighbors.

The MAC layer information is piggybacked in the RADV packet belonging to the routing layer. As the purpose of the RADV packet is to advertise new routes to every node in the neighborhood, many of the receiving nodes do not have an existing synchronization to the source node. If the advertised route has low costs, the receiver may want to synchronize to the source. However, the synchronization requires knowing the information about the access cycle of the source, thus normally requiring listening to the network channel until a network beacon from the source node is received. To avoid the energy-inefficient network scan, a RADV packet includes the MAC channel and TDMA timing information, thus providing enough information to allow synchronization to the source node.

17.3 Self-configuration with Cross-layer Information

17.3.1 Frequency and TDMA Selection

The frequency and TDMA selection is a procedure that selects a non-conflicting cluster channel and TDMA schedule. While the selection could be made solely based on the MAC layer's neighbor information, the cross-layer design with the routing layer is used to optimize the routing delay.

The basic concept of the TDMA selection is to exploit the asymmetric nature of the traffic in sensor networks in a similar way that is presented by Gang Lu et al. (2004). The TDMA schedule is selected according to the routing tree to minimize the uplink (node-to-sink) latency as shown Figure 17.3. As most of the traffic is transmitted from nodes to the sink, the low uplink latency is much more important than the downlink (sink-to-node) latency. This is especially true in measurement and surveillance networks.

To illustrate the slot selection, lets assume that a subnode is joined to node 6 in Figure 17.3. The subnode sends its data to node 6 during its active period. Although nodes 4 and 5 share the same TDMA schedule with node 6, they operate on separate channels and thus do not cause interference. Node 6 forwards the received packet to node 3, which finally forwards the packet to node 1. In total, less than one an access cycle is required to reach the sink.

The TUTWSN extends the basic TDMA selection concept in several ways. While the basic concept considers only one sink, the TUTWSN supports multiple sinks by minimizing the aggregate delay. Also, in TUTWSN a node can forward a packet to different neighbors depending the used traffic class. TUTWSN determines the delay preference between these neighbors based on the latency traffic-class metric and selects the slot that minimizes the latency in the route in question.

17.3.2 Connectivity Maintenance

The connectivity maintenance procedure selects the neighbor clusters that a node maintains synchronization with. It is not always possible to synchronize with all the neighbor nodes due to time constraints and possible overlapping cluster schedules. In addition, while a high connectivity increases robustness, it also consumes more energy as a node must

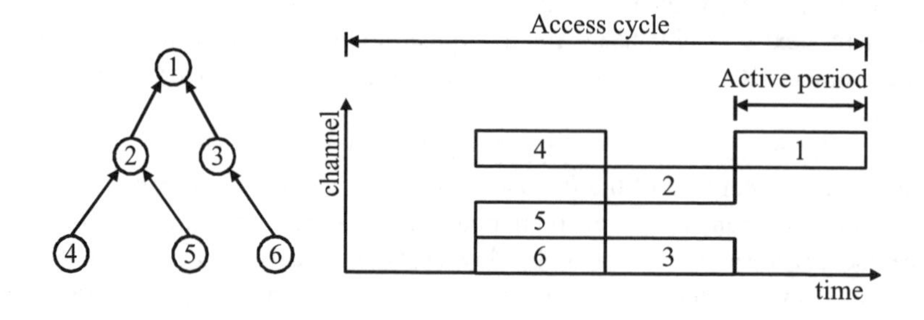

Figure 17.3 Channel and TDMA selection based on the routing tree.

receive more cluster beacons. Therefore, the connectivity maintenance procedure decides the importance of each neighbor.

The neighbor selection policy depends on the purpose of the deployment. In static measurement network, the neighbors that are nearest to the sink are selected; thus, the routing layer guides the selection. In localization, the neighbors that are situated on the different sides around a node are preferred as this allows more accurate calculation of the location. In a generic case, the nodes behave normally as in a static measurement network. However, when a node detects that it is mobile, e.g. neighbors and their RSSI change often, the neighbors that are in the direction of the movement are preferred. This way, link breaks are less probable and the amount of alternative routes can be minimized.

17.3.3 Role Selection

The role selection procedure selects between subnode and cluster head roles, which has a significant impact on several layers. In the MAC layer, a cluster head must send cluster beacons and serve superframes, thus consuming significantly more energy than a subnode. In the routing layer, a cluster head routes data from other nodes, whereas a subnode sends only its own data. In the application layer, a subnode aggregates its own data samples, whereas a cluster head aggregates the data received from other nodes. To minimize the energy consumption, the goal is to achieve full-network connectivity with a minimum amount of headnodes.

A simplified presentation of the role selection algorithm is presented in Figure 17.4. The minimum (MIN) and maximum (MAX) values control the amount of headnodes in a neighborhood, and thus the cluster density. These values are determined based on the configured cluster-density parameter. The role selection uses hysteresis by applying different MIN and MAX values, thus introducing stability into the role changes.

In addition to the cluster density, the role selection also takes into account network load, battery voltage, and available routes. If several neighbor clusters are loaded, slightly larger *MIN* and *MAX* values are used. Therefore, more clusters are allowed in the neighborhood, thus increasing the network capacity. The battery voltage information is used to balance energy consumption by preferring the subnode role on nodes with low energy. The route awareness is used to prevent voids in the network by assuming the cluster head role, when a neighbor cluster advertises sinks that the other neighbors do not know. Then, the other neighbors can obtain a route to the sink by forwarding their traffic through to node that assumed the cluster head role.

```
SELECT-ROLE()
 1: if less than MIN neighbors then
 2:    select headnode role
 3: else if more than MAX neighbors then
 4:    select subnode role
 5: else
 6:    use the old role
 7: end if
```

Figure 17.4 Simplified presentation of the role selection algorithm used in TUTWSN.

18

Protocol Analysis Models

In this chapter the principles and models for WSN performance analysis are presented. Models are closely dependent on the behavior of analyzed hardware and protocols. It is very difficult to make universal general-purpose models without over simplification, as adequate accuracy should be maintained. We will present the basic principles of analysis models.

The analysis begins with experimental measurements of steady-state power consumptions and operation-mode switch times with utilized hardware. Then, the behavior of protocols are modeled operation-specifically beginning from the PHY layer and proceeding to upper protocol layers by using basic mathematical operations. Since PHY and MAC layers have the highest effect on network performance, they are analyzed accurately, while the upper protocol layers are modeled by averaged traffic flows. The analysis models utilize averaging generally in traffic flows and network topology, thus, resulting in average network functionality. Compared to simulations, analysis results can be determined rapidly and they are easily reproduced. On the other hand, simulations can analyze the operation of individual nodes and also provide accurate information about special situations.

The performance modeling can be divided into four steps, which all increase the level of abstraction. The steps from lowest to highest are PHY power analysis, radio energy models, contention models, and node operation models.

As examples, we will next determine models for ZigBee and TUTWSN. For achieving the highest energy efficiency, IEEE 802.15.4 is configured to a synchronized low duty-cycle mode by enabling beaconing and periodic inactive time. A Beacon Order (BO) is varied from 6 to 10 resulting in a 0.98 s to 15.73 s beacon interval, corresponding to the TUTWSN access cycle length.

18.1 PHY Power Analysis

The PHY power analysis forms a basis for the performance analysis of communication protocols. It defines the steady-state power consumptions and the operation-mode switch

Table 18.1 Measured TUTWSN platform power consumptions.

Symbol	MCU	Transceiver	Power (mW)
P_{RX}	1 MIPS	RX	60.17
	1 MIPS	TX (0 dBm)	42.17
P_{TX}	1 MIPS	TX (−6 dBm)	34.67
	1 MIPS	TX (−12 dBm)	31.37
	1 MIPS	TX (−20 dBm)	29.57
P_I	1 MIPS	Idle	3.29
	1 MIPS	Sleep	3.17
P_S	Sleep	Sleep	0.037

Table 18.2 Measured TUTWSN platform operation mode switch times.

Symbol	Description	Time (μs)
t_{SI}	Sleep-to-idle	3000
t_{IA}	Idle-to-active	200
t_{RT}	Receive-to-transmit	200
t_{TR}	Transmit-to-receive	200

times of the utilized transceiver and MCU. All essential active and inactive operation modes, and operation-mode transitions are included. Hence, the power analysis result depends only on the utilized hardware. In practice, the prototype platform is configured to go sequentially through these operation modes. Power consumptions and switch times are measured at a nominal supply voltage, which typically equals to a battery voltage.

As an example, we present PHY power analysis from a 2.4 GHz TUTWSN prototype platform. The supply voltage used in the measurements is 3.0 V. The results are presented in Table 18.1 and Table 18.2. The minimum power consumption (P_I) is 37 μW, which is achieved when all components are in sleep mode. When MCU is in active mode and running at 1 MIPS speed, the power consumption increases to 3.17 mW. The transmission (TX) of a frame with active MCU consumes the power between 29.57 mW and 41.27 mW, depending on the transmission power level (−20 dBm to 0 dBm). In the following protocol analysis, we approximate that the average transmission power level is −6 dBm, which is used as the transmission mode power consumption (P_{TX}). The highest power (P_{RX}) of 60.17 mW is consumed in reception mode (RX). Depending on the operation mode, the transceiver consumes 10–20 times the power of MCU in the measured prototype platform. The operation-mode switch times from idle-to-active (transmit or receive) and between active modes are 200 μs. The wake-up time from a sleep mode is 3 ms, during which the platform consumes power P_I. Transient times from active-to-idle, and idle-to-sleep are negligibly small.

As comparison, a potential 2.4 GHz IEEE 802.15.4-platform is also analyzed. The platform is identical to the TUTWSN platform except for the transceiver, which is replaced

Table 18.3 Measured IEEE 802.15.4 platform power consumptions. © 2006 ACM. Adapted by permission.

Symbol	MCU	Transceiver	Power (mW)
P_{RX}	1 MIPS	RX	56.50
	1 MIPS	TX (0 dBm)	48.00
	1 MIPS	TX (−1 dBm)	45.00
	1 MIPS	TX (−3 dBm)	42.10
	1 MIPS	TX (−5 dBm)	39.10
P_{TX}	1 MIPS	TX (−7 dBm)	36.00
	1 MIPS	TX (−10 dBm)	32.90
	1 MIPS	TX (−15 dBm)	29.80
	1 MIPS	TX (−25 dBm)	26.60
P_I	1 MIPS	Idle	4.36
	1 MIPS	Sleep	3.17
P_S	Sleep	Sleep	0.030

Table 18.4 Measured TI CC2420 transceiver transient times. © 2006 ACM. Adapted by permission.

Symbol	Description	Time (μs)
t_{SI}	Sleep-to-idle	970
t_{IA}	Idle-to-active	192
t_{RT}	Receive-to-transmit	220
t_{TR}	Transmit-to-receive	200

with a IEEE 802.15.4 compliant Texas Instruments CC2420 transceiver having 250 kbps data rate. The power analysis of the IEEE 802.15.4 platform is presented in Table 18.3 (Kohvakka et al. 2006b) and Table 18.4 (Kohvakka et al. 2006b). The resulting power consumptions and switch times are nearly the same between the two transceivers. Compared to the nRF2401A transceiver, the lower data rate of CC2420 transceiver results in higher energy consumption per transmitted or received bit of data.

18.2 Radio Energy Models

Radio energy models increase the abstraction level from steady-state power consumptions to the energy consumptions of separate radio operations. Analyzed operations are typically the transmissions and receptions of utilized frame types, and a network scan. One of the main purposes of these models is to consider the non-ideality of PHY layer, including radio start-up times, crystal tolerances, and synchronization inaccuracy. The data processing energy of MCU is much more difficult to estimate, since it depends on the application and the types of sensors utilized. However, MCU energy consumption is very small compared to radio, as indicated by the PHY power analysis. Without a significant error, we can make an approximation that the active operation time of radio and MCU are equal.

As examples, we present the radio energy models for TUTWSN and IEEE 802.15.4. The utilized symbols and their descriptions are presented in Table 18.5.

18.2.1 TUTWSN Radio Energy Models

A frame transmission consists of a transceiver sleep-to-idle (t_{SI}) and idle-to-active (t_{IA}) transient times and actual data transmission defined as the ratio of frame length (L_F) and radio data rate (R). During a sleep-to-idle, and idle-to-active transients, the radio power consumption is approximated to be equal to the idle-mode power (P_I) and transmission-mode power (P_{TX}), respectively. Thus, the frame transmission energy E_{TX} can be modeled as

$$E_{TX} = t_{SI} P_I + \left(t_{IA} + \frac{L_F}{R} \right) P_{TX}.$$

Since all frames are transmitted in TUTWSN without carrier sensing, the energy E_{TX} is applied on all frame types.

Next, we model a frame reception energy. As TUTWSN utilizes accurate synchronization for data exchanges, the required reception time depends highly on the time drift caused by crystal tolerances. The time drift is a function of the elapsed time from the previous beacon reception. Hence, the maximum time drift is much shorter in data frames right after beacon reception compared to the reception of beacon frames, when the elapsed time from previous beacon reception is at maximum. For that reason, we model separately the reception of beacon frames and other frame types. In addition, an unsuccessful frame reception results in longer reception time than a successful reception, which also are modeled separately.

A successful beacon frame reception begins with the radio start-up transients. The radio consumes the reception-mode power P_{RX} until a frame has been received that includes an idle listening time t_I caused by synchronization inaccuracy, time drift, and possible MAC protocol delays. As synchronization is obtained by cluster beacon receptions, the time drift caused by crystal tolerance (ε) is proportional to the cluster beacon transmission interval, which equals the access cycle length (I_B). The beacon frame reception energy E_{RXB} is modeled as

$$E_{RXB} = t_{SI} P_I + \left(t_{IA} + t_I + 2\varepsilon I_B + \frac{L_F}{R} \right) P_{RX}.$$

Since data exchanges in superframes occur very soon after a cluster beacon, the time drift is negligible and can be ignored. Hence, frame reception energy E_{RX} in superframes is modeled as

$$E_{RX} = t_{SI} P_I + \left(t_{IA} + t_I + \frac{L_F}{R} \right) P_{RX}.$$

If a reception fails, a transceiver is in RX mode until the end of a data-exchange period. The data-exchange period length is sized for the worst-case scenario, when the difference between the clocks of a transmitting and a receiving node is maximal such that the clocks drift and synchronization errors occur in opposite directions. The energy consumption of an unsuccessful beacon frame reception E_{RXBU} is

$$E_{RXBU} = t_{SI} P_I + \left(t_{IA} + 2t_I + 4\varepsilon I_B + \frac{L_F}{R} \right) P_{RX}.$$

Table 18.5 Symbol descriptions.

Symbol	Description
ε	Crystal tolerance
a	Maximum number of retransmissions
b	Maximum number of CSMA backoff attempts
c	# nodes associated with a cluster head
C	# contending nodes
d	Network depth below analyzed node
D	# data transmissions during CAP
e	# transmitted beacons for signal strength estimation
f_{N}	Network beacon transmission rate
h	Probability of a hidden-node relationship
I_{B}	Cluster beacon interval
I_{D}	Downlink data transmission interval
I_{F}	Average link failure interval
I_{U}	Uplink data transmission interval
k	# maintained communication links
L_{A}	Acknowledgment frame length
L_{B}	Beacon frame length
L_{F}	Nominal frame length
L_{P}	Data frame MAC payload length
m	# neighbors associated with other clusters
n	# cluster heads in a range
n_{A}	Average number of backoff attempts
n_{B}	# beacon receptions until one with sufficient signal strength is received
n_{DL}	# nodes whose data is routed through analyzed node
n_{U}	# unsuccessful beacon receptions per access cycle
p_{C}	Probability of a clear channel by CCA
p_{D}	Collision probability due to the same backoff delay
p_{h}	Probability of a hidden-node relationship
q	Probability of the detection of a data frame transmission with the consecutive ACK by CCA
r	Radio range
R	Radio data rate
ρ	Range of a sufficient signal strength
s	Probability of a transmission attempt
S_{A}	# ALOHA slots
S_{R}	# reserved slots
t_{CAP}	CAP duration
t_{I}	Synchronization inaccuracy
u	Average # transmission attempts
v	Probability of a successful transmission

Similarly, the energy consumption of an unsuccessful frame reception in superframes E_{RXU} is

$$E_{\mathrm{RXU}} = t_{\mathrm{SI}} P_{\mathrm{I}} + \left(t_{\mathrm{IA}} + 2t_{\mathrm{I}} + \frac{L_{\mathrm{F}}}{R} \right) P_{\mathrm{RX}}.$$

Finally, we model a network-scan energy consumption. Similar to frame receptions, the network scan begins with radio start-up transients. Then, a radio is in RX mode on average duration t_{NS}. In TUTWSN, the network scan is performed until a beacon with sufficient signal strength is received. We define that the range of sufficient signal strength in proportion to maximum radio range (r) is ρ. Beacons are received from n headnodes in a range until one beacon with sufficient signal strength is received. The number of beacon receptions (n_{B}) until one beacon within the distance of ρr is successfully received is modeled by weighted average as

$$n_{\mathrm{B}} = \rho^2 + \sum_{a=2}^{n-1} \left[a \left(\prod_{b=1}^{a-1} 1 - \frac{n\rho^2}{n-(b-1)} \right) \frac{n\rho^2}{n-(a-1)} \right] + n \left(\prod_{a=1}^{n-1} 1 - \frac{n\rho^2}{n-(a-1)} \right).$$

The required network scan duration (t_{NS}) for detecting a new node with sufficient signal strength is given by

$$t_{\mathrm{NS}} = \frac{n_{\mathrm{B}}}{f_{\mathrm{N}} n}.$$

Thus, the network scan energy E_{NS} is

$$E_{\mathrm{NS}} = t_{\mathrm{SI}} P_{\mathrm{I}} + \left(t_{\mathrm{IA}} + \frac{n_{\mathrm{B}}}{f_{\mathrm{N}} n} \right) P_{\mathrm{RX}}.$$

Individual beacon receptions during the scan are negligibly small compared to the network-scan energy and are therefore ignored in the model.

18.2.2 ZigBee Radio Energy Models*

Due to the contention-based channel access of ZigBee, frame-transmission energy depends on the type of transmitted frame (Kohvakka et al. 2006b). Data frames are transmitted with CSMA-CA algorithms, while beacons and ACK frames are transmitted without carrier sensing. First, we model a beacon transmission energy (E_{TXB}). Since beacons are transmitted without carrier sensing, the energy is similar to TUTWSN frame transmission, as

$$E_{\mathrm{TXB}} = t_{\mathrm{SI}} P_{\mathrm{I}} + \left(t_{\mathrm{IA}} + \frac{L_{\mathrm{B}}}{R} \right) P_{\mathrm{TX}}.$$

An ACK frame is transmitted after a data reception without carrier sensing. The maximum allowed response time is determined by t_{AW}. We assume the node spends a half of t_{AW} for data processing, during which it consumes power P_{I}. Hence, the ACK-transmission energy (E_{TXA}) is (Kohvakka et al. 2006b)

$$E_{\mathrm{TXA}} = \frac{t_{\mathrm{AW}}}{2} P_{\mathrm{I}} + \left(t_{RT} + \frac{L_A}{R} \right) P_{\mathrm{TX}}.$$

A data frame transmission consists of a start-up transient, a backoff time due to the CSMA/CA algorithm, and the actual data transmission. We will model the backoff time (t_{BOT}) and backoff energy consumption (E_{BOT}) in detail in the following section. The energy consumption of a data frame transmission (E_{TXD}) can be presented as

$$E_{\mathrm{TXD}} = t_{\mathrm{SI}} P_{\mathrm{I}} + E_{\mathrm{BOT}} + \left(t_{\mathrm{IA}} + \frac{L_{\mathrm{F}}}{R} \right) P_{\mathrm{TX}}.$$

Next, we model frame reception energies and begin with a beacon reception. Since the beacon interval is typically several seconds, it is essential to consider the synchronization inaccuracy (t_{I}) and crystal tolerances (ε). A beacon reception begins with radio start-up transients. Due to synchronization inaccuracy, and crystal tolerances in both transmitting and receiving node, the reception of a beacon should begin at least $2\varepsilon \mathrm{I}_B + t_{\mathrm{I}}$ before a scheduled beacon moment and continue for at least the same amount of time after the scheduled beacon. On average, a beacon is received on the scheduled moment resulting in $2\varepsilon \mathrm{I}_B + t_{\mathrm{I}}$ idle-listening time. After the beacon, nodes spend in idle mode a Long Inter-Frame Space (LIFS) defined by IEEE 802.15.4. Hence, the beacon reception energy can be modeled as

$$E_{\mathrm{RXB}} = (t_{\mathrm{SI}} + t_{\mathrm{LIFS}}) P_{\mathrm{I}} + \left(t_{\mathrm{IA}} + 2\varepsilon I_{\mathrm{B}} + t_{\mathrm{I}} + \frac{L_{\mathrm{B}}}{R} \right) P_{\mathrm{RX}}.$$

After a frame transmission, a node spends in idle mode a Short Inter-Frame Space (SIFS) and waits for an ACK for at most t_{AW} during which the node must be in reception mode. Assuming that an average ACK wait time is a half of t_{AW}, the ACK-reception energy (E_{RXA})is modeled as

$$E_{\mathrm{RXA}} = t_{\mathrm{SIFS}} P_{\mathrm{I}} + \left(t_{\mathrm{TR}} + \frac{T_{\mathrm{AW}}}{2} + \frac{L_{\mathrm{A}}}{R} \right) P_{\mathrm{RX}}.$$

If ACK reception fails, the transceiver must be in reception mode until t_{AW} and the ACK-frame transmission duration are elapsed. Hence, the energy of an unsuccessful ACK reception (E_{RXAU}) is

$$E_{\mathrm{RXAU}} = t_{\mathrm{SIFS}} P_{\mathrm{I}} + \left(t_{\mathrm{TR}} + T_{\mathrm{AW}} + \frac{L_{\mathrm{A}}}{R} \right) P_{\mathrm{RX}}.$$

Since a ZigBee router (IEEE 802.15.4 coordinator) must be in reception mode for the entire contention access period, the modeling of data frame reception energy is not necessary. However, each node can request a data frame from a ZigBee router by using the indirect communication, which should be modeled.

In indirect communication, a node first transmits a data request to the ZigBee router. The maximum response time (t_{RES}) for a data request is 19.52 ms. Hence, the data reception time and energy may be significantly higher than when in direct communication. The coordinator-response time is assumed to be an average of t_{BOT} and t_{RES}. The energy (E_{RXDD}) for indirect data transmission after a data request is modeled as

$$E_{\text{RXDD}} = t_{\text{LIFS}} P_{\text{I}} + \left(t_{\text{IA}} + t_{\text{I}} + \frac{T_{\text{RES}} + t_{\text{BOT}}}{2} + \frac{L_{\text{F}}}{R} \right) P_{\text{RX}}.$$

If a node looses contact with its associated coordinator (orphans) due to mobility or a radio link failure, it may perform either an orphaned-device realignment procedure, or reset the MAC sublayer and re-associate with the network. In beacon-enabled networks the latter performs best and a node performs a passive channel scan on a single channel prior to the re-association (the operating channel of the network is known) lasting slightly over one beacon interval. The energy required for the message exchange during the association is negligible compared to the network scanning energy, and thus it is ignored in the following model. Thus, the network scanning energy (E_{NS}) is

$$E_{\text{NS}} = t_{\text{SI}} P_{\text{I}} + \left[t_{\text{IA}} + t_{\text{basesuperframe}} \times (2^{\text{BO}} + 1) \right] P_{\text{RX}}.$$

18.3 Contention Models

Contention models provide information about the behavior of a protocol in a network of nodes, which contend with each other. The results of the contention models are, for example, the probability of a successful transmission, the average number of transmission attempts per frame due to collisions, and the time and energy consumed for the backoff algorithm. It should be noted that contention models are applied only for contention-based protocols. Next, we define contention models for the ALOHA mechanism of TUTWSN and CSMA-CA algorithm of ZigBee.

18.3.1 TUTWSN Contention Models

TUTWSN utilizes ALOHA as a contention-based channel access for making network associations and slot reservations requests. Hence, the traffic in ALOHA is highly dependent on network dynamics.

The probability that a link failure (p_{F}) occurs for one link at a given access cycle is a function of a link failure interval (I_{F}). Link failures are caused by the changes in operation environment, node mobility, and node failures. We model p_{F} as

$$p_{\text{F}} = 1 - \lim_{\Delta t \to \infty} \left(1 - \frac{\Delta t}{I_{\text{F}}} \right)^{\frac{I_{\text{B}}}{\Delta t}}.$$

The average number of nodes that contend with an analyzed node in ALOHA per each superframe (C) is directly proportional to the associated nodes (c) with each headnode and the number of communication links (k) that each node maintains due to the multicluster-tree network topology. Possible collisions in ALOHA cause retransmissions and increase the number of transmission attempts (u) resulting

$$C = ckp_{\text{F}}u.$$

The probability that two nodes select the same ALOHA slots causing a collision is $1/S_{\text{A}}$, where S_{A} is the number of ALOHA slots (default: 4). Thus, the probability of a successful frame transmission (p_{s}) in ALOHA with C contenting nodes is

$$p_s = \left(1 - \frac{1}{S_A}\right)^C .$$

In the case of an unsuccessful transmission, a new ALOHA slot is randomized on a following superframe. A frame is transmitted a (default: 3) times before declaring a transmission failure. The probability of a successful transmission (v) with a attempts is modeled as

$$v = \sum_{k=1}^{a} p_s (1 - p_s)^{k-1} .$$

Finally, the average number of transmission attempts per frame (u) in ALOHA can be modeled as

$$u = (1 - v)\, a + \sum_{k=1}^{a} k p_s (1 - p_s)^{k-1} .$$

Moreover, by iterating the four last equations, u is gradually refined.

18.3.2 ZigBee Contention Models*

The contention models are used for analyzing a CSMA-CA mechanism including backoff time, collisions, and retransmissions. For simplification, the slotted structure is not considered in the equations. Hence, the following equations utilize CAP more efficiently than in reality, giving also slightly higher performance (Kohvakka et al. 2006b). The utilized parameters and their values are summarized in Table 18.5.

In contention-based protocols, the backoff delays, energies, and carrier-sensing characteristics that are required are modeled.

A hidden-node phenomenon significantly affects the performance of CCA, since a portion of the ongoing transmissions are not detected. Figure 18.1. presents a situation, where

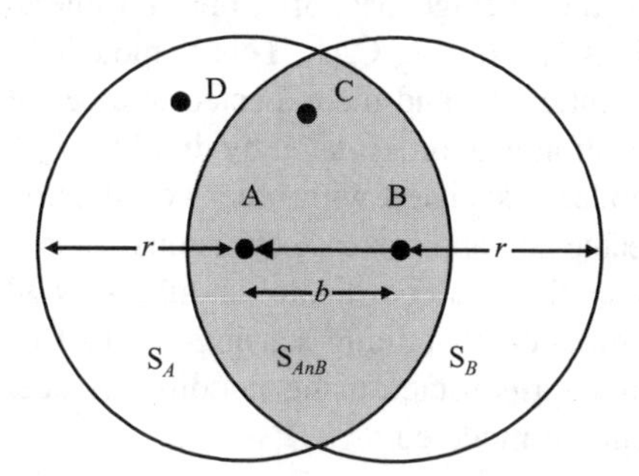

Figure 18.1 Hidden-node phenomenon. This article was published in Ad Hoc Networks, M. Kohvakka, J. Suhonen, M. Kuorilehto, V. Kaseva, M. Hännikäinen, T.D. Hämäläinen, *Energy-Efficient Neighbor Discovery Protocol for Mobile Wireless Sensor Networks*, Copyright Elsevier (2007).

a node B sends data to a node A (coordinator). At the same time, nodes C and D perform CCA, and are beginning a data transmission to the same destination (node A). The distance between nodes A and B is b, and their radio ranges form circles with a radius r. Since the node C is in the range of the both nodes A and B ($S_{A \cap B}$), it detects the ongoing data exchange, and delays its transmission. Since the node D is in S_A, but outside the area $S_{A \cap B}$, it cannot sense the transmission from node B, and begins a transmission. Due to this hidden-node phenomenon, the data transmissions from nodes B and D collide at node A. Let nodes B and D randomly locate at A's transmission range (S_A). In addition, let node B have an ongoing data transmission to the node A. Node D can detect the transmission if it is located in the intersection area $S_{A \cap B}$. The size of $S_{A \cap B}$ is calculated by the radius r and the distance b as

$$\phi(b) = 4 \int_{b/2}^{r} \sqrt{r^2 - x^2} dx.$$

Otherwise, node D is located in the area $S_A - S_{A \cap B}$, which can be calculated as $\pi r^2 - \phi(b)$. The probability of a hidden-node relationship (h) between nodes B and D can be determined by integrating the probability $(S_A - S_{A \cap B})/S_A$ over the circle of radius x centered at A for x in $[0, r]$ as

$$h = \int_0^r \frac{2\pi x \left[\pi r^2 - \phi(x)\right] / \left(\pi r^2\right)}{\pi r^2} dx \simeq 41$$

The probability (q) that a single data transmission with the consecutive ACK is detected by CCA at any time in CAP is estimated by dividing the RF transmission times of data and ACK frames by the length of CAP (t_{CAP}) as

$$q = \frac{L_{\mathrm{F}} + L_{\mathrm{A}}}{t_{\mathrm{CAP}} R}.$$

To be able to estimate the average backoff time and energy, we model the average number of data transmissions (D) during CAP. This is modeled by uplink (I_{U}) and downlink (I_{D}) data-transmission intervals, and the expected number of retransmissions (u). The effect of indirect communication is approximated by dividing I_{D} by 2. The model takes into account the neighboring nodes associated with other coordinators (m), since they may interfere with the CSMA mechanism and increase collisions when their transmissions overlap with the analyzed CAP. In addition, a coordinator should forward a relatively high amount of data from its child coordinator containing sensing values from potentially hundreds of nodes (n_{DL}). The downlink data is routed using flooding. Hence, it does not increase d in respect of n_{DL}. Thus, D can be modeled as

$$D = \left[\frac{n_{\mathrm{DL}} - c}{I_{\mathrm{U}}} + \left(\frac{1}{I_{\mathrm{U}}} + \frac{2}{I_{\mathrm{D}}}\right)\left(\frac{t_{\mathrm{CAP}}}{I_{\mathrm{B}}} m + c\right)\right] I_{\mathrm{B}} u.$$

The probability (p_{C}) that CSMA-CA detects a clear channel is determined by q, d, and h. With the two consecutive CCA analyses, p can be modeled as

$$p_{\mathrm{C}} = (1 - q)^{2D(1-h)}.$$

As the backoff algorithm is repeated at maximum b times (*macMaxCSMABackoffs*) before declaring channel-access failure, the probability of a transmission attempt (s) is

$$s = \sum_{k=1}^{b} p_{\text{C}} (1 - p_{\text{C}})^{k-1}.$$

The average number of backoff attempts (n_{A}) for each frame can be modeled as

$$n_{\text{A}} = (1 - s)\, b + \sum_{k=1}^{b} k p_{\text{C}} (1 - p_{\text{C}})^{k-1}.$$

The average time for each backoff attempt is modeled with the backoff exponent (BE), and the backoff period length (T_{BOP}) as

$$t_{\text{BO}} (BE) = \frac{2^{\text{BE}} - 1}{2} t_{\text{BOP}}.$$

After each unsuccessful backoff attempt, BE is incremented by one until $aMaxBE$ (default: 5) is reached. The total backoff time (t_{BOT}) is obtained by summing the average backoff and CCA analysis times (t_{CCA}) of each attempt. As CCA is performed twice only if the channel is assessed to be clear on a first attempt, we average the number of CCA attempts. Hence, t_{BOT} can be modeled as

$$t_{\text{BOT}} = \frac{3}{2} n_A (t_{\text{CCA}} + t_{\text{IR}}) + \sum_{k=0}^{n_{\text{A}}-1} t_{\text{BO}} \left[min\, (macMinBE + k, aMaxBE)\right].$$

Hence, the total backoff energy consumption (E_{BOT}) is

$$E_{\text{BOT}} = \frac{3}{2} n_{\text{A}} (t_{\text{CCA}} + t_{\text{IR}}) (P_{\text{CCA}} - P_{\text{I}}) + t_{\text{BOT}} P_{\text{I}},$$

which assumes that the radio is in idle mode during the backoff time. This is necessary, since the radio start-up time from sleep mode is almost 1 ms.

A transmission failure due to a collision is common in a highly loaded network. The CSMA-CA MAC is prone to two types of collisions; collisions caused by the hidden-node problem, and collisions caused by the selection of the same backoff slot with another node. Neither of these collisions can be avoided by the standard, since any type of handshaking before transmission is not used. Considering two nodes in a hidden node situation, the probability of a hidden-node collision (p_{h}) somewhere in the CAP is modeled as

$$p_{\text{h}} = \int_{0}^{t_{\text{CAP}}} q^2 dt = \frac{(t_{\text{DATA}} + t_{\text{ACK}})^2}{t_{\text{CAP}}}.$$

Due to the relatively long inactive part of the superframe, nodes most probably have gathered data during the inactive time and start a backoff procedure simultaneously at the beginning of CAP. Moreover, the nodes randomize backoff delays using a relatively low backoff exponent, since for each new transmission BE is initialized with $macMinBE$. We approximate the probability (p_{d}) that two nodes select the same backoff delay and frames collide to be

$$p_{\text{d}} = \frac{1}{2^{\text{BE}} - 1}.$$

To be able to estimate the number of nodes having data to send, instead of the number of transmissions, we slightly modify D and define the number of contending nodes (C) as

$$C = min\left(\frac{(n_{\rm DL}-c)I_{\rm B}u}{I_{\rm U}}, 1\right) + min\left[\left(\frac{I_{\rm B}}{I_{\rm U}} + \frac{2I_{\rm B}}{I_{\rm D}}\right)u, 1\right]\left(\frac{t_{\rm CAP}}{I_{\rm B}}m + c\right).$$

The probability of a successful transmission ($p_{\rm s}$) can be modeled with s, $P_{\rm h}$, and $P_{\rm d}$ as

$$p_{\rm s} = s\,(1 - p_{\rm h})^{\rm hd}\,(1 - p_{\rm d})^{C}.$$

A frame is transmitted $a = aMaxFrameRetries + 1$ times before declaring a transmission failure. Thus, the probability of a successful transmission (v) after a attempts can be modeled with $p_{\rm s}$ as

$$v = \sum_{k=1}^{a} p_{\rm s}\,(1 - p_{\rm s})^{k-1}.$$

Finally, the average number of transmission attempts per frame (u) can be calculated as

$$u = (1 - v)\,a + \sum_{k=1}^{a} kp_{\rm s}\,(1 - p_{\rm s})^{k-1}.$$

18.4 Node Operation Models

Next, the behavior of upper protocol layers is modeled by averaged traffic flows. The transmitted and received data and signaling frames per time unit are estimated. For example, beacon exchanges, data exchanges, association requests, and network scans are included. Node operation models are further divided into throughput models and power-consumption models. To demonstrate, these models are determined for TUTWSN and ZigBee.

The performance of the network is analyzed in a large-scale, low data-rate WSN application, where each node transmits 16-byte MAC payloads ($L_{\rm P}$) of sensing items at uplink direction at $I_{\rm U}$ intervals. These frames are multihop routed to a single sink node via a cluster-tree (ZigBee) or multicluster-tree (TUTWSN) topology. Control frames are broadcast at downlink direction in reserved slots at intervals of $I_{\rm D}$. Each cluster head has c child nodes associated with consisting of 12 child devices ($n_{\rm D}$) and 3 child cluster heads ($n_{\rm C}$). Network depth is 4. Due to the multicluster-tree topology, TUTWSN nodes maintain synchronization with k parent cluster heads (headnodes), while ZigBee nodes maintain synchronization only with one parent.

For estimating the amount of routed data, we first model the number $n_{\rm DL}$ of child and grandchild nodes, whose data is routed through an analyzed cluster head. By assuming an uniform network topology, $n_{\rm DL}$ equals to (Kohvakka et al. 2006b)

$$n_{\rm DL} = -1 + \sum_{i=0}^{d-1} n_{\rm C}^{i}\,(1 + n_{\rm D}),$$

where d is the network depth below the analyzed cluster head. Both cluster heads and devices are included. As d increases from 1 to 4, $n_{\rm DL}$ get values from 12 to 519. Next, we model the throughput and power consumptions for TUTWSN and ZigBee nodes in this network.

18.4.1 TUTWSN Throughput Models

In this subsection, we model the throughput required by an application and determine achieved goodputs for TUTWSN, measured from a MAC layer payload. A requested throughput (T_{REQ}) is determined according to the 16-B data payloads L_P routed from n_{DL} nodes at I_U intervals. Control frames are broadcast in the downlink direction in reserved slots at intervals of I_D. Hence, T_{REQ} is modeled as

$$T_{REQ} = \left(\frac{n_{DL}}{I_U} + \frac{1}{I_D}\right) L_P.$$

We model the throughput that an analyzed headnode is able receive and transmit on its superframes. For fulfilling the requested throughput, a headnode needs to assign R_R reserved slots for each superframe. Although a link failure causes unsuccessful data reception and a redundant reserved slot, a missed data payload is retransmitted in an ALOHA slot piggybacked with the frame containing association and slot reservation requests. Thus, link failures do not increase R_R. R_R can be modeled as

$$R_R = \frac{T_{REQ} I_B}{L_P}.$$

The actual utilization of reserved slots (U_R) per each superframe is R_R limited by the maximum number of reserved slots (S_R) (default: 8). U_R can be formulated as

$$U_R = min\,(R_R, S_R)\,.$$

Goodput (G) is defined as the achieved throughput in the headnode's own superframes including the incoming and outgoing data in ALOHA and reserved slots. Occasional link failures reduce the goodput of reserved slots resulting

$$G = \left[(1 - p_F)\, U_R + U_A\right] \frac{L_P}{I_B}.$$

It should be noted that ACK frames can carry the same goodput to downlink direction. In the given application, however, downlink data is very low data-rate broadcast traffic, which utilizes its own slot reservations. Hence, the goodput obtained from ACK frames is unexploited and not considered in this analysis.

The headnode buffers all the received and measured sensing items, and forwards the data to a parent headnode. If the requested bandwidth exceeds the network capacity, we assume that the parent headnode limits assigned bandwidth equally from all nodes in proportion to their bandwidth requests. We model that a parent headnode allocates on average (R_U) slots for the analyzed headnode in each superframe, which is

$$R_U = \frac{(n_{DL} + 1)\, I_B}{I_U} \frac{U_R\,(d+1)}{R_R\,(d+1)}.$$

18.4.2 ZigBee Throughput Models*

The requested throughput (T_{REQ}) for ZigBee differs due to an indirect data transmission mechanism in the downlink direction. This means that a device must request data from its

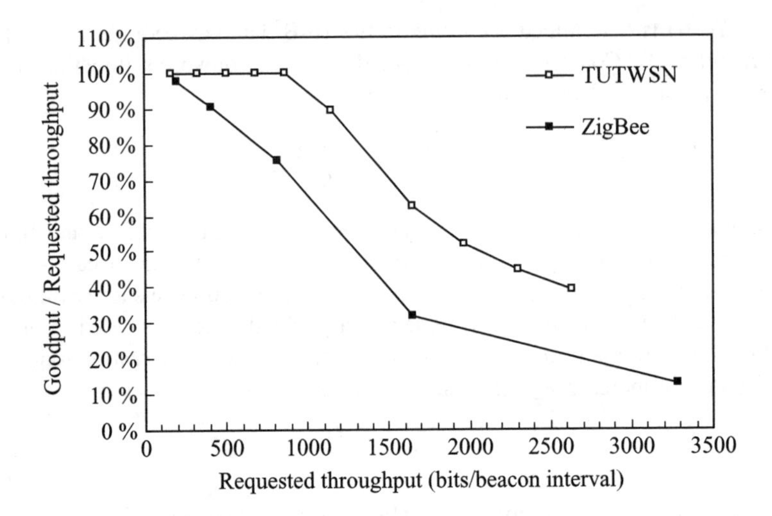

Figure 18.2 Achieved goodput versus the requested throughput in TUTWSN and ZigBee cluster heads.

coordinator, which approximately doubles the downlink frame exchanges in a superframe compared to the direct transmissions used in TUTWSN. Hence, the requested throughput for ZigBee differs slightly from TUTWSN. T_{REQ} is modeled as

$$T_{REQ} = \left(\frac{n_{DL}}{I_U} + \frac{2c}{I_D}\right) L_P.$$

The achieved goodput (G) to a coordinator can be modeled with the requested throughput and the probability of a successful transmission (υ) as (Kohvakka et al. 2006b)

$$G = T_{REQ}\upsilon.$$

For comparison, the goodputs of TUTWSN and ZigBee are compared. The SO of IEEE 802.15.4 is set to 3, equaling to 123 ms superframe length. For comparability, both ZigBee and TUTWSN utilize 16-B MAC payload lengths. The resulted goodput curve is plotted in Figure 18.2. As shown in the figure, TUTWSN clearly outperforms ZigBee with these settings. ZigBee would perform significantly better if the number of child nodes is decreased, due to reduced collisions.

18.4.3 TUTWSN Power Consumption Models

In this section, we determine the average power consumptions for a TUTWSN subnode and a headnode. Due to the multicluster-tree topology, a subnode receives cluster beacons from k headnodes on each access cycle. We define that the signal strength estimation utilizes two consecutive beacons (e) in each access cycle. Since beacons are transmitted with various power levels, some of the beacon receptions fails. We assume that the average number of a unsuccessful beacon receptions per access cycle (n_U) is 0.5. Hence, the beacon reception power (P_{SB}) is

$$P_{SB} = \frac{n_U E_{RXBU} + E_{RXB}}{I_B} k.$$

A subnode transmit 16-B data payloads to some of its k parents at I_U intervals. After each data-frame transmission, the subnode receives an acknowledgment. In addition, a control frame is received at I_D intervals. A link failure occurs at probability p_F resulting in an unsuccessful reception. Since the data frame is piggybacked with an association-request frame, a retransmission is typically not necessary. Thus, the data exchange power consumption P_{SS} is

$$P_{SS} = \frac{E_{TX}}{I_U} + \left(\frac{1}{I_U} + \frac{1}{I_D}\right)\left[(1 - p_F)\,E_{RX} + p_F E_{RXU}\right].$$

A link failure occurs on average at I_F intervals for each of k links causing a network scan, u transmission attempts, and $(u-1)$ unsuccessful acknowledgment receptions in ALOHA. The power consumption caused by link failures (P_{LF}) is

$$P_{LF} = \frac{k}{I_F}\left(E_{NS} + uE_{TX} + (u-1)E_{RXU} + E_{RX}\right).$$

Finally, subnode power consumption (P_S) is obtained as

$$P_S = P_{SB} + P_{SS} + P_{LF} + P_S.$$

The power consumption of a subnode is plotted as the function of a cluster beacon interval in Figure 18.3. In addition, the separate power consumptions for beacon receptions, data exchanges, sleep mode, and link failures are presented. The uplink data transmission interval (I_U) is 40 s. The average link failure interval is set to 3 hours, which corresponds to a fairly stable network. Subnode power consumption ranges from 56 μW to 144 μW, depending on the utilized beacon interval. At access cycle lengths shorter than 4 s, beacon-reception power consumption is dominating. At longer access cycles, the sleep mode dominates the power drain. Data exchanges consume only 1.8 μW power.

Next, the power consumption is determined for a headnode. Beacons are transmitted in the cluster channel at I_B intervals and in the network channel at rate f_N. Each beacon

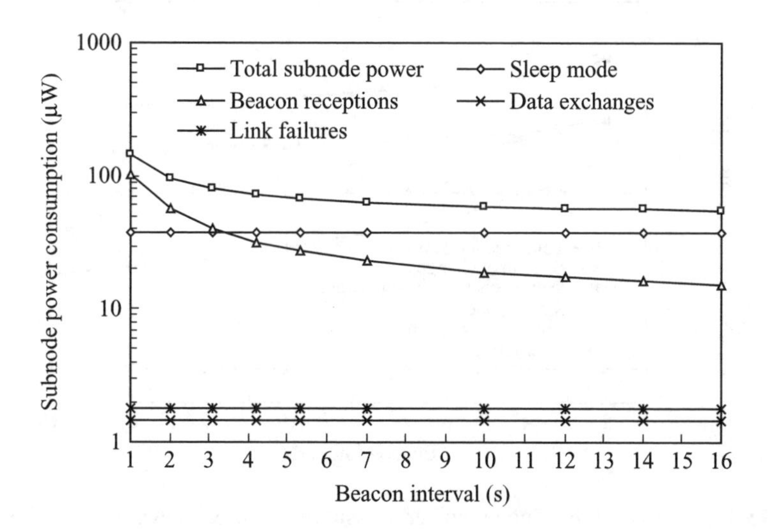

Figure 18.3 TUTWSN subnode power consumption breakdown ($I_U = 40$ s, $I_F = 3$ h).

is transmitted e times due to the signal strength estimation. Cluster beacons are received from k parent headnodes. Hence, the beacon-exchange power (P_{HB}) is

$$P_{HB} = \frac{k\,(n_U E_{RXBU} + E_{RXB}) + eE_{TX}}{I_B} + eE_{TX} f_N.$$

The amount of data exchanges in the superframe depends on the network depth below the analyzed node, and the data transmitting interval. We determine the power consumption of data exchanges according to the average utilization of reserved (U_R) and ALOHA (U_A) slots and a beacon interval. Thus, (P_{HS}) is

$$P_{HS} = (S_A - U_A + p_F U_R)\,\frac{E_{RXU}}{I_B} + \left[U_A + (1 - p_F)\,U_R\right]\frac{E_{RX} + E_{TX}}{I_B}.$$

In addition, the headnode consumes a power (P_{HP}) in data exchanges with its parent headnode. Averagely R_U reserved slots are used for data exchanges. Thus, P_{HP} is

$$P_{HP} = \frac{(R_U)E_{TX}}{I_B} + \left(\frac{R_U}{I_B} + \frac{1}{I_D}\right)\left[(1 - p_F)\,E_{RX} + p_F E_{RXU}\right].$$

Finally, the average power consumption of a headnode (P_H) is

$$P_H = P_{HB} + P_{HS} + P_{HP} + P_{LF} + P_S.$$

The power consumption of a headnode as the function of the beacon interval is presented in Figure 18.4. The average link failure interval is set to 3, while each node generates a data frame and transmits it at the uplink direction (I_U) at 40 s intervals. The network

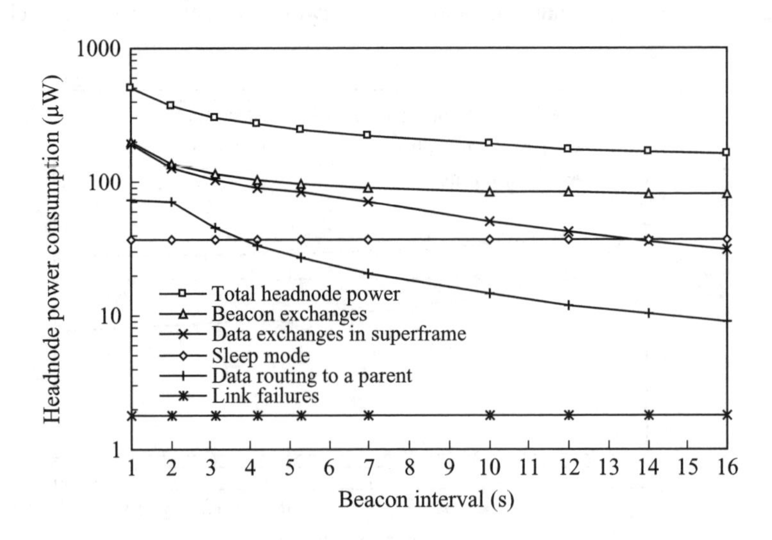

Figure 18.4 TUTWSN headnode power consumption breakdown ($I_U = 40$ s, $d = 2$, $I_F = 3$ h).

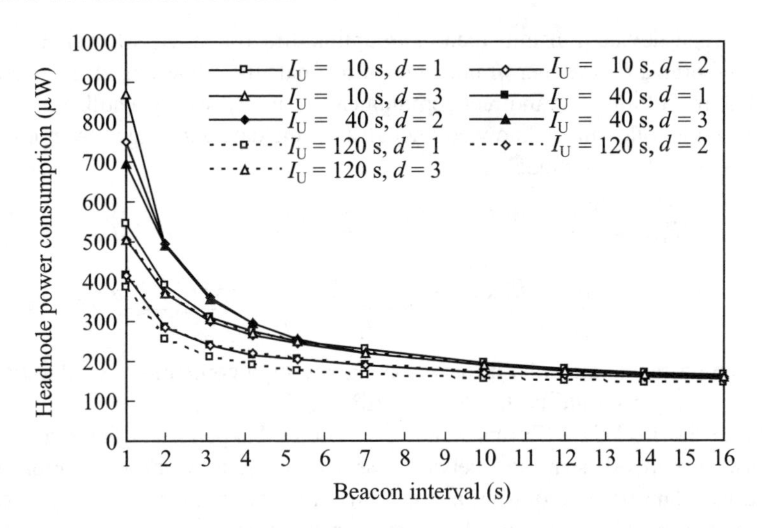

Figure 18.5 TUTWSN headnode power consumption as the function of the cluster beacon interval, as the uplink data transmission interval, and the network depth varies.

has two levels of hierarchy (d) below the analyzed headnode equaling to 51 nodes. The resulted power consumption is between 160 μW and 505 μW. At shorter access cycles, the data exchanges in superframes dominate the power consumption. At longer superframes, the frequency of ALOHA slots reduces and the network throughput may saturate, which reduces the data-exchange power consumption. Beacon exchanges consume around 100 μW. The effect of traffic load to headnode power consumption is presented in Figure 18.5. Traffic load is varied by adjusting the sensor sampling interval (I_U) and network depth (d) below the analyzed node. The results indicate that the headnode power consumption can be roughly divided into a static power caused by network management operations, which is around 150 μW, and a dynamic, traffic-dependent power, which is from zero to 700 μW.

18.4.4 ZigBee Power Consumption Models*

The power consumption of a ZigBee end-device and ZigBee router are modeled. To be able to operate in a beacon-enabled network, an end-device receives beacons and performs data exchanges with its cluster head (ZigBee router). If a communication link to the coordinator is lost then a device performs a network scan. The rest of the time a device is in sleep mode.

As a ZigBee end-device receives beacons from a parent router at I_B intervals, the beacon-exchange power consumption (P_{DB}) is determined simply by

$$P_{DB} = \frac{E_{RXB}}{I_B}.$$

A ZigBee end-device transmits data in an uplink direction at I_U intervals using average on u transmission attempts per frame. Downlink data is requested at I_D intervals. By assuming that most failures in indirect communication are caused by a collision in either the data request or the following acknowledgment frame, the data-exchange power consumption for an end-device (P_{DS}) is modeled as

$$P_{DS} = \frac{E_{TXD}u + E_{RXAU}(u-1) + E_{RXA}}{I_U} + \frac{E_{TXD}u + E_{RXAU}(u-1) + E_{RXA} + E_{RXDD} + E_{TXA}}{I_D}.$$

For obtaining realistic results in networks having high contention, the duration of data transmission attempts should be limited to CAP length.

Similar to the in TUTWSN analysis, link failures happen at I_F intervals. Each link failure causes a network scan and network association. A network-association request is transmitted in a data frame requiring average on u retransmissions and $u - 1$ unsuccessful acknowledgment receptions. Hence, the power consumption caused by link failures (P_{LF}) is

$$P_{LF} = \frac{1}{I_F}\left[E_{NS} + uE_{TXD} + (u-1)E_{RXAU} + E_{RXA}\right].$$

Finally, the average power consumption for a ZigBee end-device (P_D) is

$$P_D = P_{DB} + \min\left(P_{DS}, \frac{t_{CAP}P_{RX}}{I_B}\right) + P_{LF} + P_S.$$

The power consumption of an end-device is plotted as the function of the access cycle length, as shown in Figure 18.6. The separate power consumptions for beacon receptions, data exchanges, sleep mode, and link failures are also presented. For comparability with the TUTWSN results, the uplink data transmission interval is set to 40 s and the average link

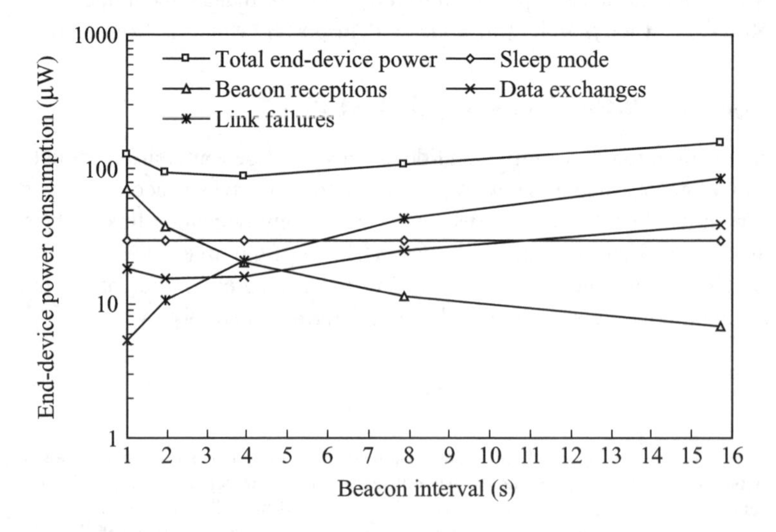

Figure 18.6 ZigBee end-device power consumption breakdown ($I_U = 40$ s, $I_F = 3$ h).

failure is 3 hours. The power consumption of an end-device ranges from 87 μW to 160 μW, depending on the beacon interval. At beacon intervals shorter than 4 s, beacon receptions dominate the power consumption. At longer beacon intervals, link failures dominate the power drain. Data exchanges consume around 20 μW power.

Next, we model the power consumption of a ZigBee router. First, the beacon-exchange power consumption is determined. As beacons are transmitted and received at I_B intervals, beacon-exchange power consumption (P_{CB}) is

$$P_{CB} = \frac{E_{TXB} + E_{RXB}}{I_B}.$$

A ZigBee router is active during the entire CAP. Since the transceiver power consumption is nearly the same in transmission and reception mode, we can approximate that a ZigBee router consumes power P_{RX} during the entire CAP. Hence, the average data-exchange power consumption (P_{CS}) of a ZigBee router in its own superframes can be modeled as

$$P_{CS} = \frac{t_{CAP} P_{RX}}{I_B}.$$

In addition, a ZigBee router consumes power P_{CP} in data exchanges with its parent router. Besides the data originated from the router itself, data is routed from n_{DL} nodes. Hence, P_{CP} is

$$P_{CP} = \frac{E_{TXD}u + E_{RXAU}(u-1) + E_{RXA}}{I_U}(n_{DL}+1)$$
$$+ \frac{E_{TXD}u + E_{RXAU}(u-1) + E_{RXA} + E_{RXDD} + E_{TXA}}{I_D}.$$

In addition, P_{CP} should be limited to P_{CS} corresponding to a situation that the router is active during the entire CAP. Finally, the average power consumption (P_C) of a ZigBee router is

$$P_C = P_{CB} + P_{CS} + \min(P_{CP}, P_{CS}) + P_{LF} + P_S.$$

The resulted ZigBee router power consumption as the function of a beacon interval is presented in Figure 18.7. For comparability, network parameters are similar with TUTWSN analysis. The resulted power consumption is between 1.01 mW and 7.50 mW. At shorter beacon intervals, data exchanges with child nodes dominate the power consumption. At longer beacon intervals, the increase of contention in CAP raises the power consumption of data exchanges with a parent. The results also indicate that the ZigBee router consumes one order of magnitude more power than a TUTWSN headnode in a similar application. The effect of traffic load to the power consumption of a ZigBee router is presented in Figure 18.8. Similar to the TUTWSN analysis, the traffic load is varied by adjusting the sensor sampling interval (I_U) and network depth (d) below the analyzed router. The results indicate that router power consumption is highly dependent on the traffic load. Yet, the CAP length and beacon interval limit the maximum duty cycle and the power consumption of data exchanges.

18.5 Summary

Performance analysis is a feasible approach for determining the performance of physical and data link layers. As the variation of protocol parameters is easy, the behavior of protocol can

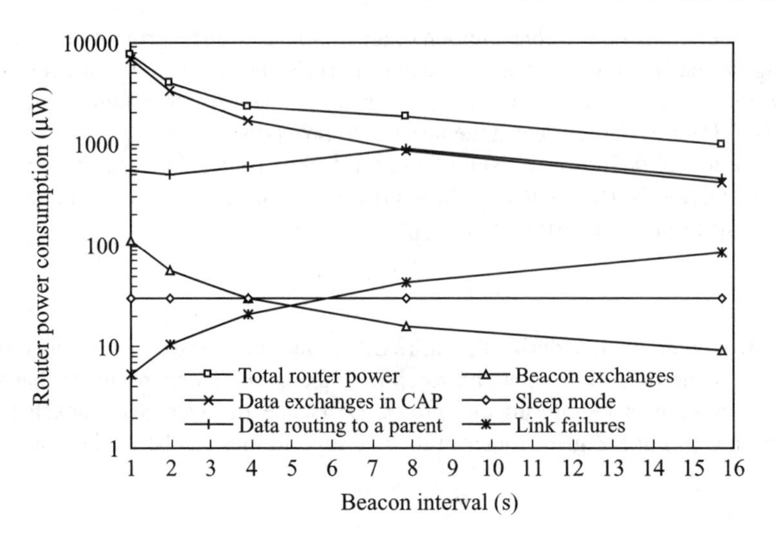

Figure 18.7 ZigBee router power consumption breakdown ($I_U = 40$ s, $d = 2$, $I_F = 3$ h).

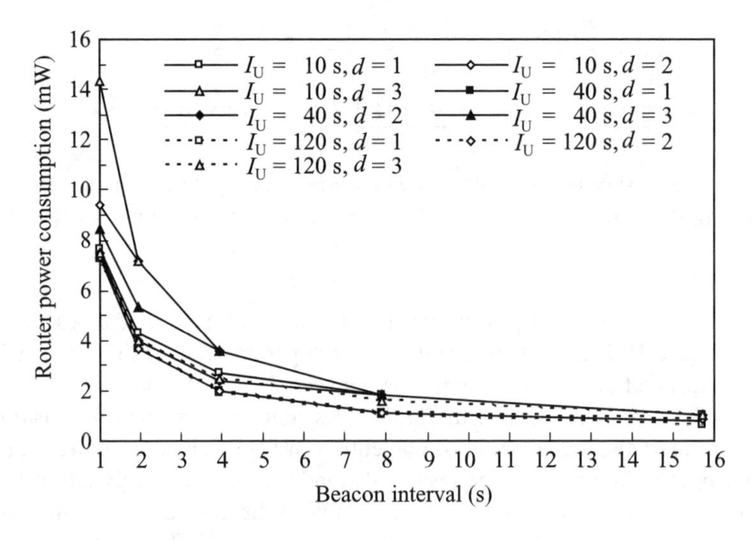

Figure 18.8 ZigBee router power consumption as the function of beacon interval, as uplink data transmission interval and network depth below the analyzed node vary.

be evaluated much more rapidly and extensively than using simulations or experimental measurements. Performance analysis is an effective tool for determining the effect of non-ideal characteristics of the operation environment and PHY layer on the achieved performance. For maintaining the modeling rapid, only the most essential non-idealities having a significant effect on performance results are included. Analysis focuses on the average network operation, while special situations are very difficult to model or even detect. For these purposes, the utilization of a simulator is more feasible, as presented in the next chapter.

19

WISENES Design and Evaluation Environment

One of the key challenges for future WSNs technologies is the tailoring of networks for different applications. A hindrance is that the low-level optimization and prototyping of WSNs is time consuming and can generate misleading results. Furthermore, the estimation of performance and robustness of large networks is difficult, if the operation is tested only on a small scale.

WISENES framework is aimed at easing the development of application-specific WSNs. WISENES enables the design, simulation, implementation, and evaluation of WSN protocols and applications in one framework with measured back-annotated information. The difference between WISENES and other WSN evaluation tools is that there is no need to carry out a separate high-abstraction WSN modeling and evaluation project and a separate development project for the actual implementation. However, if preferred, WISENES can also be used for plain simulations like other WSN simulators. In all cases, WISENES eases the assessment of the protocol and application interoperability, and their applicability to different sensor node platforms (Kuorilehto et al. 2007b).

19.1 Features

The key benefit of WISENES is that the evaluation of protocols, applications, and their different configurations is carried out starting from the design phase. The framework defines rules and interfaces for the protocol stack and application implementation. WSN nodes, transmission medium, and inspected phenomena are modeled separately under the same WISENES framework. The functionality of WSN protocols and WSN applications are implemented in Specification and Description Language (SDL) (ITU 2002). The models of high-abstraction level SDL are compiled to executables used for both simulation and final implementation.

Compared to other WSN simulators, especially SenQ (Varshney et al. 2007) and SensorSim (Park et al. 2001), the battery and sensing models of WISENES are currently

Ultra-Low Energy Wireless Sensor Networks in Practice: Theory, Realization and Deployment

more inaccurate. The distinctive features for WISENES are the complete design flow from high-abstraction level graphical models to the final node implementation, accurate full-scale simulations with configurable protocol stack and node platform models, and the back-annotation of performance information from real platforms.

19.2 WSN Design with WISENES

The design flow of a WISENES simulation is presented in Figure 19.1. First, the designer chooses the protocol stack and configuration parameters that fit the application requirements. Then, after ensuring an initial performance with the simulator, WISENES framework is used to generate C source code for real platform nodes. As the real-life environment rarely corresponds to the envisioned conditions in the simulation, the WISENES uses the back-annotation mechanism to match the configuration parameters to the actual measurement results. Thus, the source code is iteratively optimized for the application-specific scenario.

The design of WSNs in WISENES is based on the abstraction of the main functional components. In WISENES the abstraction is realized by four models that are defined by a designer. These are *application model*, *communication model*, *node model*, and *environment model*. The hierarchy and main properties of the models are depicted in Figure 19.2 (Kuorilehto et al. 2007c).

The environment model defines the parameters for wireless communication (signal propagation, noise), describes overall characteristics (average values, natural variation) for different phenomena, and specifies separate target areas (e.g. buildings) and objects (e.g. humans, animals, vehicles) and their characteristics. The environment model also defines the mobility of nodes and target objects.

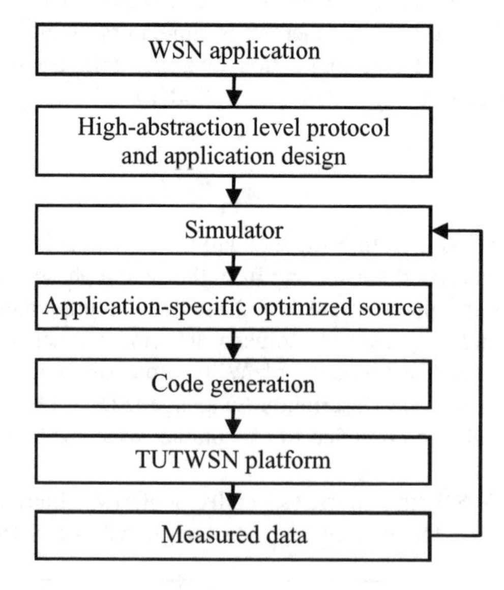

Figure 19.1 Design flow of a WISENES simulation.

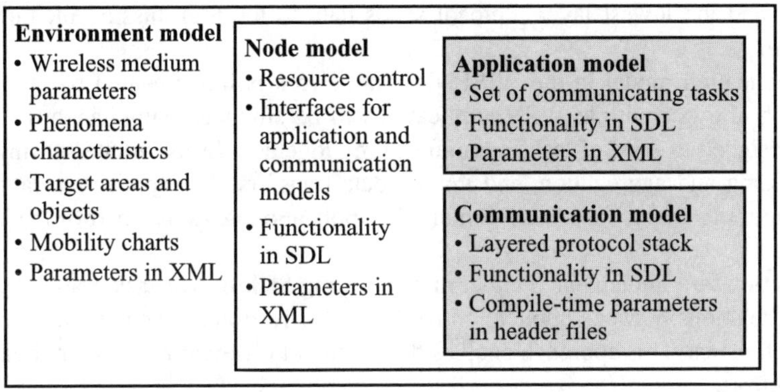

Environment model
- Wireless medium parameters
- Phenomena characteristics
- Target areas and objects
- Mobility charts
- Parameters in XML

Node model
- Resource control
- Interfaces for application and communication models
- Functionality in SDL
- Parameters in XML

Application model
- Set of communicating tasks
- Functionality in SDL
- Parameters in XML

Communication model
- Layered protocol stack
- Functionality in SDL
- Compile-time parameters in header files

Figure 19.2 WISENES models for the designer and their main characteristics. With kind permission of Springer Science and Business Media.

The WISENES application model allows a designer to describe the functionality and requirements of an application separately. This eases the exploration of the performance and suitability of application configurations for different kinds of networks and environments. The functionality of an application is divided into tasks that are implemented in SDL. The operational parameters and requirements of the application are specified in XML configuration files. These parameters define the dependencies between the application tasks, task activation patterns, and QoS requirements for the applications. The QoS parameters define task priorities, networking requirements in terms of data reliability and urgency, and an overall network optimization target. The optimization target is used to steer the communication model and it can be, for example, load balancing, high performance, or a maximal network lifetime.

The communication model specifies the networking for WSN applications. The WISENES communication model consists of a protocol stack implemented in SDL and a set of configuration parameters. Protocol configuration parameters are set at design time, while application-specific requirements, such as the network optimization target, are input from the application model during runtime.

The node model describes the characteristics and capabilities of physical node platforms. The node model is parameterized in XML, which defines the node resources, peripherals and transceivers. The functionality of the node model is realized by an OS-type interface provided to application and communication models for resource management and execution control.

19.3 WISENES Framework

The WISENES framework defines rules and interfaces for the model design and provides a library that contains existing communication model implementations and a set of implementations for known functions. The implementation of the design models is mapped straightforwardly to corresponding WISENES framework components.

The composition of the protocol stacks implementing communication models is network and application dependent. The transceiver unit at the physical layer is part of the WISENES framework and the lowest layer protocol sends data to the transmission medium through its interface.

The application model in the WISENES framework is implemented by an application layer, which allows multiple WSN applications to be simulated simultaneously. Applications are designed as a set of tasks communicating together. Tasks initiate sensing, perform data processing and aggregation, and initiate data transfers. The application layer component, which implements a host environment for application tasks, is a part of the WISENES framework.

Applications are implemented either in detail using SDL or by a task graph. A task graph defines the relations between application tasks, task activation frequencies, and sensing and data characteristics. This approach enables the testing of different types of applications with minimum effort.

The input parameter and the output result groups of WISENES are summarized in Figure 19.3. WISENES input parameters are defined using XML, each parameter set having a dedicated file with a predefined structure.

Node, transceiver unit, and peripheral parameters define the node model by setting the capabilities of node platforms. Physical node parameters are given in two separate files. The first defines the capabilities of node platforms, and the other per node platform type and node coordinates. Protocol-related XML configuration parameters define the static memory consumption for each protocol, while other characteristics are specified by the communication model implementation.

The environment model is parameterized by transmission medium, sensing and mobility parameters. The first one defines the characteristics for the signal attenuation and wireless communication related aspects, while the last two specify active phenomena, target objects, and the mobility of nodes and target objects.

WISENES outputs information in two forms. Detailed information about the simulated WSN and nodes is collected to log files. Each event during a simulation is logged with parameters that define the cause and the consequence of the event. Log data are written to per-node directories, each protocol layer, application, and control instance modeling OS

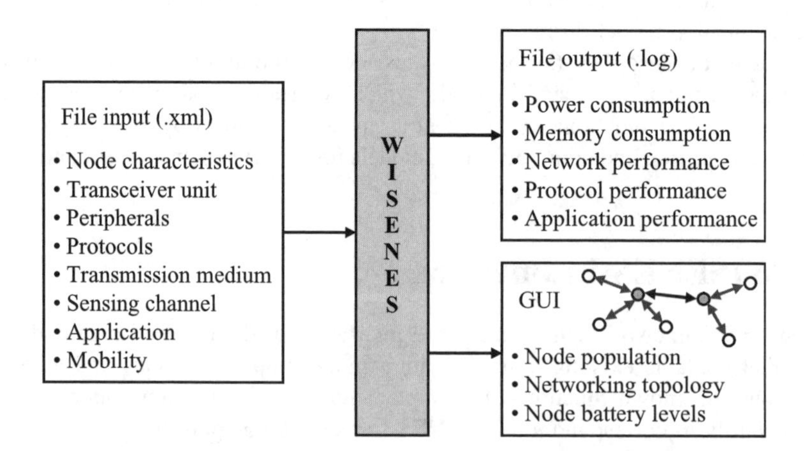

Figure 19.3 WISENES input and output parameters.

routines having a dedicated ".log-file". During an active simulation run, the progress of a simulation is illustrated through GUI that presents the node population, network topology, and node energy levels.

WISENES has two UIs for controlling and monitoring simulation runs. Simulations are started and controlled through a command-line interface. The progress of the simulation and the topology of a simulated WSN are visualized in WISENES GUI. GUI is implemented in Java using Java foundation classes Swing packages (Sun Microsystems 2006a). The communication between GUI and WISENES is implemented by a socket interface.

The WISENES framework consists of SDL models for transmission medium, sensing channel, sensor node, and a central simulation control that manages simulations and handles simulator I/O. Networking protocols and power consumption emulation are embedded to sensor node model.

In addition to the application tasks and protocols described by the designer, SDL is used for the implementation of the WISENES framework. The tool used for SDL development is Telelogic TAU SDL Suite (Telelogic 2006), version 4.5. The SDL suite uses a graphical notation for SDL design, and provides tools for simulation, integration, and implementation.

19.3.1 Short Introduction to SDL

The MoC used in SDL is parallel communicating Extended Finite State Machines (EFSMs). SDL hierarchy has multiple levels, of which the system level consists of a number of blocks that clarify the representation. They can be recursively divided into sub-blocks. The behavior of a block is implemented in processes described by EFSMs. Processes can be simplified by implementing a part of the functionality in a procedure. Blocks and processes can be implemented as "types" that describe the functionality. For execution, multiple instances of these types can be created. These type definitions can be included with other type definitions to SDL packages that facilitate modular system design. The maximum number of instantiated blocks must be defined at a compile time, whereas processes can be created dynamically during runtime (ITU 2002).

Processes communicate by asynchronous signals that can carry any number of parameters. Each process has an infinite First-In-First-Out (FIFO) buffer for incoming signals. Signal routes define which type of signals a process can send and to which processes. Communication between processes can also be executed synchronously by calling remote procedures, which are exported on a process interface (ITU 2002).

Due to its formality, SDL can be automatically converted, for example, to C source code, which can then be used to make an executable application or simulation. The Telelogic TAU SDL simulation engine supports discrete-event simulations and realtime simulations. In WISENES we utilize discrete-event simulation, in which events are processed and handled in the order of occurrence. This makes the time concept fully parallel and avoids active waiting during idle times.

Environment functions implement the interaction between SDL and its execution environment. Dedicated functions are defined for environment initialization, unloading, signal output, and signal input. The output function is called when a signal is sent from the SDL system to the environment. Because a method for interrupting is absent in SDL, the input function must be polled for receiving signals from the environment. In the Telelogic TAU SDL simulation engine, the input function is called after every transaction, which is an execution flow from a state to another triggered by an incoming signal. An SDL procedure can be substituted by an external function, in a case where SDL lacks expressivity or more efficient implementation is desired. Both environment and external functions are implemented in C for WISENES.

19.3.2 WISENES Instantiation

The instantiation of WISENES is depicted in Figure 19.4. The designer selects the protocols from the library or implements new ones in SDL and integrates them to the WISENES framework. The upper and lower interfaces of the protocol stack are defined interfaces

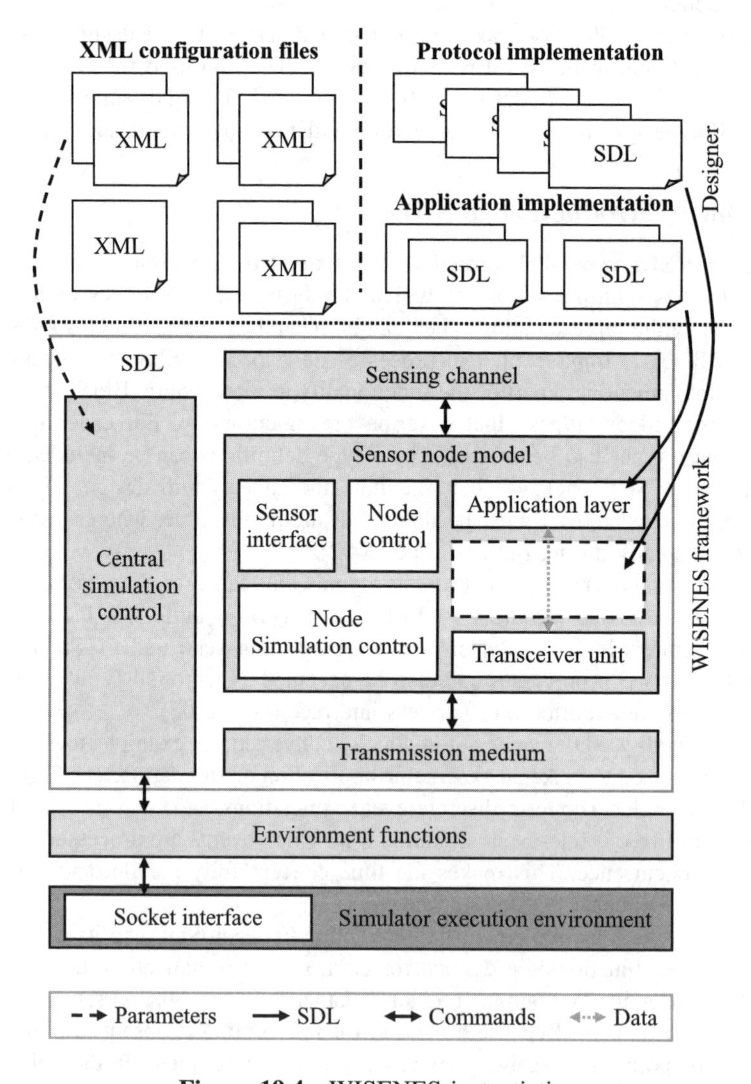

Figure 19.4 WISENES instantiation.

of the application layer and the transceiver unit, respectively. Application functionality is either implemented as SDL procedures or described by a task graph (Kuorilehto et al. 2007b).

The composition of the protocol stack can be freely selected by the designer, but a typical protocol stack consists of data link, network, and middleware layers. The protocol stack composition is specified by an SDL package containing the SDL implementation. The block instantiates the defined protocol layer implementations that are also SDL blocks implemented in SDL packages. The internal implementation of layers is not restricted in any way.

Node platforms are parameterized in XML configuration files that are parsed by *central simulation control*. The parameters are passed to *sensor node* SDL model. Node coordinates are relayed also to *transmission medium* and *sensing channel*, both of which have also dedicated parameters.

The interfacing of WISENES GUI is implemented in environment functions that maintain the socket connection. Information to GUI is updated only periodically in order to lessen communication. A data structure that defines sensor node parameters is sent to the environment functions as a signal parameter and parsed to the socket.

19.3.3 Central Simulation Control

The central simulation control initiates the WISENES framework, controls active simulation runs, and performs event and statistics logging. The information gathering for the control and logging is implemented in remote procedures. Separate procedures are implemented for log data gathering, GUI information updating, and predetermined end-condition signaling. The end condition is set by the designer and it defines a simulation time limit, a percentage of dead nodes, or e.g. a limit for an application task activation count.

Distinct remote procedures for event and data packet logging are exported for each protocol layer, application tasks, and for framework components. Their parameters vary depending on the layer. Logs are stored in dedicated data structures and written to files either periodically when the WISENES memory consumption exceeds a predefined limit, or at the end of the simulation.

19.3.4 Transmission Medium

The transmission medium provides the connectivity between sensor nodes. It is implemented as an SDL process that redirects signals from a source to destination sensor node SDL blocks. Sensor nodes register their node identifier and transceiver unit PID to the transmission medium for enabling the data redirecting. Due to the nature of SDL data typing, transmitted data are separately copied for each destination node.

The signal propagation in the transmission medium follows a selected algorithm. Currently supported algorithms are free-space loss, two-ray ground, and a transceiver unit dependent signal attenuation curve. The transmission medium takes into account collisions and bit errors, which allows a realistic modeling of, for example, packet losses and hidden-node problem.

A delay during a transmission is calculated by dividing the packet length by the transceiver unit throughput. The delay of signal propagation in the medium is omitted. Thus, a packet is relayed to the destinations immediately after the calculated transfer delay.

19.3.5 Sensing Channel

The sensing channel simulates physical phenomena. Similar to the transmission medium, the sensing channel utilizes node coordinates. Each phenomenon is modeled separately with individual propagation characteristics. The propagation depends also on the media in the vicinity.

Our current sensing channel implementation generates random stimuli for phenomena. The upper and lower limits are defined for each phenomenon. Currently simulated phenomena are temperature, humidity, vibration, sound, and luminance. The selected approach is applicable for environmental monitoring applications, but a more detailed sensing channel must be implemented for example object tracking applications.

19.3.6 Sensor Node

The sensor node SDL block consisting of node, communication, and application model implementations is depicted in Figure 19.5. On the sensor node SDL block, *physical layer, sensor interface, application layer,* and *node control* blocks are part of the WISENES framework and implement the node model, while the instantiated protocol layers defining

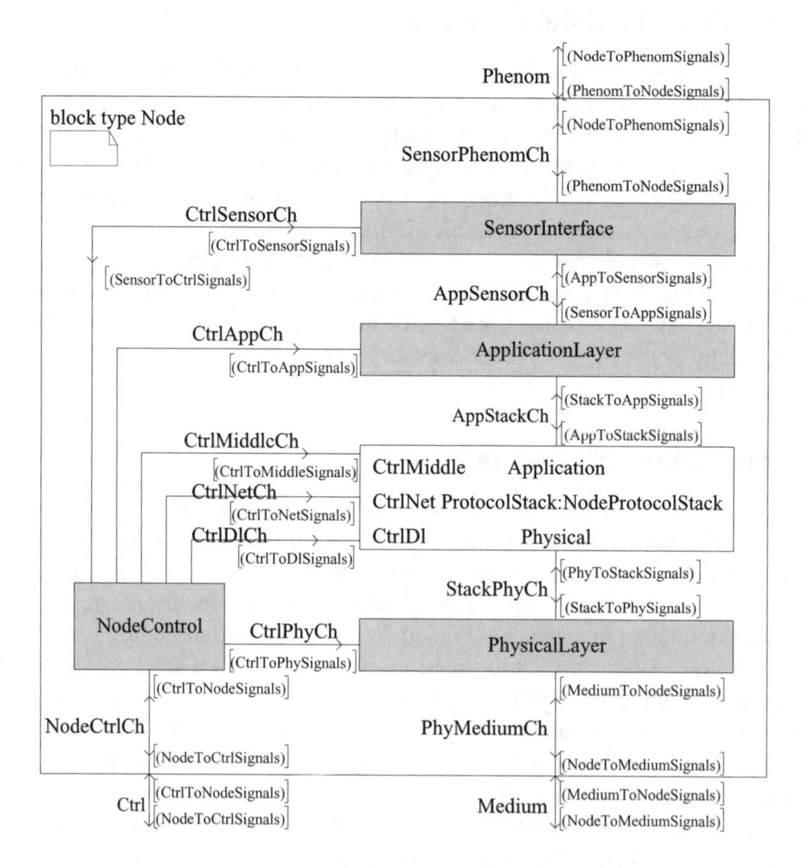

Figure 19.5 WISENES sensor node SDL block.

the communication model are selected by the designer. The signal routes between protocol layers are for data communications, whereas the signals sent from the node control to the other blocks are for initiation and shutdown.

Physical layer

A transceiver unit process at the physical layer models the hardware and its device driver. The process implements the interface to the transmission medium, performs collision detection, and models the internal delay in a transceiver unit. Depending on the modeled hardware, additional hardware-accelerated features, such as CRC or encryption, are incorporated to the model. The upper interface of the transceiver unit defines a set of signals for the lowest protocol layer. Dedicated signals are declared for transceiver unit control, transmitter or receiver enabling, carrier sensing, data sending, send confirmation, and data reception indication.

Transferring data to or from the transceiver unit causes delay, because a transceiver unit is typically a distinct hardware component on a node platform. A delay of data length in bits divided by the interface bit rate is generated in data loading.

Sensor interface

Sensor interface SDL block consists of a process that models ADC and sensor operation. When an application initiates a sensing task, it sends a signal to the sensor interface process. The process activates a sensor for that phenomenon and ADC, which samples the analog sensor output. The sensor interface process acquires a value from the sensing channel by signal exchange. The delay in operation depends on the sampling frequency and associated sensor. These values are defined in input parameters.

Application layer

The application layer implements the scheduling of application tasks. This approach is selected to facilitate the task scheduling when implemented as SDL procedures. When an application is described as a task graph, the application layer process emulates the execution of tasks. In this case, no real functionality apart from sensing and data-transfer initiation is implemented.

The functionality of application task in SDL procedures is implemented by the designer. The task-state control and scheduling are implemented in the application layer process. Task state is *running* when it is executed, *ready* when it is ready for execution but another task is running, or *wait* when the task requires an event to occur before running. Supported scheduling algorithms are round-robin and static priority scheduling.

When a task is ready and scheduled for processing, the application layer process calls the procedure that implements the task. The current state and the event that moved the task to the ready state are given as parameters to the procedure. The task completes its next transition and enters to the wait state. The waited event is returned to the application layer process. In occurrence of an event, all tasks waiting for it are set to the ready state.

The application initiates a data transfer by sending a signal to the protocol stack. The signal defines the destination task, optimization target for the data transfer, and the data content itself. The actual data routing is left to the protocol stack. Similarly, a received data event is indicated to the application layer through a signal that defines the source task and data.

Node control

The node control block consists of two processes, *node control* and *node simulation control*. Similar to central simulation control, both processes use remote procedures for communication with the protocols. The node control process implements OS routines in WISENES. Separate procedures are exported for execution and scheduling, sleep-state activation, and memory management.

The node simulation control implements a per-node interface to the central simulation control and models the power consumption of node platform. In initiation, node simulation control relays node parameters in signals to the node control and protocol layers. During an active simulation run, the node simulation control gathers GUI-related information from the node and passes it to central simulation control.

The node power consumption modeling is implemented in the node simulation control by a linear battery model, in which the component power consumption is independent of battery discharge rate. Power consumption-related remote procedures are called only from the SDL processes that are part of the WISENES framework.

19.4 Existing WISENES Designs

Currently, the WISENES environment incorporates two full-feature protocol stacks: TUTWSN and ZigBee (Zig 2004). A widely adopted reference stack in WSN domain namely S-MAC (Ye et al. 2004) and directed diffusion (Intanangonwiwat et al. 2003) is under development. The latter protocol stack is omitted in here because the scalability of the current WISENES S-MAC implementation is not sufficient for large-scale WSNs. Each stack is implemented in the WISENES environment in SDL.

As stated, the implementation of protocol stacks in WISENES is done by utilizing SDL packages. The structure, namely the protocol layers and the signal routes between them, are defined in the top-level protocol stack package. The internals of each protocol layer are also implemented as separate SDL packages that define the processes and their communication separately for each layer. Such an approach allows a modular exchange of protocol layer implementations requiring only the interface adaptation.

The statistics of TUTWSN and ZigBee WISENES implementations are presented in Table 19.1. The statistics are gathered by a complexity measurement tool incorporated into the Telelogic TAU SDL Suite. The columns give statistical information about the SDL-specific implementation details.

The *declarations* column includes all procedures, data types, variables, and other similar entity declarations in the process. The number of distinct *states* in the process and its procedures are showed in the next column. A *transition* starts from a state with a reception of an input signal and ends when the process enters to another state. Number of *procedures* visualize the hierarchy of the implementation. Note that procedures are also included in the declarations.

In the graphical SDL design environment used in WISENES the implementation is divided into *symbols*. A symbol may be, for example, a task containing several *statements*,

Table 19.1 Statistics of WISENES SDL implementations for TUTWSN and ZigBee.

SDL process	Declarations	States	Transitions	Procedures	Symbols	Statements	Execution paths
TUTWSN MAC							
TutwsnMacChannelAccess	1122	20	194	129	2079	3141	3072
TutwsnMacManagementEntity	195	9	32	13	297	381	130
TUTWSN Routing							
TutwsnRouteManagement	479	10	81	43	838	1290	1276
TutwsnRouteDataService	158	3	41	21	314	373	125
TUTWSN API							
TutwsnApiService	150	4	25	10	180	283	71
LR-WPAN MAC							
WpanChannelAccess	834	34	163	75	2202	2732	2205
ZigBee NWK layer							
ZigbeeNWKManagementEntity	327	12	55	30	574	899	366
ZigbeeNWKDataEntity	277	4	47	26	393	519	184
Adaptation middleware							
DataRedirectingProcess	72	4	20	4	167	185	105

a procedure call, or output signal. A statement is, for example, an assignment within a task symbol, i.e. its is closer to C statements. The *execution paths* column counts the possible different paths from a transition start to the transition end. The used analysis tool does not guarantee that all possible paths are included but it should give a good estimate.

19.4.1 TUTWSN Stack

The TUTWSN protocol stack SDL implementation in WISENES consists of three layers as depicted in Figure 19.6. A full-feature TUTWSN MAC protocol is implemented at the data link layer and TUTWSN routing at the network layer. TUTWSN API implementation at the middleware layer does not support profiles, but all networking and data-related functions are covered.

The task division of the TUTWSN protocol stack in WISENES follows that presented in Chapters 13–15. Yet, in WISENES some of the MAC protocol functions are implemented

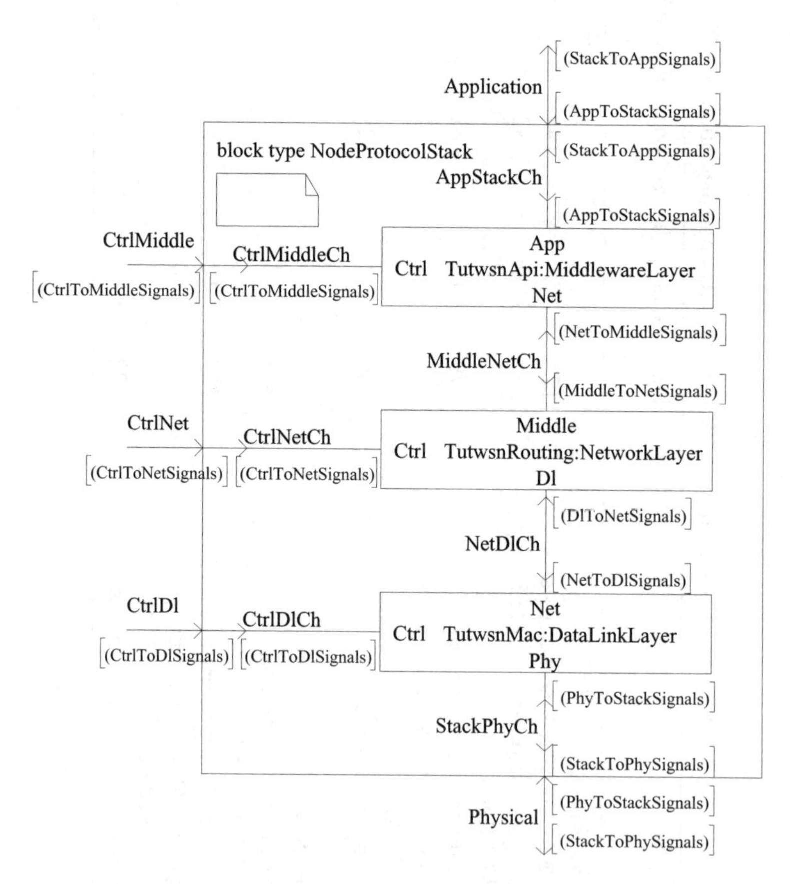

Figure 19.6 TUTWSN node protocol stack SDL block.

at the network layer, since the sharing of data between layers is complicated in SDL. Therefore, association to neighbor clusters and role selection are controlled by the network layer, as most of the data required for the decision making are hosted by the TUTWSN routing protocol.

The interfaces between layers consist of signals and remote procedures. Signals are used for passing data for processing and for requesting control operations. Remote procedures are mainly utilized for getting "shared" data from another layer. Control signals from the WISENES framework, mainly for initialization and shutdown, are passed separately to each layer.

TUTWSN data link layer

The functionality of the TUTWSN data link layer is implemented by two SDL processes, `TutwsnMacChannelAccess` and `TutwsnMacManagementEntity`, as depicted in Figure 19.7. The sharing of internal data between the processes is performed by signals that pass global variables between them.

`TutwsnMacChannelAccess` implements TUTWSN subnode and headnode functionality, neighbor discovery, and data buffering. Further, it performs channel access through the transceiver unit. Most of the algorithms and functionality are implemented in SDL procedures, as shown in Table 19.1. The actual SDL process mainly incorporates the control for the operation.

`TutwsnMacManagementEntity` is responsible for runtime adaptation of channel access parameters (e.g. access cycle length as well as ALOHA and reservation slot counts), and selection of the frequency channel and time slot for a cluster's active period. The neighbor information needed by the algorithms is provided by the `TutwsnMacChannelAccess` and routing protocol.

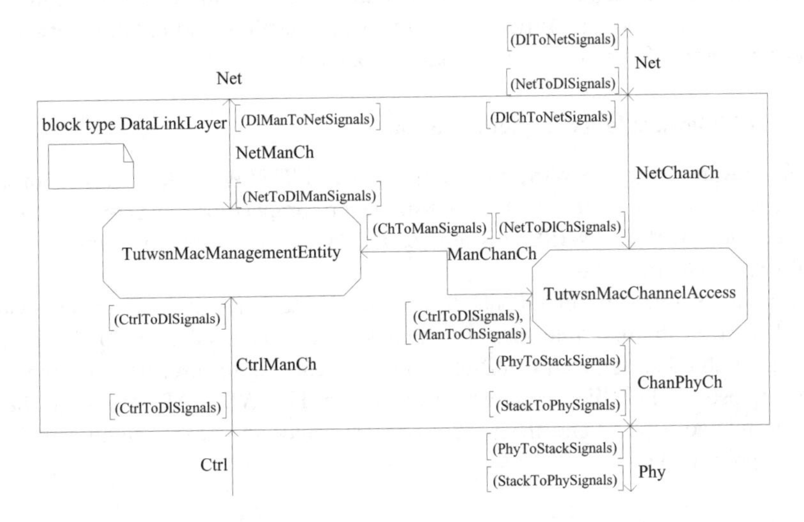

Figure 19.7 SDL block describing TUTWSN MAC protocol implementation.

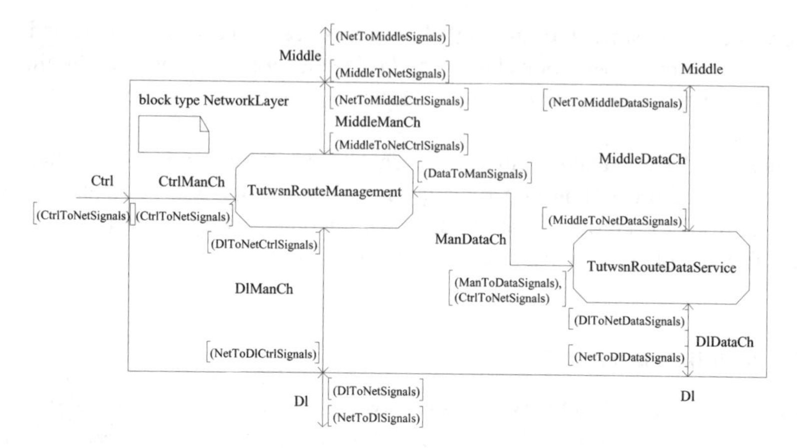

Figure 19.8 SDL block describing TUTWSN routing protocol implementation.

TUTWSN network layer

Similar to the data link layer, the TUTWSN network layer consists of two SDL processes. These are `TutwsnRouteDataService` and `TutwsnRouteManagementEntity` as shown in Figure 19.8. Compared to TUTWSN MAC, the division of the routing protocol to SDL processes is more straightforward.

`TutwsnRouteDataService` is responsible for data buffering. It handles data-related lower and upper interfaces and offers a service for transferring internal routing control messages. A route for data is queried from the `TutwsnRouteManagementEntity`.

All the routing decisions are made by `TutwsnRouteManagementEntity`. It maintains route tables and information about the available sinks within the network. In addition, `TutwsnRouteManagementEntity` controls role selection, and initiates network scans and other neighbor information-related control operations.

TUTWSN middleware layer implementation

At the middleware layer, the WISENES SDL model of TUTWSN API consists of a single SDL process; see Figure 19.9, which shows the layer composition. `TutwsnApiService` process interfaces with the WISENES application layer and adapts the communication to the TUTWSN protocol stack.

`TutwsnApiService` implements data redirection and handling of network-wide data interests. Based on the data interests, it activates or deactivates defined application tasks. The API maps the data targeted to different tasks according to information provided with the data interests. WISENES implementation of the TUTWSN API omits the hardware information-related aspects and the profile concept, since they are insignificant from the networking point of view.

19.4.2 ZigBee Stack

Similarly to TUTWSN, the ZigBee protocol stack in WISENES consists of three layers, as shown in Figure 19.10. At the data link layer, the IEEE 802.15.4 LR-WPAN MAC

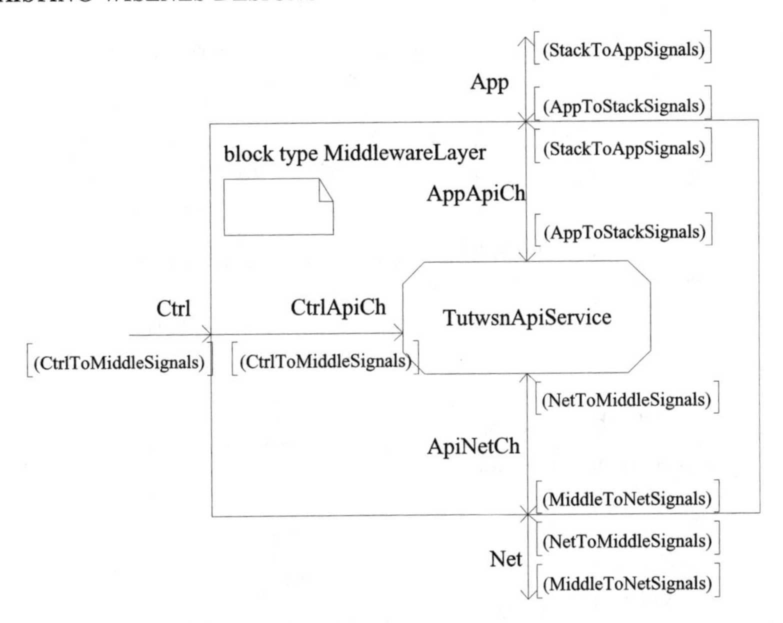

Figure 19.9 SDL block describing TUTWSN API implementation.

protocol is implemented according to the standard (IEE 2003b). The ZigBee NWK layer does not conform to all features of the ZigBee specification (Zig 2004) but supports only cluster-tree topology. The middleware layer adapts the WISENES application layer to the lower layers.

The LR-WPAN MAC protocol implements the signal interface specified in the standard at its upper interface (IEE 2003b) but the ZigBee NWK layer provides only data access and simple control interfaces for the middleware layer. Due to the tightly layered architecture, all inter-layer communications are done via signals, thus remote procedures are not used.

ZigBee data link layer

The LR-WPAN standard includes an appendix having an SDL description of the protocol. However, this description is not utilized in WISENES due to its superficial nature and poor presentation. Instead, a complete implementation is done in a single SDL process, `WpanChannelAccess`, as depicted in Figure 19.11.

`WpanChannelAccess` implements the beacon-enabled mode of LR-WPAN standard, data buffering, and the interface for the upper layers. Currently, non-beacon mode is not fully supported, since it is not required by the upper layers. The implementation approach is quite similar to TUTWSN MAC, i.e. most of the functionality is implemented in SDL procedures.

ZigBee network layer

The ZigBee NWK layer in WISENES has two processes; `ZigbeeNWKDataEntity` and `ZigbeeNWKManagementEntity`. The structure of the network layer SDL block is

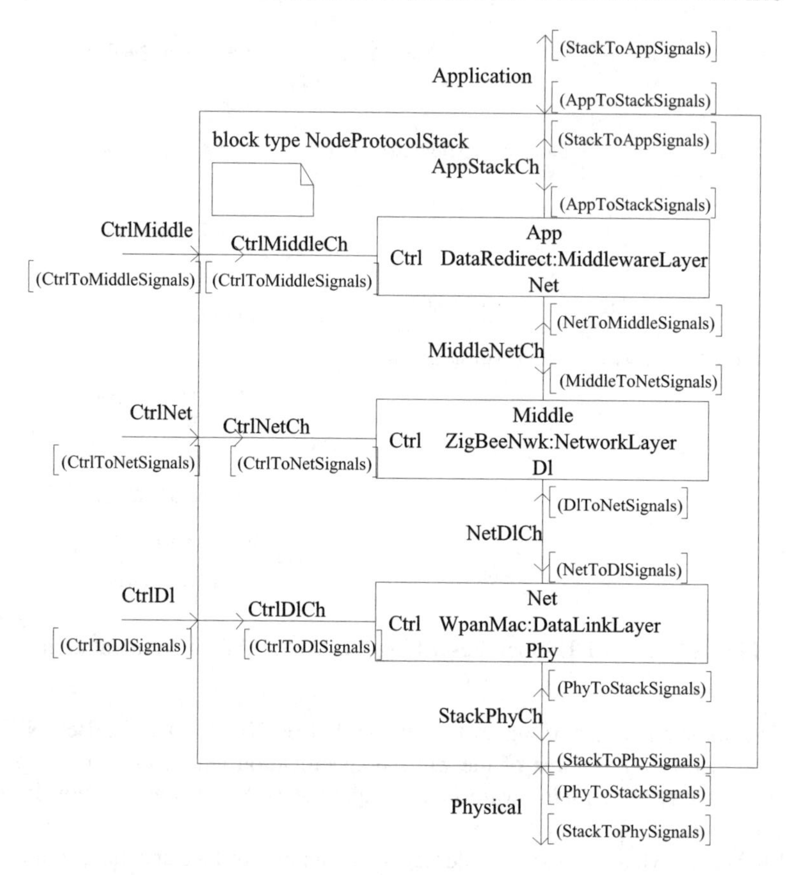

Figure 19.10 ZigBee node protocol stack SDL block.

shown in Figure 19.12. The interdependencies between the two processes are kept to a minimum, which makes it possible to use only signals at the internal interface.

`ZigbeeNWKDataEntity` is purely for data buffering and relaying between the middleware and data link layers. Since the NWK layer does not incorporate any internal communication primitives, the only dependency to the `ZigbeeNWKManagementEntity` is the request for route information.

`ZigbeeNWKManagementEntity` implements the hierarchical addressing and routing scheme defined in the specification (Zig 2004). Thus, a ZigBee coordinator sets the address spaces for its children and routes are formed according to the addresses. Contrary to the specification, the WISENES ZigBee NWK layer also contains the logic for selecting the parent for association. Thus, this task is not left to the upper layers.

ZigBee middleware layer implementation

The ZigBee middleware layer in WISENES does not have any correspondence to the application layer of the specification. Instead, the WISENES implementation consists of a simple process, `DataRedirectingProcess`, as depicted in Figure 19.13.

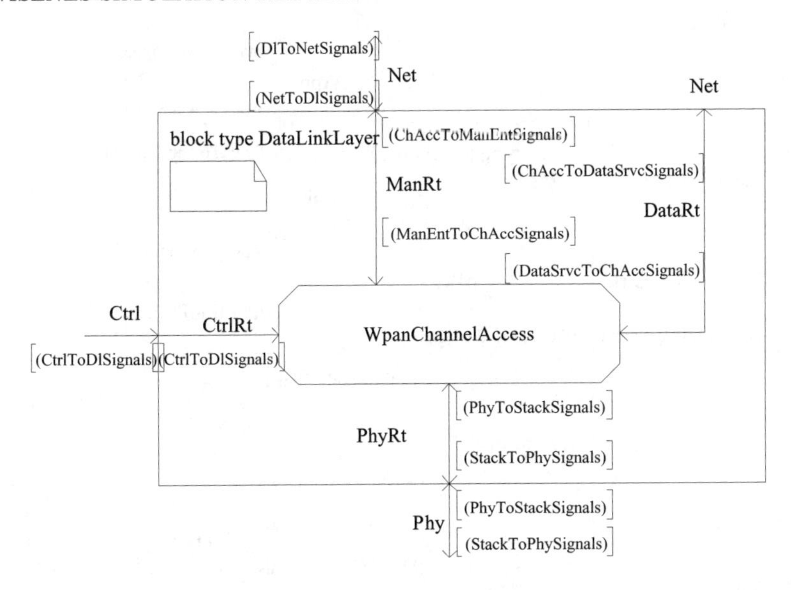

Figure 19.11 SDL block describing LR-WPAN MAC protocol implementation.

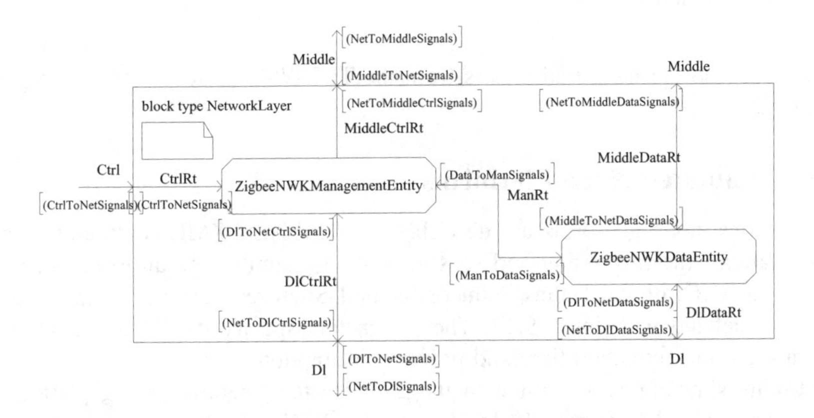

Figure 19.12 SDL block describing ZigBee NWK layer implementation.

`DataRedirectingProcess` does not include any control logic, but it simply adapts the WISENES application layer to the rest of the stack by modifying and mapping WISENES specific signaling to that of ZigBee. The layer implementation is straightforward as illustrated by the complexity figures in Table 19.1.

19.5 WISENES Simulation Results

The purpose of simulations is to evaluate the performance of the TUTWSN and ZigBee protocols with different protocol configurations. The results are comparable to the analysis

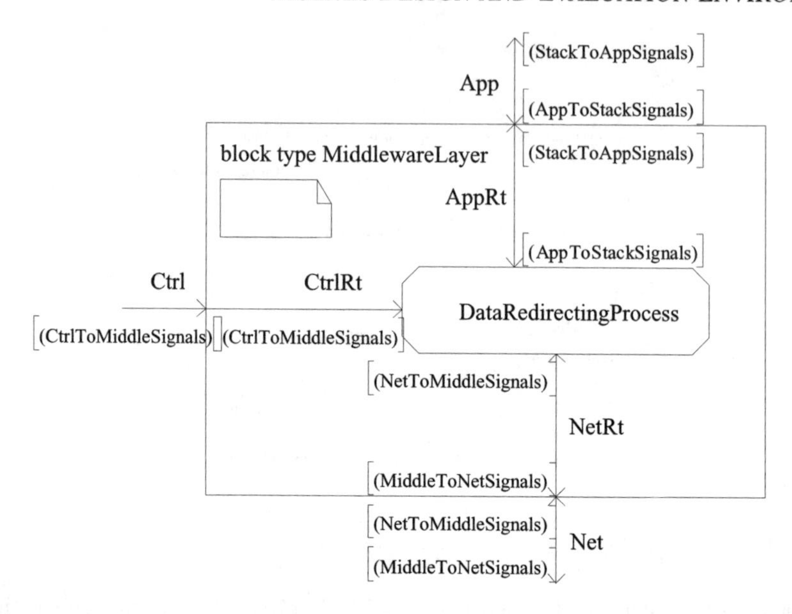

Figure 19.13 SDL block describing data redirecting and adaptation middleware for ZigBee protocol stack implementation.

in Chapter 18 and the measured results from the TUTWSN prototypes that are presented in Chapter 23.

19.5.1 Simulated Node Platforms

For simulations, node platforms are described in WISENES XML configuration files. In both simulations, the processing and sensing units are similar. As an example, peripherals including ADC, digital Dallas Semiconductor DS620 temperature sensor, and a GPS receiver are defined in Figure 19.14. The parameters specify the dependencies to other peripherals and the activation time and power consumption.

TUTWSN simulations are run with the 2.4 GHz temperature-sensing platform introduced in Section 4.3.1. Figure 19.15 depicts the XML for the Nordic Semiconductor nRF2401 transceiver unit on the platform. The parameters define the transceiver and host-side capabilities and power consumption in different operation modes. IEEE 802.15.4 standard (IEE 2003b) compliant 2.4 GHz Texas Instruments CC2420 transceiver used in the ZigBee simulations is similarly specified in Figure 19.16. This transceiver has carrier-sensing capability. The capabilities of the transceiver unit are measured using a Chipcon SmartRF CC2420DK Development Kit.

The components of the nodes are composed in an XML configuration file that defines node types, as shown in Figure 19.17. The file defines the processing and memory capacities of the MCU as well as its power-saving modes. The crystal accuracy is specified in ppm. Battery capacity and its operational characteristics are defined, including also possible

```
<peripheral id="1" name="ADC">
  <phenomenon>VOLTAGE</phenomenon>
  <dependency id="0"/>
  <active_period ms="0.5" uW="15"/>
</peripheral>
<peripheral id="2" name="Temperature sensor">
  <phenomenon>TEMPERATURE</phenomenon>
  <dependency id="0"/>
  <active_period ms="1" uW="10"/>
</peripheral>
<peripheral id="3" name="GPS receiver">
  <phenomenon>GPS</phenomenon>
  <dependency id="0"/>
  <active_period ms="500" uW="100"/>
</peripheral>
```

Figure 19.14 WISENES XML configuration file defining peripherals.

```
<transceiver_unit id="1" name="nRF2401">
  <throughput bps="1000000"/>
  <rssi capability="NO"/>
  <data mode="BURST"max_length="32"/>
  <data_load_info bps="1000000" uW="1310"/>
  <rxtx_change capability="NO"/>
  <receiver_info>
    <rx_sensitivity dBm="-85"/>
    <idle_transient ms="0.25" uJ="10.9"/>
    <tx_transient ms="0" uJ="0"/>
    <active_power uW="57600"/>
  </receiver_info>
  <transmitter_info>
    <idle_transient ms="0.25" uJ="8.2"/>
    <tx_transient ms="0" uJ="0"/>
    <tx_power_levels>
      <tx_power dBm="-20" uW="26400"/>
      <tx_power dBm="-10" uW="28500"/>
      <tx_power dBm="-5" uW="31800"/>
      <tx_power dBm="0" uW="39000"/>
    </tx_power_levels>
  </transmitter_info>
  <carrier_sense capability="NO" uW="0"/>
</transceiver_unit>
```

Figure 19.15 WISENES XML configuration file defining nRF2401 transceiver unit.

```
<transceiver_unit id="2" name="CC2420">
  <throughput bps="250000"/>
  <rssi capability="YES"/>
  <data mode="DIRECT" max_length="128"/>
  <data_load_info bps="1000000"uW="1280"/>
  <rxtx_change capability="YES"/>
  <receiver_info>
    <rx_sensitivity dBm="-85"/>
    <idle_transient ms="0.192"uJ="7.6"/>
    <tx_transient ms="0.2"uJ="8.6"/>
    <active_power uW="55000"/>
  </receiver_info>
  <transmitter_info>
    <idle_transient ms="0.192"uJ="7.6"/>
    <tx_transient ms="0.2"uJ="8.6"/>
    <tx_power_levels>
      <tx_power dBm="-15" uW="28300"/>
      <tx_power dBm="-10" uW="31400"/>
      <tx_power dBm="-5" uW="37600"/>
      <tx_power dBm="0" uW="46400"/>
    </tx_power_levels>
  </transmitter_info>
  <carrier_sense capability="YES" uW="54300"/>
</transceiver_unit>
```

Figure 19.16 WISENES XML configuration file defining Chipcon CC2420 transceiver unit.

energy scavenging. The peripherals and transceiver unit information is included by setting corresponding IDs. Although the ZigBee node does not describe a real prototype, the combination of the components in WISENES results in a realistic platform model.

19.5.2 Accuracy of Simulation Results

The accuracy of the WISENES results is evaluated by comparing simulation results to those measured from real TUTWSN indoor-temperature node platforms. For measuring the power consumption of a subnode and headnode, the nodes are powered by super capacitors. The average power consumption of a node is then derived from the slopes of capacitor terminal voltages during a 10-minute measurement period.

The evaluation is performed with an indoor-surveillance system application. The application is described in Section 23.1 together with the measurement case details. Since the TUTWSN protocols are highly dynamic by their nature, the creation of an identical network configuration is difficult. Thus, for the WISENES simulations, the objective is to construct as similar an environment as possible.

In the evaluation of the accuracy of the results, power consumption and data-routing delay are considered. The communication environment has a significant impact on delay measurements, since the delay is mainly caused by buffering due to retransmissions and congestion. In the case of power consumption, the effect is not as considerable.

```
<node_type id="1" name="TUTWSN PIC node">
  <type>FFU</type>
  <cpu MHz="2000000"/>
  <memory code="65536" data="3986"/>
  <max_ppm>20</max_ppm>
  <battery>
    <voltage initial="3.0" limit="2.4"/>
    <capacity mAh="2100" efficiency="0.91"/>
    <harvest uW="0"/>
  </battery>
  <state_info>
    <state name="active" ms="0" uW="3170"/>
    <state name="sleep" ms ="3" uW="37"/>
  </state_info>
  <transceiver_unit id="1"/>
  <peripheral_info>
    <peripheral id="1"/>
    <peripheral id="2"/>
  </peripheral_info>
</node_type>
<node_type id="2" name="ZigBee PIC node">
  <type>FFU</type>
  <cpu MHz="2000000"/>
  <memory code="65536" data="3986"/>
  <max_ppm>20</max_ppm>
  <battery>
    <voltage initial="3.0" limit="2.4"/>
    <capacity mAh="2100" efficiency="0.91"/>
    <harvest uW="0"/>
  </battery>
  <state_info>
    <state name="active" ms="0" uW="3170"/>
    <state name="sleep" ms ="3" uW="37"/>
  </state_info>
  <transceiver_unit id="2"/>
  <peripheral_info>
    <peripheral id="1"/>
    <peripheral id="2"/>
  </peripheral_info>
</node_type>
```

Figure 19.17 WISENES XML configuration file defining basic node types for TUTWSN and ZigBee simulations.

The results presented in Section 23.1 show that the variation between the measured and simulated power consumption is 6.6 % on average. Similarly, the average variation in delay is 9.5 %. These results indicate that WISENES models the node platforms and environment very accurately, and it can be used to obtain realistic performance estimations already during design time.

19.5.3 Protocol Comparison Simulations

WISENES is intended for large-scale simulations for evaluating the functionality and performance of protocols and applications to the scale required by the final application. In order to show WISENES performance in the simulations of WSNs with hundreds of nodes, TUTWSN and ZigBee protocol stacks are simulated with the temperature-sensing application.

The temperature-sensing application measures temperature and forwards the result towards a sink node. Temperature is measured within one minute intervals. A reliable data transfer is requested for the measurement results. Thus, data is retransmitted for the protocol-dependent number of times if an acknowledgment is not received.

For the simulations, a 250-node network is deployed to a 150 m $\times$ 150 m area. The data gathering point, a sink node in TUTWSN or ZigBee coordinator, is located at the middle of the area while the rest of the nodes are randomly deployed. The node locations are the same in every simulation. The length of the access cycle is altered between 1, 2, 4, and 8 seconds.

In each simulation case the simulated time is one week. For realistic results, nodes with completely erroneous operation are filtered away. Thus, if, for example, in ZigBee simulations a node has not been able to associate to any neighbor router, its performance is not considered in the results.

TUTWSN simulations

In TUTWSN simulations, 99 headnode-capable devices and 150 subnodes are deployed in addition to the sink node. Yet, the role selection algorithm of TUTWSN is used for determining the role of the headnode-capable devices. The most significant configuration parameters of the TUTWSN protocols set for the simulations are shown and described in Table 19.2 and Table 19.3.

The simulated power consumption of the TUTWSN headnode, node, and subnode as a function of the access cycle length are depicted in Figure 19.18. A "node" refers TUTWSN nodes that select their role dynamically depending on the surrounding neighbors. The results are averaged over a randomly selected 20 nodes. The averaged headnode power consumptions are 1.29 mW, 0.877 mW, 0.505 mW, and 0.327 mW with access cycle lengths of 1, 2, 4, and 8 seconds. For the subnode, the power consumptions are 0.542 mW, 0.322 mW, 0.211 mW, and 0.119 mW. The power consumptions of nodes changing their role dynamically depends greatly on the surroundings and deployed network configuration. In the simulated case with different access cycle lengths, the averaged power consumptions are 0.949 mW, 0.662 mW, 0.420 mW, and 0.271 mW.

Compared to the results given for analysis and measurements, the obtained simulation power consumptions are a bit larger. This is caused mainly due to the more dynamic and error robust operation. Thus, in simulations the reception margins are longer and the preferred number of synchronized clusters are larger. The network topology is also more dynamic, which causes sporadic network scans and other neighbor-discovery operations. The results also show the benefits of dynamic role selection over statically configured headnode roles.

The data-routing delay as the function of number of hops is presented in Figure 19.19. The delay is shown separately for different access cycle lengths. The access cycle length

Table 19.2 Configuration parameters for TUTWSN MAC protocol used in WISENES simulations.

Parameter	Value	Description
TUTWSN MAC parameters		
ac	*(varied)* s	Access cycle length
idleBeaconInterval	500 ms	Time between idle network beacons
slotLength	20 ms	Length of a slot in superframe
scanLength	750 ms	Length of a network scan
clusterBeaconCount	2	Number of cluster beacons in superframe
alohaSlots	4	Number of ALOHA slots in superframe
maxReservedSlots	4	Maximum number of reservation slots in superframe
memberAssocTimeout	30 *ac*	Idle interval, after which a member needs to renew association
reservationCycleLength	30 *ac*	Number of access cycles to which reservation slots are divided
reservationMinBandwidth	10	Minimum number of slots for member in reservation cycle
maxRetries	3	Maximum number of retries for failed operation
maxQueueLength	16	Maximum data buffer size in MAC layer
maxQueueTime	60 s	Maximum time data is stored in buffer
tdmaSlotLength	500 ms	Time slot length used in the selection of own active period time
channelSpacing	2	Number of free channels between used frequency channels

Table 19.3 Configuration parameters for TUTWSN routing and API used in WISENES simulations.

Parameter	Value	Description
TUTWSN routing parameters		
maxSyncClusters	5	Maximum number of synchronized clusters
minSyncClusters	1	Maximum number of synchronized clusters
preferredSyncClusters	3	Preferred number of synchronized clusters
maxMembers	14	Maximum number of member nodes allowed in a cluster
routePendingTimeout	6 *ac*	Time waited before a lost route is renewed
routeCostTimeout	6 *ac*	Wait time before a route is changed because of increased cost
routeSeqTimeout	3 *ac*	Wait time before a route is changed because of increased sequence
scanIntervalShort	7 s	Time between scans if no neighbors found
scanIntervalLongSub	32000 s	Time between periodic scans in subnodes
scanIntervalLongHead	1800 s	Time between periodic scans in headnodes
nborQueryInterval	5 *ac*	Period for querying neighbor information
nborSyncInterval	12 *ac*	Time between synchronization attempts for neighbors
roleCheckInterval	100 s	Maximum time between consecutive role checks
dataMcastTimeout	6 *ac*	Access cycles after which multicast is cancelled
dataMaxHoldTime	60 s	Maximum storage time for data with no routes
TUTWSN API parameters		
interestRefreshInterval	300 s	Time between data interests updates

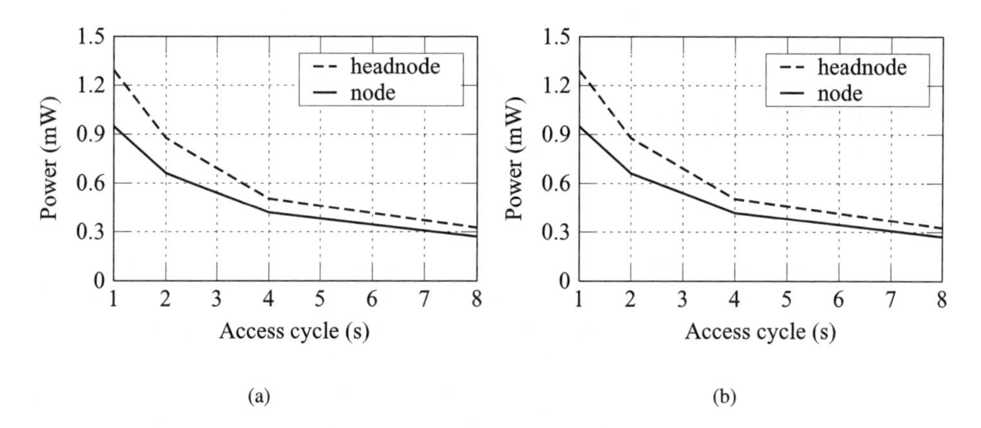

Figure 19.18 TUTWSN power consumption for (a) headnodes and nodes that select role dynamically, and (b) subnodes.

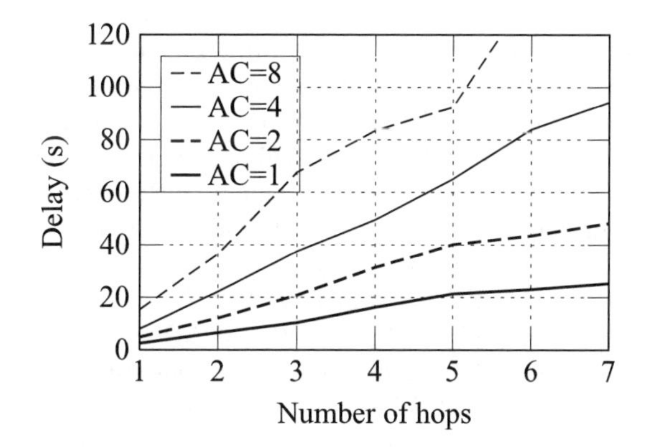

Figure 19.19 TUTWSN data-routing delay with different access cycle lengths as the function of number of hops from sink.

has a considerable effect on the delay since the temperature data is not considered delay critical. Therefore, reserved data slots are preferred with the cost of increased delay. The average delays per hop are 3.58 s, 6.82 s, 12.1 s, and 21.2 s with access cycle lengths of 1, 2, 4, and 8 seconds, respectively. As shown in the figure, the variances in average per hop delay are quite small.

According to the simulations, TUTWSN scales well to 250-node networks, and with moderate access cycle lengths, the network bandwidth is adequate for simulated temperature-sensing application. With an 8 second access cycle, the data-routing delays are considerable. Further, data losses in the network are quite common due to the 60-second limit set in *maxQueueTime*. If a node loses a route, a new route is searched for after a 48 second timeout (*routePendingTimeout*).

In the simulated scenario, the main challenge for the TUTWSN protocols was the frequency and time slot selection. Sometimes, nodes tended to select overlapping slots because of the insufficient neighbor information. However, the TUTWSN protocols are able to recover from the situation by re-triggering the frequency and time slot selection algorithm after detecting poor channel conditions.

ZigBee simulations

In 250-node ZigBee simulations, 29 of the nodes are ZigBee routers and 220 ZigBee end-devices. In ZigBee, the BO values corresponding to the simulated access cycle lengths are 6, 7, 8, and 9. In addition to BO, SO is set to values 1 and 2 in simulations. Since the data payload is considerably larger than in the case of TUTWSN, each ZigBee router aggregates the data it receives from its child end-devices before forwarding them. The key protocol parameters in ZigBee simulations are depicted in Table 19.4. The parameters that are not included in the table are set during runtime or are not used because of the selected configuration.

In the simulations, there were on average 6–8 child routers and end-devices associated per ZigBee router. The routers nearer to the ZigBee coordinator were considerably more loaded than those further away. Since the network formation of ZigBee in cluster-tree topology with hierarchical addressing is a bit deficient in completely random deployments, node placement and start-up in simulations is performed in a coordinated manner.

The power consumptions for ZigBee router and end-devices are shown in Figure 19.20 as the function of access cycle and superframe lengths. ZigBee-router power consumptions with BO varying from 6 to 9 are 2.14 mW, 1.10 mW, 0.584 mW, and 0.322 mW with SO=1, and 4.12 mW, 2.09 mW, 1.08 mW, and 0.578 mW with SO=2. For end-devices, the power consumptions are 0.164 mW, 0.104 mW, 0.080 mW, and 0.069 mW with SO=1, and 0.164 mW, 0.107 mW, 0.081 mW, and 0.072 mW with SO=2.

As depicted in the figure, the ZigBee-router power consumption is considerably larger than that of the ZigBee end-device. While a router needs to listen actively for an entire CAP, a device receives a beacon and sends data periodically. Thus, varying SO between

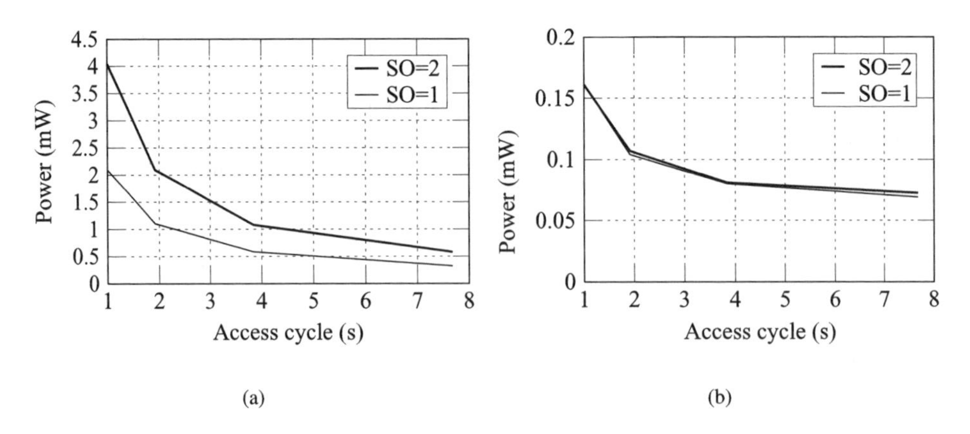

Figure 19.20 ZigBee power consumption for (a) routers, and (b) devices.

Table 19.4 Configuration parameters for TUTWSN routing and API used in WISENES simulations.

Parameter	Value	Description
IEEE 802.15.4 MAC parameters		
macAckWaitDuration	54	Number of symbols, for which an acknowledgment is waited
macAutoRequest	TRUE	Whether a device sends an automatic data request for pending data
macBattLifeExt	FALSE	Whether battery-life extension mode is used
macBeaconOrder	*(varied)*	Used BO
macSuperframeOrder	*(varied)*	Used SO
macGTSPermit	FALSE	Whether coordinator accepts GTS requests
macMaxCSMABackoffs	4	Number of backoff attempts before channel access failure
macMinBE	3	Minimum value for backoff exponent in CSMA algorithm
macPromiscuousMode	FALSE	Whether MAC accepts all packets from PHY
macRxOnWhenIdle	FALSE	Whether receiver is enabled when MAC layer is on idle state
macSecurityMode	0	Used security mode (0: none)
ZigBee NWK layer parameters		
nwkMaxChildren	12	Maximum number of child devices a router can have
nwkMaxDepth	4	Maximum depth (number of hops from coordinator) to the network
nwkMaxRouters	4	Maximum number of child routers a router can have
nwkUseTreeAddrAlloc	TRUE	Whether hierarchical addressing scheme is used
nwkUseTreeRouting	TRUE	Whether hierarchical routing scheme is used
nwkAddressIncrement	1	Amount of address increment in a router
nwkTransactionPersistenceTime	$16 * 2^{(BO-SO)}$	Superframe periods a transaction is stored

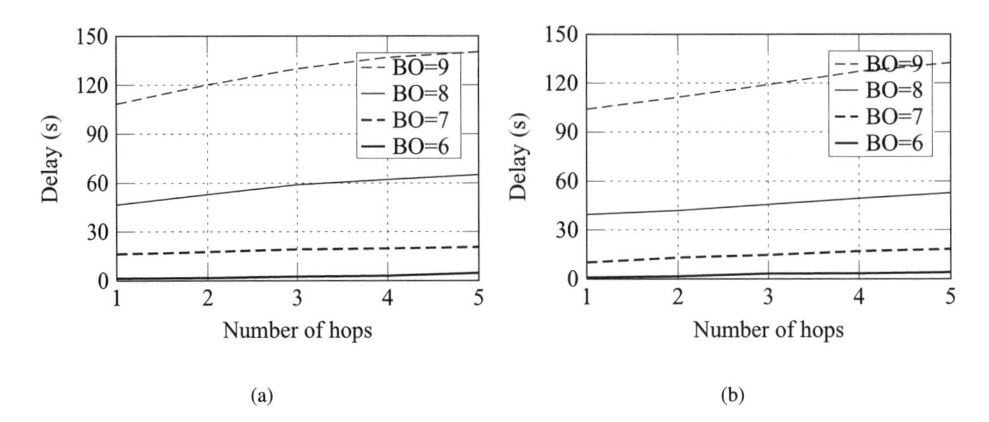

Figure 19.21 Data-routing delay in ZigBee with (a) SO 1, and (b) SO 2.

1 and 2 has a significant effect on ZigBee-router power consumption, while in the case of devices the power consumptions are quite identical. Compared to the analysis results given in Chapter 18, the simulated results are a bit smaller. The difference is caused mainly due to the lesser dynamics resulting from the coordinated network formation.

The delay of individual measurements depends on the data-routing delay but also on aggregation. For fair comparison with the TUTWSN results, only the routing delay is considered in the results. The results are shown in Figure 19.21 as the function of number of hops for different access cycle lengths.

In case of ZigBee, the number of hops is smaller due to the coordinated network formation, and used transmission power (−5 dBm). As shown, the SO value has, especially with longer access cycle lengths, a significant effect on the routing delay. A per hop delay increases with the larger number of total hops. This is caused by congestion and buffering in the nodes near to the ZigBee coordinator.

As the results show, with the given data interval, only BO=6 leads to a good performance. With other BO values, the routers nearest to the ZigBee coordinator are congested, which is visualized by the longer delay in the first hop. Further more, the packet-loss rate in these routers is quite high at larger BO values. The effect of SO is not considerable, since the end part of the CAP is less utilized. In general, routers succeed to send several frames during CAP whereas end-devices need more attempts and CCAs. Routers benefit from the longer frames that reserve the channel and enable additional sends during the rest of the CAP.

The simulations show that the performance of ZigBee in the simulated application scenario depends on selected BO and SO values. These parameters balance the power consumption and bandwidth of the network. Once the ZigBee network is formed, its operation is quite reliable assuming that there are no external interference sources.

In general, the hierarchical addressing and routing schemes are applicable only for very static deployments, just as they are intended. A random error causes a reconstruction of a complete sub-tree, which is typically very energy consuming. Further, since there are no timeouts for child association, the address space is exhausted if nodes periodically lose the connection to their parents due to communication errors or interference. If this is prevented by removing node associations, duplicate addresses may occur. If a coordinator removes the neighbor table entry of a device on failed association, the device may still have successfully received the association response.

The difficulties in network formation are caused by the time slot selection when a coordinator is started. A time slot has to be decided based on the information obtained during a passive scan. However, in dense networks the delay between the scan and the start of the router operation may be considerable, because parent association is needed before a router starts its own superframe; thus, a coordinator may not know the time slots of its neighbors. This causes problems especially if the number of available slots is limited (smaller BO and larger SO).

Comparison

The power consumption of TUTWSN subnodes and ZigBee end-devices are in the same order of magnitude. ZigBee end-devices are a bit more energy efficient, since they maintain only one connection to a parent, while a TUTWSN subnode prefers three active parent connections. Compared to TUTWSN headnodes, ZigBee routers consume significantly more power because of the CAP serving. In addition, TUTWSN benefits from the ability of the nodes to dynamically change their role.

As the results show, TUTWSN does not saturate as easily as ZigBee due to the guaranteed bandwidth provided by reservation slots. With longer access cycle lengths ZigBee routing delays increase as the ZigBee routers near the coordinator are congested. Further, each TUTWSN node maintains several alternative routes to a sink node. Compared to ZigBee, this increases reliability and robustness of the network in error situations.

Part V

DEPLOYMENT

20

TUTWSN Deployments

In the case of WSNs, the term *deployment* is typically understood as the placing of nodes (either manually or randomly) to their target locations. Yet, the operation of a WSN also requires other aspects that need to be initialized and integrated with the nodes. These include gateway nodes, backbone network infrastructure, and client SW components that store, relay, or process data gathered from the network. Thus, in this chapter, the term deployment is considered to include all these aspects that are required to make the WSN functional from a human user point of view.

The purpose of a WSN deployment is defined by its target application. As stated, the applications for WSNs are diverse and span to different domains. Thus, a single deployment may not be suitable for other types of applications due to the design choices and configurations of protocols and nodes. If the application of a deployed WSN changes, the network may need to be reconfigured or even re-deployed.

In order to obtain the required knowledge regarding the suitability of a WSN and the possible configurations for different applications, the WSNs must be tested in real deployments.

The need for realistic deployments during the development phase is also evident. Only a realistic deployment of the network and its support infrastructure reveal the challenges for the system and the interferences and errors caused by the environment. Thus, the complexity of a system cannot be accounted for in simulations or prototype testing in a laboratory environment unless it has been tested in the final destination environment.

This part of the book presents several application scenarios for WSNs that possess differing requirements and characteristics. An example solution is given for each application. The solutions are implemented with TUTWSN by configuring the protocols and selecting or assembling nodes. The presented solutions may not be the best ones and are definitely not the only ones, but their main purpose is to show the tools, methods, and building blocks required for successful WSN deployment. Further, the examples show the feasibility of TUTWSN for different kinds of applications and usage.

Adaptive MAC and routing protocols together with the rich set of configuration parameters enables the application-specific tailoring of TUTWSN protocols. Further, the wide family of node platforms based on COTS components makes it possible to either utilize

Ultra-Low Energy Wireless Sensor Networks in Practice: Theory, Realization and Deployment

existing node platforms or to design an application-specific platform. The deployment architecture, including gateways, servers, and UIs, is flexible for different kinds of applications. Both Web-based WSN data presentation and Java UI can be easily adopted to visualize new types of application data. These features make the utilization of TUTWSN possible in different kinds of use cases.

20.1 TUTWSN Deployment Architecture

A typical TUTWSN deployment consists of the network itself and a support infrastructure. Since TUTWSN does not allow direct Internet access to the data of individual sensors, the support infrastructure consists of components that make the data access, storing, manipulation, and visualization feasible.

The logical topology of a TUTWSN deployment is presented in Figure 20.1. The main entity in the topology is a *WSN Server* that is responsible for managing connections and relaying data between other entities. A *WSN* is connected to the WSN Server through a *Gateway*(GW) that abstracts the heterogeneities of WSN architectures. The number of Gateways connecting WSNs to the WSN Server is not limited. It should also be noted that many gateways can exist in one WSN.

The main purpose of a centralized WSN Server architecture is to allow scalable implementation of data storing and visualization entities. Through the WSN Server, the gathered data can be relayed either to a single *database* or several databases that may store data from one or more WSNs. Similarly, several *UIs* can simultaneously visualize the data of a single WSN. On the other hand, a single UI can visualize the data of several networks.

In addition to the scalability, the WSN Server abstracts the locations of the other entities. Gateways, the WSN Server, databases, and UIs can reside in different PCs, or even in different continents considering that they are connected through the Internet.

The entities in the logical topology are mapped to physical components in an example deployment scenario depicted in Figure 20.2. In the illustrated case, a deployed TUTWSN is accessed through an external Gateway device. The Gateway and UIs communicate with the WSN Server through the Internet.

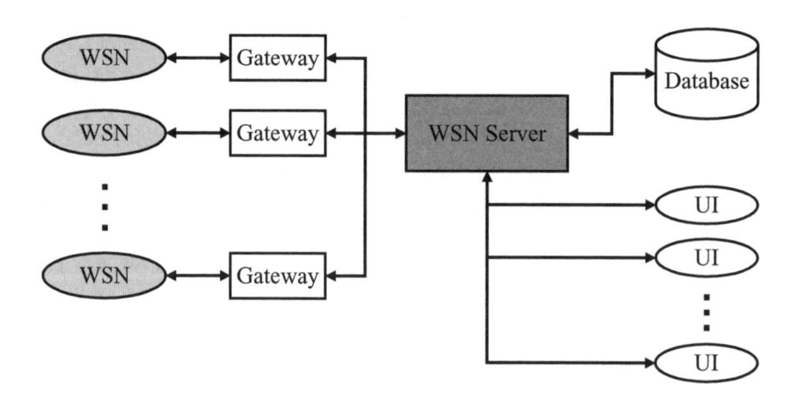

Figure 20.1 The overview of logical components and their relations in a TUTWSN deployment.

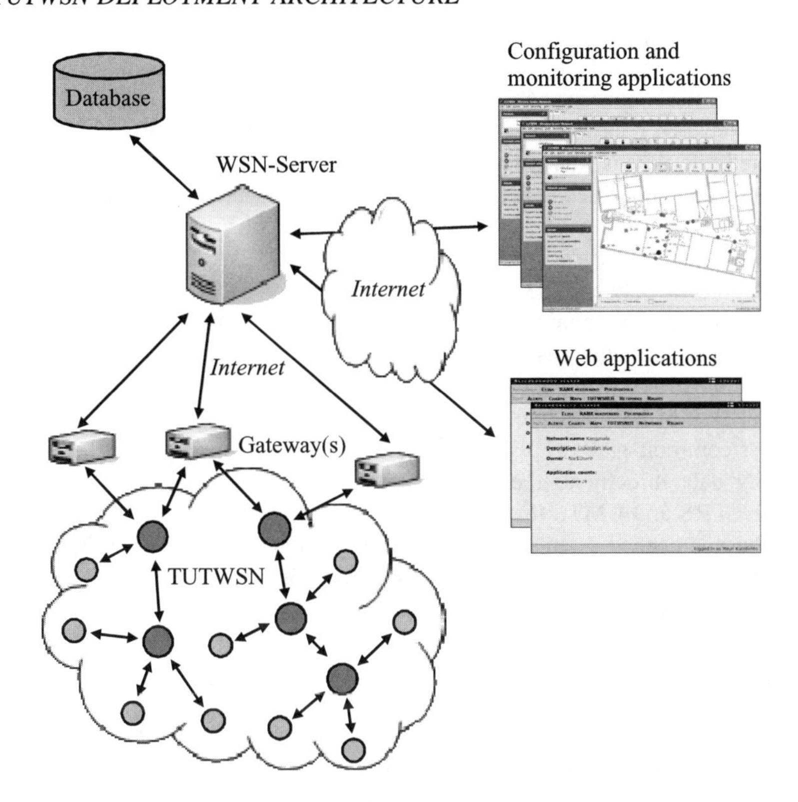

Figure 20.2 An example component architecture with one WSNs Server, three gateways, a database, and five different kinds of UI applications for data visualization.

20.1.1 WSN Server

The WSN Server manages the communication between other components. The communication is implemented with Java Message Service (JMS) (Sun Microsystems 2002). The basic JMS architecture consists of a *JMS provider* implementing a Message Oriented Middleware (MOM) interface and *JMS clients* communicating with *JMS messages*. A JMS client can be either a *JMS producer* sending messages or a *JMS consumer* that receives the messages. A JMS message is a Java object with internal implementation and data content.

In the deployment architecture, the WSN Server acts as a JMS provider utilizing a publish/subscribe model. In the model, messages can be published to different *topics*. Each client that has subscribed to a topic receives all messages published on that topic. The WSN Server maintains a separate topic for each physical WSN. Thus, a gateway publishes WSN data to the network-specific topic, and databases and UIs listening for that topic receive the message. The same architecture is applied also for sending data to the network.

Since JMS messages are sent through the Internet, security issues need to be taken into account. Therefore, WSN Server requires authentication prior to a client publishing the messages or subscribing to a specific topic. A gateway authenticates using the credentials of the network owner, while UIs use the credentials of the user running the application. The communication privacy is implemented with Secure Sockets Layer (SSL).

20.1.2 WSN and Gateway

A deployed TUTWSN consists of an arbitrary number of nodes that are tailored application specifically. Network data are accessed and operation configured through the TUTWSN API. As stated, the API is implemented in gateway nodes. Depending on its computation and communication capacities, a gateway node can be connected to the WSN Server either directly or through a more powerful gateway device.

The purpose of the gateway is to parse and interpret TUTWSN API messages and relay the message content between the WSN and WSN Server. Depending on the capabilities of a gateway, it may communicate with the WSN Server using JMS messages or native TCP/IP packets. In the latter case, a JMS message is created either at the WSN Server or at an intermediate proxy. High-end gateways may incorporate a local database for avoiding data losses in case of backbone network failures.

The most common gateway used in TUTWSN deployments is an Ethernet GW that bridges WSN data directly to the Ethernet using native packets. Similar gateways are available for GPRS and UMTS that allow a wireless connection to the WSN server. When a local database is needed as an intermediate storage, a gateway node can be connected to a PC-class device through a serial port. The PC runs an interface component that composes JMS messages and manages local database access.

20.1.3 Database

All data gathered from TUTWSNs are stored to a database for post-processing and for viewing long-term statistics and reports. The stored data include node-specific information, such as profile information, raw sensory data, and preprocessed application results.

The database access is implemented by a dedicated component referred to as *DBhandler*. Each database has an own DBhandler that subscribes to a set of network topics. Thus, the DBhandler receives all JMS messages concerning subscribed networks. The data are parsed from the messages and inserted in to the database through a Java Database Connectivity (JDBC) API (Sun Microsystems 2006b). DBhandler is also responsible for handling and responding to client database queries. Thus, all database accesses are performed through DBhandler.

20.1.4 User Interfaces

TUTWSN has several different types of UIs. In general, the two main alternatives are a configuration and monitoring UI implemented in Java and a Web-based interface. Other UIs include tailored interfaces for different types of mobile devices and pure textual data outputs for printing unprocessed data.

Configuration and monitoring UIs

The configuration and monitoring UI has two operation modes. *TUTWSN Control Panel* supports several network configuration features such as service selection, runtime protocol parameterizing, and network reprogramming. Instead, *TUTWSN Application Panel* is purely for the visualization of processed application results and does not support any management features. The former is intended only for the network administrators.

The configuration and monitoring UI is implemented in Java using Swing packages (Sun Microsystems 2006a). It has several views for WSN data visualization as depicted later in this and followings chapter. These include a map-based overview, a detailed textual listing, and different kinds of application status views, long-term reports and charts. The underlying implementation architecture is designed to be easily extensible for new applications and features.

Web interface

While the configuration and monitoring UI allows a realtime presentation of the application results, the Web interface is mainly for static queries. Its main objective is to provide a simple overview of application results in situations where the TUTWSN Application Panel is not available. The interface is implemented using Hypertext Pre-Processor (PHP) and static Hypertext Markup Language (HTML) pages in order to enable the operation on a wide variety of client devices.

The web interface implements a map view showing the last measurement results, and supports the presentation of long-term measurement reports as graphs and textual listings. Further, it includes methods for generating Short Message Service (SMS) and email alerts from WSN data. When a monitored value crosses a limit, an indication is sent to a user that requested the alert.

The Web interface also implements a management interface for user and access control. Through the interface, network administrators are able to create new users and grant them with monitoring or configuration privileges for the networks.

20.2 Network Self-diagnostics

The self-diagnostics facility is not a "real" WSN application but a network development tool. However, its design and implementation possess similar characteristics to those of other applications. The main purpose of the self-diagnostics application is to instruct the network protocols and nodes to gather performance information and to visualize this data in UI so that it can be benefit in WSN development. The self-diagnostics tool helps a customer in network maintenance by detecting depleted nodes and erroneous behavior. Also, it is used in the network design and deployment phases, making it the most important tool for a developer. The runtime monitoring of the deployed WSN provides insight to the designer, which aids in the detection of possible flaws and network bottlenecks.

20.2.1 Problem Statement

Monitoring the performance of a deployed WSN is problematic because of the numerous possible sources of errors and bottlenecks in the network. Therefore, making correct decisions based solely on the application traffic profile received at the sink node is impossible. As node debugging and instrumenting through cables is inconvenient in large-scale deployments, network self-diagnostics is needed.

Unfortunately, the limited resources in a sensor node set restrictions on how the diagnostics can be collected. The transferred diagnostic messages must be small and sent infrequently, as they would otherwise interfere with other network traffic and cause unnecessarily large energy consumption. Also, the limited memory capacity requires choosing the stored information selectively and calculating aggregate values.

20.2.2 Implementation

The self-diagnostic application is naturally implemented in nodes. It does not need any extra components and the monitoring is purely done with special algorithms. A microcontroller is used to observe the different parameters of nodes, e.g. route information, access cycle timing, and battery voltage. This information is sent to the user interface via a gateway. The user interface visualizes data to an easily comprehendable form.

Implementation on nodes

Diagnostics information is collected on different protocol layers as shown in Figure 20.3. Each layer maintains information about its state and operation. The self-diagnostics control module collects the data from different layers and combines it to get extensive information about the state of the node. In addition, the control module maintains statistics from individually measured parameters, which is used quantify resources status, e.g. average link reliability. Protocol layers use the diagnostics data internally to make operational decisions. For example, link reliability information is crucial when selecting the highest reliability route. The self-diagnostics application composes reports from the self-diagnostics data and generates diagnostic PDUs for a sink.

The self-diagnostics information is not only collected by counters on each protocol layer, but also by gathering operation and function level information. The operation level information records occurred events consisting of network scans, synchronization losses, route and role changes, and boots. Each event is accompanied by a reason and neighbor reference. The reason is a simple integer value with a predetermined meaning that describes the cause for an event. The neighbor reference identifies the neighbor address relating to the event, e.g. when synchronization is lost.

Function level information is intended mainly for debugging the sensor code. It is essential in complex sensor network development because the debugging of an embedded software is often hard. As the programming errors may surface only on specific situations, getting information during normal network operation is important. A programmer uses the

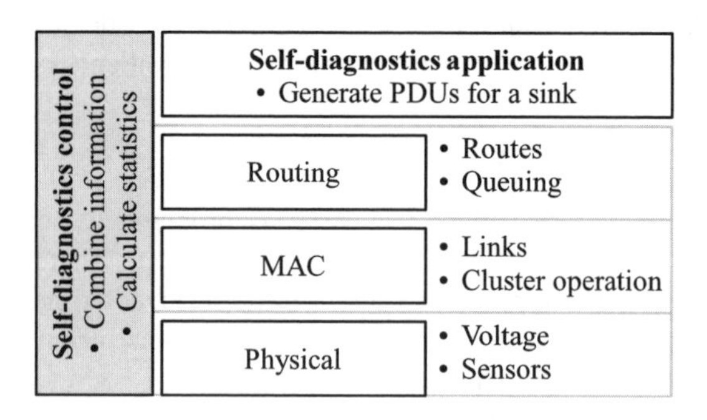

Figure 20.3 The implementation of self-diagnostics on different protocol layers and the collected information.

function level debugging by putting assertions to the source code. When an assertion is triggered, diagnostics record the call stack and the place in the code. This information is then transmitted to the sink node.

As all the collected diagnostics information may not be needed at the same time, the diagnostics data is categorized and only the requested categories are the ones that are transmitted. A category is defined in an interest that also determines the interval used to send the diagnostics' packets. Different intervals may be used by defining several interests.

Self-diagnostics information comprises five packet types that are summarized in Table 20.1. In TUTWSN, each of these packets are fitted into short, 18 B payloads. The payload size is limited by the maximum transfer unit of the transceiver hardware. The self-diagnostics'packets and their usage are described as follows:

- **Node information packet**: The node information packet is used to get overall impression of the node performance. It contains the operational-level events, the amount of networks scans, battery voltage, and queuing information. This data allows for

Table 20.1 Contents of the diagnostics packets.

Information	Description
Node information packet	
Voltage	Latest voltage measurement
Network scans	Network scan counter
Queue delay	Average delay per traffic class
Queue usage	Average and maximum queue usage
Events	A list of occurred events
Cluster traffic packet	
Access cycles	The number of served access cycles
Channel	The channel in which the cluster operates
ALOHA RXs	Successfully received frames
Reserved RXs	Successfully received frames and receive failures
Data RXs	Received unique and duplicated data frames
Broadcast TXs	Broadcast frame transmission attempts
Unicast TXs	Unicast frame transmission attempts and failures
Max slot usage	Maximum slot usage, separately for ALOHA and reserved
Neighbor mapping packet	
Address map	Maps network addresses (24-bit) to short identifiers (8-bit)
Neighbor information packet	
Neighbor type	Indicates member, neighbor cluster, and/or neighbor's neighbor
Channel	The channel in which the neighbor operates
Access cycle	Time difference between own and neighbor's access cycle
RSSI	Signal strength required to communicate with the neighbor
Neighbor traffic packet	
ALOHA TXs	Frame transmission attempts and failures on ALOHA
Reserved TXs	Frame transmission attempts and failures on reserved

detecting the symptoms of misbehavior and thus, the need for declaring interests that start the collection of other diagnostics categories.

Voltage information gives the rate of depletion and is used to determine when to replace a node. Too high a depletion rate might also indicate performance problems and a requirement for deeper examination of the node.

Queue delay and usage information is used to detect forwarding problems. High queue usage increases delays and is usually an indication of a performance bottleneck. However, if the delays are high while queue usage is low, the next-hop link may not be reliable, which causes several retransmission attempts.

- **Cluster traffic packet**: The cluster traffic packet describes the activity of maintained access cycles. As the information contained in this packet is cluster-head specific, the packet is not transmitted if the node is acting as a subnode.

  The reliability of the traffic is calculated based on the packet counters. As the cluster head knows exactly in which reserved slot a member node is expected to transmit, slot reliability can be calculated by comparing the successfully received frame counter against the expected reception counter. However, this method is not applicable for ALOHA slots because the cluster head cannot know when the member is trying to transmit. Therefore, only the successfully received frame counter is maintained for the ALOHA.

  Interference and hidden-node problems may cause asymmetric unreliability. That is, the cluster head may be able to receive its member transmissions but the member cannot receive acknowledgments. As this happens, the member node unnecessarily retransmits its frame, which the cluster head considers as duplicate. For detecting this situation, the received frames are further divided into *unique* and *duplicate* counters.

  Slot-usage counters are used to analyze bottlenecks. The cluster traffic packet contains maximum slot usage information, whereas the average slot usage is calculated based on the number of elapsed access cycles and the total number of transmissions. A high slot usage indicates a bottleneck and thus the requirement to add more cluster heads to balance the load.

- **Neighbor mapping packet**: Node addresses are required in both neighbor information and neighbor traffic packets. As the used address is relatively long (24-bit), neighbor mapping packet assigns a shorter (8-bit) identifier for each neighbor. Thus, the mapping allows fitting more neighbors to the neighbor information and neighbor traffic packets. Furthermore, as the mapping does not to change often, it can be send more infrequently than other packets.

- **Neighbor information packet**: The neighbor information packet contains a list of source node's neighbors. A neighbor information contains address, the type of neighbor, RSSI, channel, and access cycle timing. Based on this information, a sink node can construct a view of the network topology.

The neighbors are classified as members, cluster heads, or neighbors' neighbors. The classification allows determining the direction of link connectivity, and thus route directions between nodes. A member is a neighbor that is joined to the cluster maintained by the source node, whereas the connectivity is the other way around with a cluster head neighbor. A node does not communicate with a neighbor's neighbor but receives the channel and access cycle information indirectly from its other neighbors.

- **Neighbor traffic packet**: The neighbor traffic packet contains counters for ALOHA and reserved transmission attempts and failures, thus allowing determining the reliability of a link. The reasons for unreliability can be concluded together with the neighbor information, as bad RSSI is a probable reason for unreliability. However, if link has a good RSSI, the unreliability might be caused by interference.

The information from different self-diagnostics categories can be combined to get more accurate reason for possible problems. Figure 20.4 shows how the reason for link unreliability can be determined. A high ALOHA usage increases collision probability, thus decreasing reliability. If ALOHA is congested, an additional node should be added near the cluster head to allow load balancing. Low RSSI causes unreliability, as a node may be just barely within the communication range. In this case, a forwarding node should be added in between a cluster head and its member to allow the member node to communicate with a shorter and more reliable link. If the unreliability cannot be explained by ALOHA congestion or low RSSI, interference is assumed. The reasons for interference are other wireless networks (e.g. WLAN) operating on the same frequency band. Also, network configuration errors may cause requiring more channels than are available, thus forcing channels to overlap.

UI implementation

As the collected numeric data is not intuitive, the diagnostics UI is used to visualize the diagnostics information in an easily comprehended form to make network analyzation easier. The statistics of a specific counter can be explicitly requested, but the UI also incorporates automatic reasoning and suggestions similar to the deduction presented in

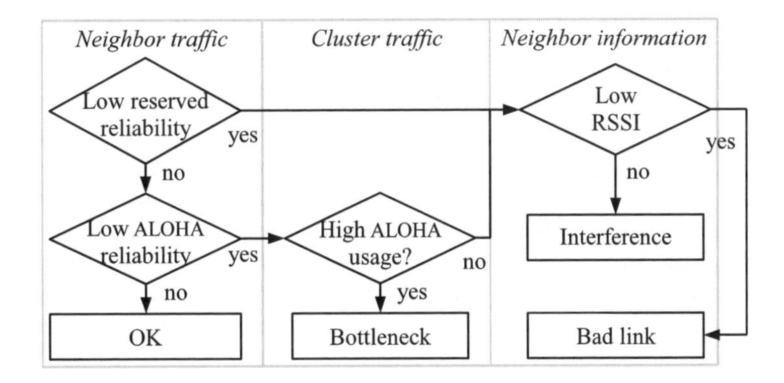

Figure 20.4 Using the self-diagnostics information to determine the reason for link unreliability.

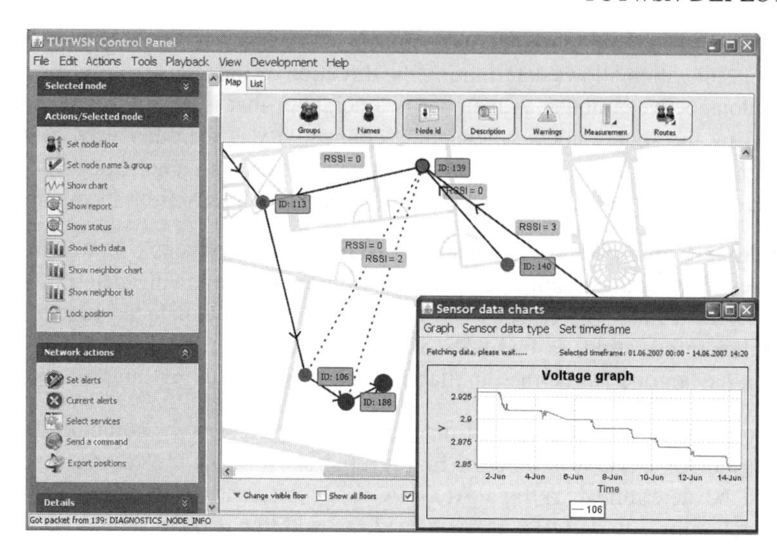

Figure 20.5 Visualization of diagnostics information in map and graph views.

Figure 20.4. For example, when a bad link is detected, the UI displays a warning message identifying the node and the link in question.

The diagnostics UI comprises map and graph views. The map view shows topology information corresponding to the real-world deployment, thus making the replacement of an erroneous sensor node easier. Node placements are set manually, but can be determined automatically if nodes have GPS sensors. An overview of the diagnostics UI is presented in Figure 20.5. The map view shows a latest route through which the last packet from each node has traveled. Alternatively, the UI can show the average bandwidth, usage, and reliability of each link based on the neighbor traffic information. The graph view shown in the figure visualizes the development of voltage levels in a selected node.

The graph view is used to show straight forward diagnostic counters, but can also combine the data for more expressive charts. Figure 20.6 shows the dialog that is used to display new charts. The main window is showing the buffering delay of low-priority traffic class packets in two selected nodes.

A momentary network connectivity can be examined from the map view that has the ability to show the signal strength of each link. For example, the links and neighbors of node 139 are visualized in Figure 20.5. While the momentary connectivity allows locating bad links, a longer connectivity history is used for detecting interference and for visualizing bad links. The link changes of a selected node are examined with the graph view as shown in Figure 20.7. The size of the circle indicates the RSSI of a link. While the examined node (106) uses momentarily up to four links, it is usually connected to three neighbors. The diagnostics chart indicates that two of these long-term links have varying link quality, which is attributed to the environmental changes.

Finding potential problems by simply going through the collected counters could be tedious. For allowing a rapid examination of the network functionality, the transmitted packets that are received by a sink are visualized in Figure 20.8. As each node is configured to send

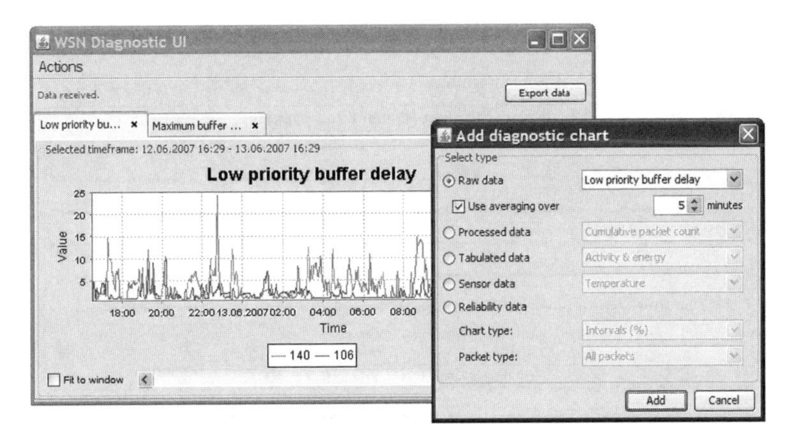

Figure 20.6 Graph view used to examine collected diagnostic counters.

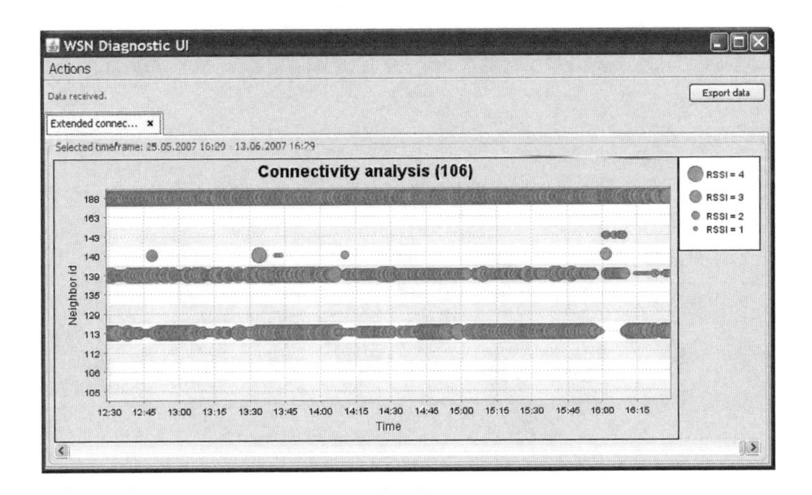

Figure 20.7 Visualization of connectivity and the RSSI of active links.

data periodically, the packet reception interval should equal that of the configured period. Thus, any hole in the transmitted packet graph indicates that packets have not been received from a node. For example, in Figure 20.8, the node with identifier 4022 had reliability problems between 14:00 and 14:15. Because the time when the node did not function as expected is seen precisely, the problems can be tracked easier based on the other diagnostics' data.

The distribution of packet intervals allows for determining which nodes have not functioned as expected. Figure 20.9 shows the fractiles of intervals between the arrival times of packets received by a sink. The average interval (50% fractile) should be equal to that of the configured interest interval. Interval variation is caused by retransmissions, packet losses, and buffering delays, and therefore, indicate problems on a routing path. Thus, high variation or too high interval values suggest that node-specific diagnostics data should be examined to determine the cause.

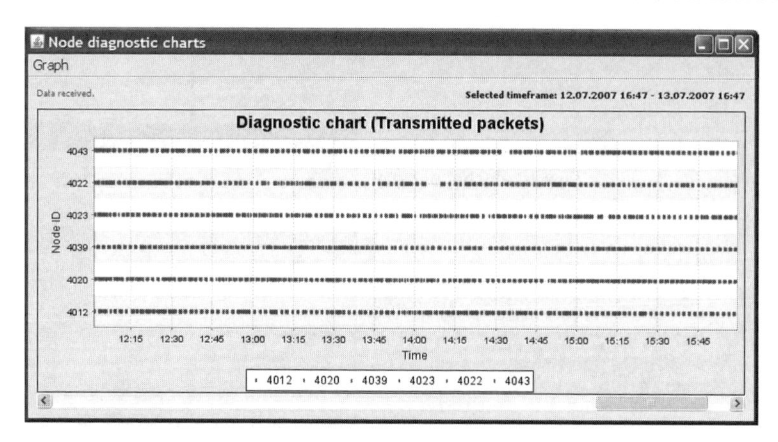

Figure 20.8 Visualization of the reception times of transmitted packets. Long pauses indicate problems.

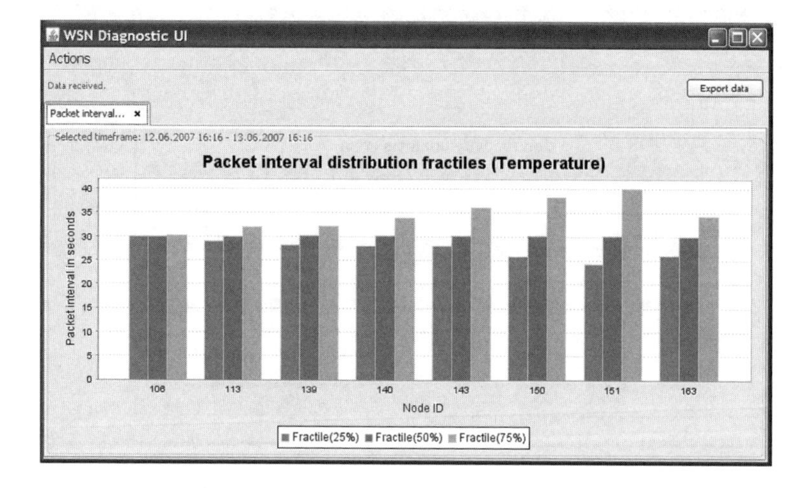

Figure 20.9 Distribution of received-packet intervals. A high interval variation indicates potential problems on a routing path.

20.3 Security Experiments

Similar to self-diagnostics, the security experiment use case is not a pure WSN application. Instead, it presents a centralized authentication and key distribution design and implementation, which can be utilized in WSN applications requiring such security functionalities.

A large part of WSN security research is concentrated on probabilistic key pre-distribution techniques (Section 12.3) as well as on the efficient implementation of cryptographic algorithms. In order to extend the evaluation of KDC-based solutions in practice, a centralized WSN authentication service has been developed using the TUTWSN technology

(Hämäläinen et al. 2006). The service can be used for both network access control and key distribution using only symmetric-key cryptography.

20.3.1 Experimental KDC-based Key Distribution and Authentication Scheme

The developed experimental key distribution and authentication scheme uses a similar authentication architecture and the same type of a protocol as Kerberos (Kohl and Neuman 1993). In contrast to Kerberos, the freshness of messages is ensured with *nonces* instead of timing information. Hence, nodes do not have to maintain clock synchronization but they require a random number generator.

The scheme is derived from the KDC proposal of Perrig et al. (2002), which does not report an implementation. Instead of direct communication with the KDC in Perrig et al. (2002), in the version of this work all messages are communicated through a headnode, supporting the clustered TUTWSN topology. Also, authentication with the newly distributed key is included. In the presented scheme, the key agreement and authentication procedure can be performed between a headnode and a subnode or between two headnodes.

The scheme operates as follows. During the WSN deployment, each node is allocated a symmetric key K_X (consisting of an encryption key and a MIC key), which they share with the KDC (S). For authenticating and establishing a symmetric key K_{AB} between two nodes, the node A and the headnode B, the following protocol is executed:

1. $A \rightarrow B : N_A \mid A$
2. $B \rightarrow S : N_A \mid N_B \mid A \mid B \mid \mathsf{MIC}(K_B, N_A|N_B|A|B|S)$
3. $S \rightarrow B : A \mid \mathsf{E}(K_B, K_{AB}) \mid T \mid \mathsf{MIC}(K_B, A|\mathsf{E}(K_B, K_{AB})|T|N_B|S|B)$
4. $B \rightarrow A : N'_B \mid S \mid T \mid \mathsf{MIC}(K_{AB}, N_A|N'_B|S|T|B|A)$
5. $A \rightarrow B : \mathsf{MIC}(K_{AB}, N'_B|A|B)$

where $T = \mathsf{E}(K_A, K_{AB}) \mid \mathsf{MIC}(K_A, \mathsf{E}(K_A, K_{AB})|S|N_A|A|B)$ and $|$ stands for concatenation. Above N_X is a nonce created by X, $\mathsf{MIC}(K, Y)$ is a MIC computed with the key K over the data Y, and $\mathsf{E}(K, Y)$ is the encryption of Y with the key K. K_{AB} is chosen by S, which is trusted by the nodes.

Message 1 of the protocol serves as a challenge from A to B and S. With Message 2, B requests K_{AB} from S, including N_B into the message for challenging S to prove its authenticity in Message 3. With Message 4, B proves A its knowledge of K_{AB} as it has been able to decrypt the key from Message 3 with K_B and compute the MIC over N_A. Finally, A proves B its knowledge of K_{AB}, which it has obtained from T, by transmitting the MIC computed over N'_B.

20.3.2 Implementation Experiments

The TUTWSN prototype node shown in Figure 4.7 was used in the experimental implementation of the key distribution scheme. For the security experiments, the payload size of a network packet was 22 B in the chosen TUTWSN configuration. As a result, the maximum payload data rate for a link was 1.4 kbit/s in both directions.

In the implementation, the AES block cipher (NIS 2001) with 128-bit keys was used for encryption as well as for MIC computations. For efficiency, the encryption of K_{AB} was chosen to be carried out in the Electronic Codebook (ECB) mode. MICs are computed in the Cipher-based Message Authentication Code (CMAC) mode (Song et al. 2006). The sizes of the protocol message fields were chosen so that each message fits at maximum into two packets of the used TUTWSN configuration.

The implementation was carried out as software (C language) in the TUTWSN nodes. For AES the assembly implementation of (Mic 2005) was utilized. The experimental implementation supports authentication of nodes that are at maximum two hops away from the KDC since multihop routing for the protocol messages was not implemented. For example, for covering the network of Figure 21.4, an increase in transmission power is required.

With the used TUTWSN configuration the protocol run takes about 2 s, in which the channel access and communication latencies dominate. For instance, the processing time for the largest message, Message 3, is below 100 ms in the TUTWSN prototype node. In networks of large number of hops and key requests, the increase of latencies can be reduced by allocating more processing power to the KDC and accessing it through uniformly deployed gateway nodes. Furthermore, the distance of nodes from the KDC (as the number of hops) can be shortened by increasing the transmission power. Also, the KDC approach can be combined with a certain amount of pre-distributed pairwise keys.

The experiments suggest that KDC-based key distribution mechanisms can be feasible solutions in certain WSN configurations and applications. This is the case especially when a centralized, possibly physically protected location for WSN access control is desired.

21

Sensing Applications

In a sensing application, WSN nodes simply collect sensor data and send it to a sink node. This chapter presents three sensing applications: linear-position metering, indoor-temperature sensing, and environmental monitoring. The linear-position metering and the indoor-temperature sensing applications are implemented with the first versions of TUTWSN protocols and platforms. The linear-position metering is an industrial monitoring application that replaces traditional cabling with a WSN. Due to the harsh industrial environment, the nodes are designed to withstand rough operating conditions. The indoor-temperature sensing is targeted at simpler operating conditions for long-term temperature measuring in a home environment. The environmental monitoring use case contains a large number of nodes that are deployed for monitoring outdoor environmental conditions. The main objectives of the network are long network lifetime and autonomous operation without maintenance.

21.1 Linear-position Metering*

The deployment described in this section is one of the first deployments of TUTWSN carried out in 2004. It demonstrates the applicability of WSN technology in an industrial environment. In this deployment, the WSN nodes are equipped with linear-position sensors, which are widely used in industry for measuring the displacements of moving parts. The utilization of WSN eliminates the need for extensive cabling, thus reducing both material and installation costs (Kohvakka et al. 2005a).

21.1.1 Problem Statement

The linear-position sensing application design contains a complete linear-position metering system from a sensor to data analysis on a workstation. The design utilizes TUTWSN as a base technology, including its protocol stack and network architecture.

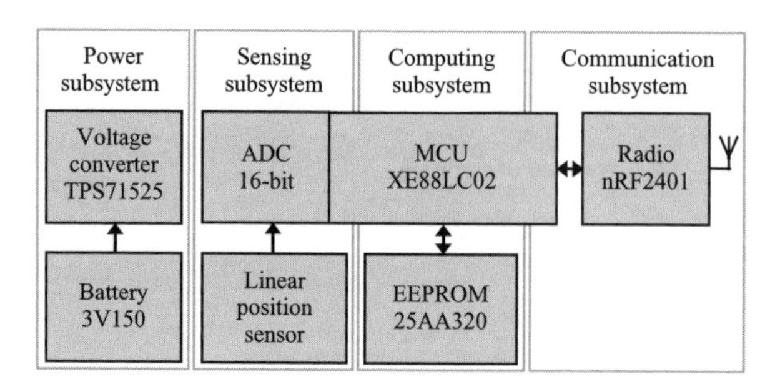

Figure 21.1 Wireless linear-position sensor node prototype architecture. © 2005 IEEE. Reprinted, with permission, from *Proceedings of the 8th Euromicro Conference on Digital system Design.*

The linear-position metering system measures low-frequency movements, where data is gathered from sensors at 1 Hz sample rate and 16-bit resolution. All samples are time stamped. The expected number of sensors is 50, but this should not be limited. The network range is one hundred meters, where sensors are at most ten meters apart from each other. WSN data is multihop routed in the network while the targeted latency per hop is 1 s. The required network throughput is 512 bits/s. There are no strict delay requirements for the data collection, but the reliability of the WSN is critical. Therefore, the node must be able to buffer and retransmit samples (Kohvakka et al. 2005a).

A wireless linear-position sensor is assembled by two parts, as a WSN node is attached on the physical connector of a linear position sensor. Thus, the WSN nodes should not significantly increase the size or weight of the existing physical sensor. The sensors will operate in harsh physical environment containing fluids, dust, vibration, and electromagnetic interference. Hence, WSN nodes should be sealed in waterproof metal containers. The sensors should be powered by rechargeable batteries having at least a one month lifetime.

21.1.2 Implementation

The hardware architecture for the linear-position sensing WSN node is presented in Figure 21.1 (Kohvakka et al. 2005a). The communication subsystem consists of a Nordic Semiconductor nRF2401A (Nordic Semiconductor ASA 2006) radio transceiver operating at the 2.4 GHz ISM band. The 2.4 GHz frequency band is high enough for avoiding interference from machinery (Rappaport 1989) and for allowing physically small and efficient antennas. The antenna is implemented with a quarter-wave external monopole antenna due to adequate efficiency and omnidirectional radiation. Measured radio range is about 30 m.

The computing subsystem consists of a Semtech (formerly Xemics) XE88LC02 (Semtech Corp. 2006) MCU, which integrates a CoolRisc 816 processor core with a 16-bit ADC, 22 kB program memory and 1 kB data memory. The maximum instruction execution speed is 2 MIPS. The MCU is connected to an 8 kB Microchip 25AA320 EEPROM for nonvolatile data memory. The MCU is selected due to very high 1344 MIPS/W energy efficiency, and accurate 16-bit ADC. Compared to 8-bit PIC MCUs, the energy efficiency of Semtech is exceptionally high due to efficient 16-bit instruction. Moreover, low 5 μW sleep-mode power consumption and versatile timers are the advances of the MCU. The small memory resources were adequate for the development version of TUTWSN in the year 2004.

The sensing subsystem gathers requested data from node environment. The subsystem consists of a linear-position sensor and the ADC integrated with MCU. The sensor utilizes a 5 kΩ(Kilo ohms) linear potentiometer. During a measurement, a constant voltage is applied over the potentiometer and the deviation is determined by the wiper voltage. ADC has pre-amplification and offset compensation stages, which increase sampling accuracy. The maximum sample rate using 16-bit resolution is 2 kHz.

The power subsystem utilizes three serially connected 150 mAh(milliamperes per hour) Varta V150H NiMH batteries achieving a 3.6 V nominal voltage. The NiMH chemistry is selected due to good energy density, cheap price, and better robustness against high temperatures and currents than lithium chemistries. The battery voltage is converted down to 2.5 V supply voltage by a Texas Instruments TPS71525 linear voltage regulator. The converter is selected due to low electromagnetic interferences and quiescent current. These are important, since proximity to other components is short and the average current consumption is low. Moreover, the converter is small and cheap. For the required one month WSN node lifetime, the power subsystem can deliver 208 μA average current drain equaling to 624 μW at 3.0 V supply voltage.

The WSN node prototype is implemented in a four-layer FR-4 PCB, presented in Figure 21.2 (Kohvakka et al. 2005a). The top side of the prototype contains radio, battery, programming connector, antenna, and two diagnostics LEDs. The bottom side contains MCU, voltage regulator, and a connector for the linear-position sensor. The prototype is enclosed in an aluminum casing presented in Figure 21.3 (Kohvakka et al. 2005a). The size of the prototype is 120 × 25 × 25 mm. The connector for the linear position sensor is also used for battery recharging.

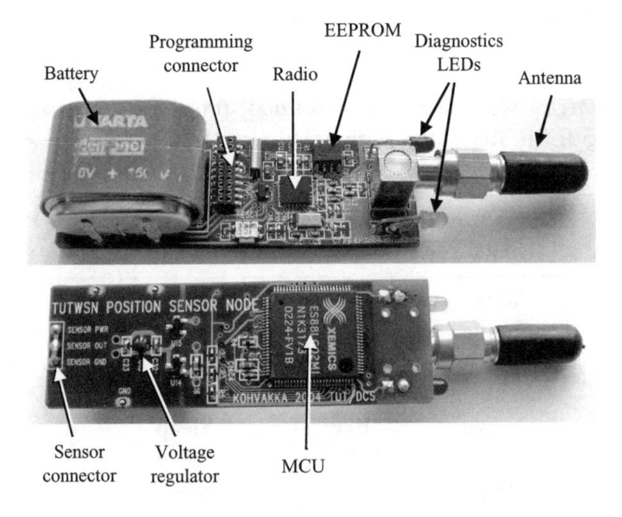

Figure 21.2 WSN node prototype. © 2005 IEEE. Reprinted, with permission, from *Proceedings of the 8th Euromicro Conference on Digital System Design.*

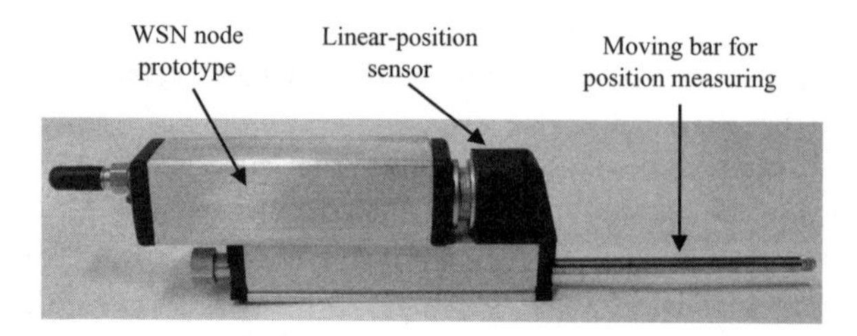

Figure 21.3 Complete wireless linear-position sensor prototype consisting of a WSN node attached to a position sensor. © 2005 IEEE. Reprinted, with permission, from *Proceedings of the 8th Euromicro Conference on Digital System Design.*

21.1.3 Results

The static power consumptions of the radio, MCU, ADC, and the linear-position sensor are measured at different operating modes. The MCU power consumption is measured in an active mode, when the core is running on 1.8 MIPS speed, and in a sleep mode, when only a wake-up timer is active. The ADC and sensor power consumptions are measured in active and shut down (off) modes. The radio power consumption is measured in various operation modes, including data loading between MCU and the radio buffer (Kohvakka et al. 2005a).

Measurements were done at 3 V supply voltage including power dissipation in the power subsystem. The analysis results are presented in Table 21.1. The minimum power consumption is 21 μW, which is achieved when all components are inactive. The maximum required power is 57.60 mW, when all components are active and radio is in receive mode. In the implemented prototype, the radio clearly dominates the momentary power consumption, radio receiving time is particularly, costly. The node power consumption doubles during reception compared to transmission with the lowest −20 dBm transmission power. The linear-position sensor power consumption is 960 μW.

Table 21.1 Wireless linear sensor node prototype power analysis. © 2005 IEEE. Reprinted, with permission, from *Proceedings of the 8th Euromicro conference on Digital System Design.*

MCU mode	ADC mode	Sensor mode	Radio mode	Power[a] (mW)
Active	Active	Active	RX	57.60
Active	Active	Active	TX (0 dBm)	39.73
Active	Active	Active	TX (−20 dBm)	26.46
Active	Active	Active	Data loading	4.70
Active	Active	Active	Sleep	3.06
Active	Active	Off	Sleep	3.01
Active	Off	Off	Sleep	1.69
Sleep	Off	Off	Sleep	0.021

[a] Measured at 3.0V supply voltage

Measured average power consumption of the platform in the linear-position sensing application is presented in Section 13.8 (Kohvakka et al. 2005a). The average power consumption of a network consisting of five times more subnodes than headnodes was 299 μW.

21.2 Indoor-temperature Sensing*

Conditional monitoring in buildings, including temperature, humidity, and gas measurements, is straightforward and one of the most potential applications for WSNs. Compared to wired sensors, the utilization of WSN technology reduces cabling and installation costs and makes the system easily extendable, as the maximum number of measurement points is extensive. The data rate and latency requirements are quite low, but the required node lifetime should be years without the need for maintenance (Kohvakka et al. 2005c).

In this section, an indoor-temperature sensing deployment carried out in 2004 is presented. This is one of the first deployments of TUTWSN targeting for the verification of energy efficiency and applicability of developed TUTWSN protocols.

The deployment is targeted to verify that TUTWSN can achieve significantly longer network lifetimes with smaller node size compared to current proposals. Desired network lifetime is at least one year from a couple of AA-type primary batteries.

The measurement network consists of sensor nodes and a gateway to connect the network to a PC for data analysis. WSN nodes are mounted on the ceilings and walls of a building as depicted in Figure 21.4 (Kohvakka et al. 2005c). The nodes acquire temperature

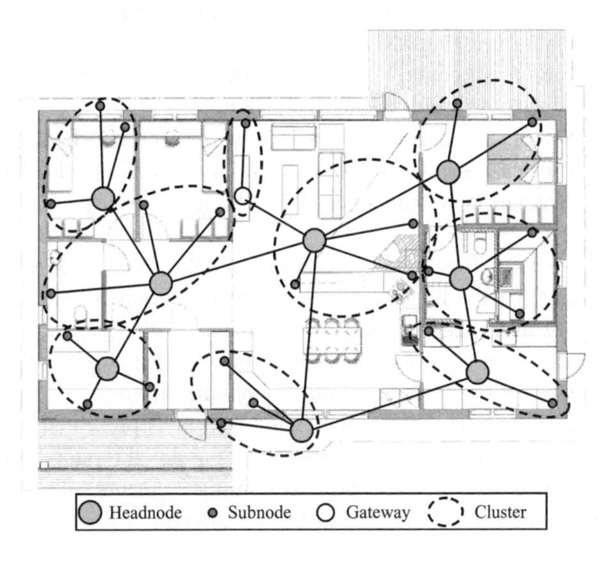

Figure 21.4 TUTWSN topology in temperature monitoring application. © 2005 IEEE. Reprinted, with permission, from *Proceedings of 2005 IEEE 16th International Symposium on Personal, Indoor and Mobile Radio Communications.*

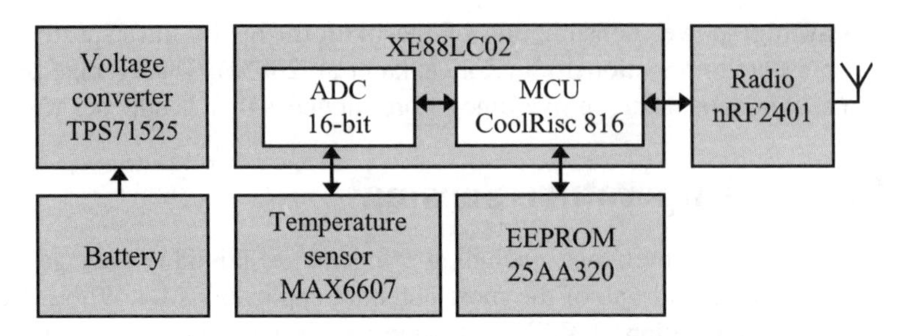

Figure 21.5 TUTWSN node prototype hardware architecture. © 2005 IEEE. Reprinted, with permission, from *Proceedings of 2005 IEEE 16th International Symposium on Personal, Indoor and Mobile Radio Communications.*

values at 5 s intervals with at least 1°C accuracy. The number of nodes is 100, but this is not limited by TUTWSN. The network range is up to 100 m, where the proximity between the nodes is at most 10 m. There are no strict delay requirements for the data collection, but the latency per hop should be below 5 s.

21.2.1 WSN Node Design

The node prototype hardware architecture is presented in Figure 21.5 (Kohvakka et al. 2005c). As the linear-position metering platform has high energy efficiency, adequate computing, and communication resources, and its operation has been verified in experimental tests, a similar hardware architecture is utilized. However, the sensor element, antenna, and battery are changed due to different deployment requirements.

The sensing subsystem consists of a Nordic Semiconductor nRF2401 transceiver and a quarter-wave GigaAnt Rufa SMD antenna. The SMA antenna is selected due to its small size, omnidirectional radiation, and adequate 68% efficiency. Measured radio range is about 15 m.

The temperature-sensing function is implemented with a Maxim MAX6607 temperature sensor, a Maxim MAX6018 voltage reference, and the integrated ADC. The performance of thc ADC integrated in MCU has adequate accuracy for the application.

The prototype has a connector for an external battery that can be selected according to measured power consumptions and the required network lifetime.

The TUTWSN node prototype is implemented in a PCB presented in Figure 21.6 (Kohvakka et al. 2005c). The top side of the prototype contains the radio, antenna, EEPROM, temperature sensor, some test pins, and a programming connector. The underside contains MCU and the voltage regulator. The size of the prototype is 31 × 23 × 5 mm.

21.2.2 Results

Power analysis measures the static power consumption of the main components of the prototype. The measurements are carried out in states where the radio, MCU, ADC, and sensor are programmed to enter different operation modes. The MCU power consumption is measured in an active mode when the MCU core is executing the sensor application, and in a sleep mode only when a wake-up timer is active. The ADC and sensor power

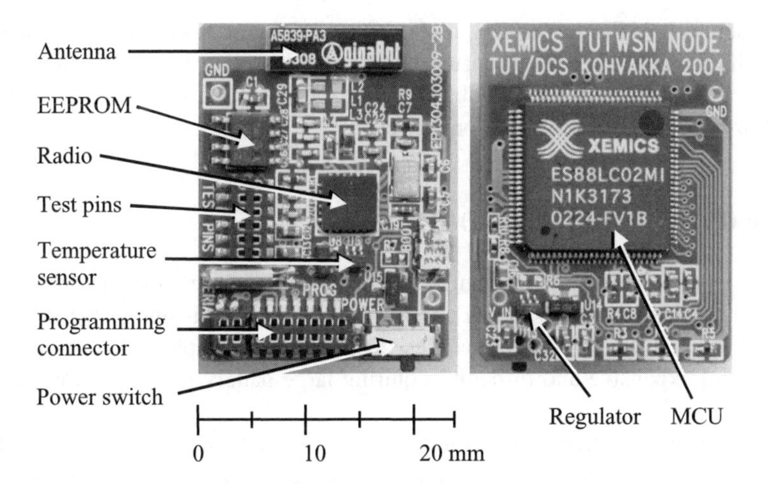

Figure 21.6 TUTWSN node prototype. © 2005 IEEE. Reprinted, with permission, from *Proceedings of 2005 IEEE 16th International Symposium on Personal, Indoor and Mobile Radio Communications*.

consumptions are measured in active and shut down (off) modes. The radio power consumption is measured in sleep, reception, and transmission modes with a minimum and maximum transmission power level.

The results are presented in Table 21.2 (Kohvakka et al. 2005c) and include the power dissipation in the voltage converter, while the prototypes are supplied with 3 V supply voltage. The minimum power consumption is 24 μW, when all components are inactive. The maximum power consumption is achieved when all components are in active mode and the radio is in receive mode; then, the power consumption is 59.55 mW. The analysis

Table 21.2 The power analysis of the TUTWSN node prototype. © 2005 IEEE. Reprinted, with permission, from *Proceedings of 2005 IEEE 16th International Symposium on Personal, Indoor and Mobile Radio Communications*.

MCU mode	ADC mode	Sensor mode	Radio mode	Power[a] (mW)
Active	Active	Active	RX	59.55
Active	Active	Active	TX (0 dBm)	37.70
Active	Active	Active	TX (−20 dBm)	26.60
Active	Active	Active	Sleep	3.04
Active	Active	Off	Sleep	3.01
Active	Off	Off	Sleep	1.69
Sleep	Off	Off	Sleep	0.024

[a] Measured with 3.0 V supply voltage

shows the Semtech MCU has a very high energy efficiency. Moreover, the radio clearly dominates the momentary power consumption; radio receiving time is particularly costly. The power consumption at least doubles during the reception compared to transmission with the lowest −20 dBm transmission power.

21.3 Environmental Monitoring

An environmental sensor network can be deployed in hostile environments or over large geographical areas to provide accurate and localized data. This is one of the most potential WSN applications. By conventional wireless technology the implementation of such functionality is very expensive and difficult, requiring large battery packs and manual network configuration.

The deployment described has been our first large-scale and long-term field test. The deployment has been active since autumn 2005. During the deployment the network has been upgraded and tested with a varying amount of nodes, ranging up to 50 nodes. The target of the deployment is the demonstration and verification of the TUTWSN operability in harsh outdoor conditions. Additionally, the Web interface is offered for local residents providing accurate regional temperature information. The service has been especially useful in farming and gardening as it can determine the right time to begin planting in spring and provides frost alerts in autumn.

21.3.1 Problem Statement

The deployed network should operate in arctic outdoor conditions for at least half a year. Dozens of WSN nodes are deployed in an area of a few square kilometers resulting in a very sparse network. Required hop distances should be in the order of 1 km in non-line-of-sight conditions. Transmit powers can be raised somewhat, while considering the network lifetime requirement. The size of nodes may be larger than previous platforms, which allows improving antenna efficiency.

21.3.2 Implementation

In this section, the hardware implementations of the environmental monitoring network is first discussed. Then, the utilized network architecture and user interfaces are presented.

Hardware platform design

The large geographic area is covered by multihop networking and long-range node platforms designed especially for outdoor use. The hardware platform is based on the verified architecture of the TUTWSN temperature-sensing platform presented in Section 4.3.1. However, due to different communication range requirements, the transceiver is replaced with a long-range Nordic Semiconductor nRF905 transceiver operating at the 433 MHz frequency band. The hardware architecture of the long-range TUTWSN prototype is presented in Figure 21.7. All the components have an extended temperature range starting from (−40 °C) to allow operation in arctic conditions.

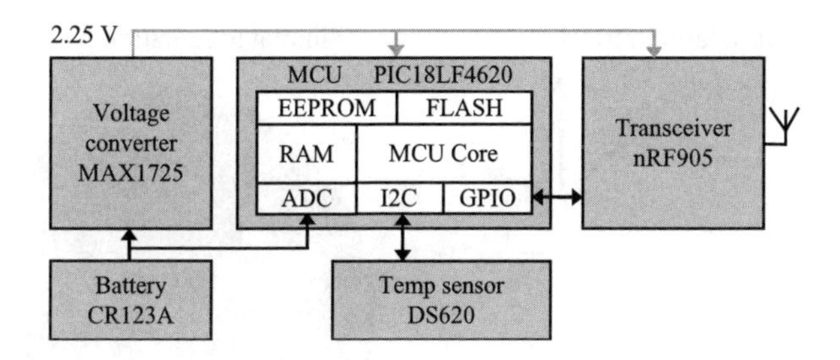

Figure 21.7 Long-range TUTWSN prototype architecture.

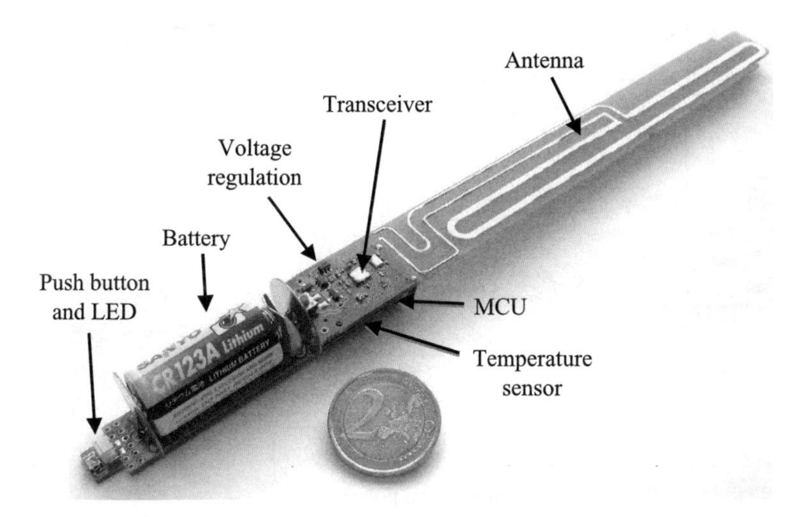

Figure 21.8 Long-range TUTWSN prototype. With kind permission of Springer Science and Business Media.

The implemented long-range TUTWSN prototype is presented in Figure 21.8. The prototype is 255 × 21 mm in size, and encapsulated in a plastic waterproof enclosure. The prototype consists of two separate boards, one for MCU, radio, voltage regulation, and temperature sensor, and other extension board is for the battery, push button, LED and I/O connector. Also, other types of sensors and energy-scavenging circuits can be easily implemented into the extension board increasing flexibility for various applications.

The antenna is designed according to the requirements of omnidirectional radiation and high efficiency. In addition, it has to be fitted to a 21 mm diameter pipe. For achieving these requirements, the antenna is constructed as a multiple folded dipole that is implemented directly onto a printed circuit board. The antenna impedance is also quite near to the transceiver output impedance requiring only a minimum impedance matching network. The antenna properties are presented in Figure 21.9.

The platform uses Dallas DS620U temperature as described in Section 4.3.1. However, in addition to the one on-board temperature sensor, a few nodes are modified to include additional temperature sensors that are connected to the node with a 30 cm to 1 m wire.

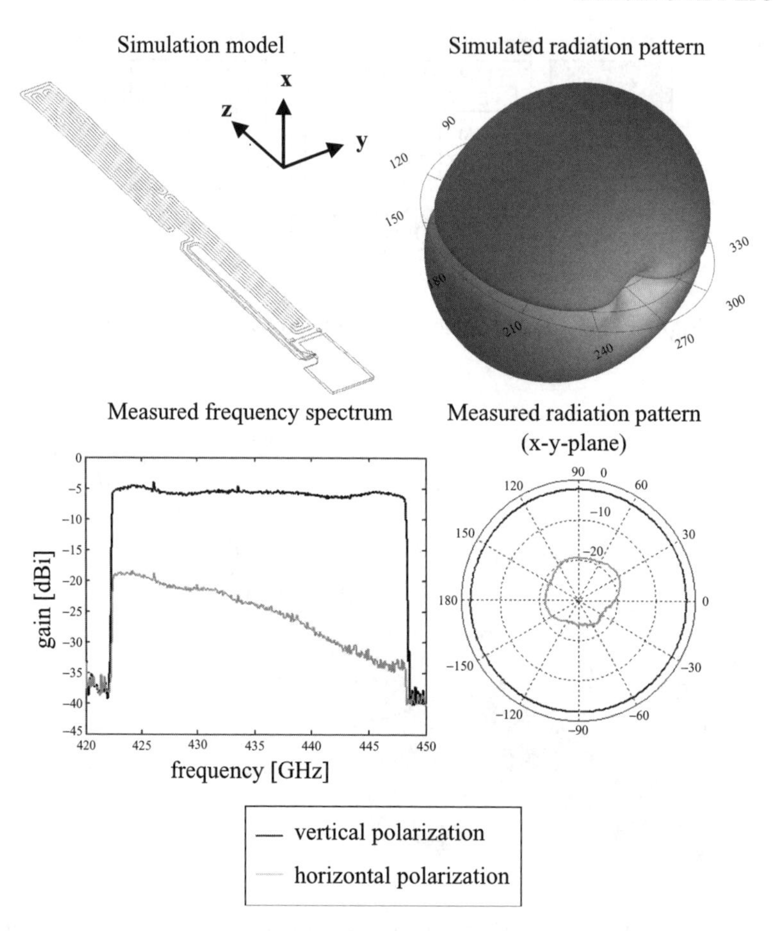

Figure 21.9 Radiation properties of the 433 MHz long-range TUTWSN prototype with multifolded dipole antenna.

The temperature sensors operate via Serial Peripheral Interface Bus (SPI) in which each sensor can be individually addressed. By having several sensors, a node can measure air, water, and earth temperatures simultaneously.

Static power consumption

The peak power consumption of the long-range TUTWSN prototype is measured in different operation modes at 3.0 V supply voltage, as presented in Table 21.3. The minimum achievable power consumption is 31 μW when the system is in a sleep mode. As MCU is in active mode and running at 1 MIPS speed, prototype power consumption increases to 3.17 mW. Transceiver reception and active MCU, but inactive ADC and sensor, consume 40.6 mW of power. The maximum power of 95.9 mW is consumed as all components are in active mode and radio is in transmission mode at 10 dBm transmission power.

Table 21.3 The static power consumption of the long-range TUTWSN prototype.

MCU mode	ACD mode	Sensor mode	Radio mode	Power[a] (mW)
Active	Active	Active	TX (10 dBm)	95.9
Active	Active	Active	TX (6 dBm)	65.9
Active	Active	Active	TX (−2 dBm)	47.9
Active	Active	Active	TX (−10 dBm)	32.9
Active	Active	Active	RX	43.3
Active	Active	Active	Sleep	5.91
Active	Active	Off	Sleep	3.68
Active	Off	Off	Sleep	3.17
Sleep	Off	Off	Sleep	0.031

[a] Measured at 3.0V supply voltage.

According to the measured power consumptions, transceiver peak power is one order of magnitude higher than the power of the rest of prototype components. Transceiver reception mode consumes 12.8 times the power of the MCU. Data transmission at 10 dBm transmission power consumes 29.4 times the power of the MCU. Thus, the transmission of 1 bit of data at 50 kbps data rate consumes the energy of the execution of 588 instructions on MCU. For energy efficiency, both the transmission and reception time should be minimized.

Network architecture

The environmental monitoring uses the TUTWSN deployment architecture that was presented earlier in this chapter. In this deployment, the sink node is connected to a PC gateway via a serial port interface. The gateway forwards measurements to the WSN server. In case of a connection break, the PC gateway stores measurements to a local database. The cached measurements are sent to the server when the connection is restored. The gateway laptop computer is depicted in Figure 21.10 together with UI and a deployed sensor node.

Sensor nodes have three embedded sensor applications: sensor control, temperature, and WSN self-diagnostics. These sensor applications are activated by defining related interests to the TUTWSN control panel that is shown in Figure 21.11. The sensor control application handles received control messages, thus allowing the remote configuration of a node. The temperature application performs sensing on a digital sensor with the interval that is defined in the related interest. In addition, the temperature interest defines the measurement range that causes transmission of a packet only when the measured values are within a certain threshold. The self-diagnostics application is used to monitor and analyze the network.

The Web interface for the environmental monitoring deployment is targeted at end-users and can be used with any device having a Web browser. The data is processed completely on the server side with Java Servlets, which eases the requirements of the device using the service. A map view on the Web interface showing the last measured temperatures on a sensor field is presented in Figure 21.12. The selected node has multiple temperature sensors that separately measure air and water temperatures. A user can also examine the

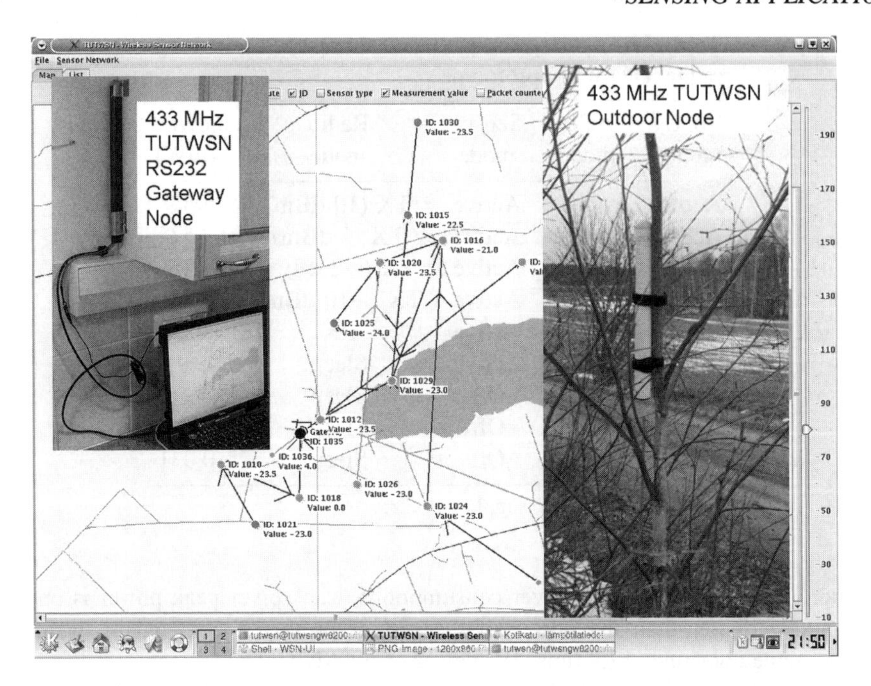

Figure 21.10 Environmental monitoring system overview. Nodes are deployed to the ground and the measurement data is forwarded through a gateway computer to the UI.

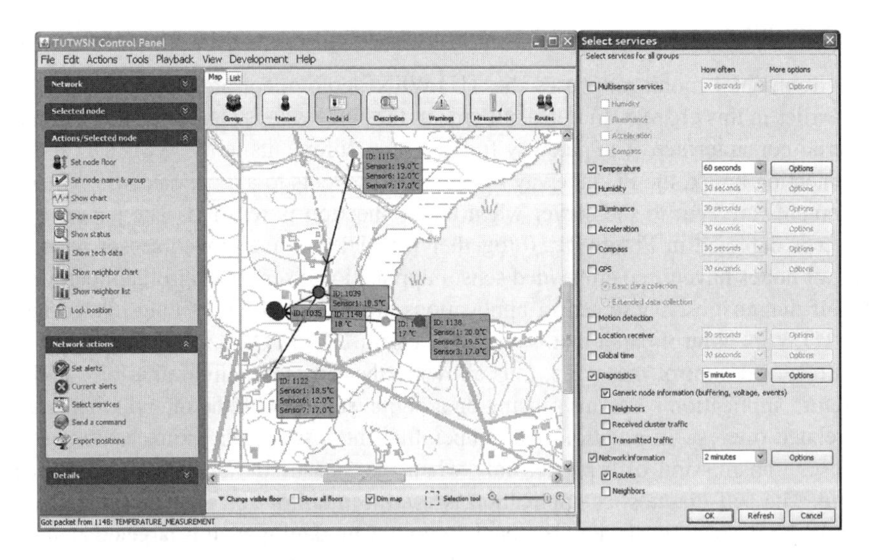

Figure 21.11 TUTWSN control panel showing the realtime status and active routes in the network. A dialog for setting interests is presented on the right.

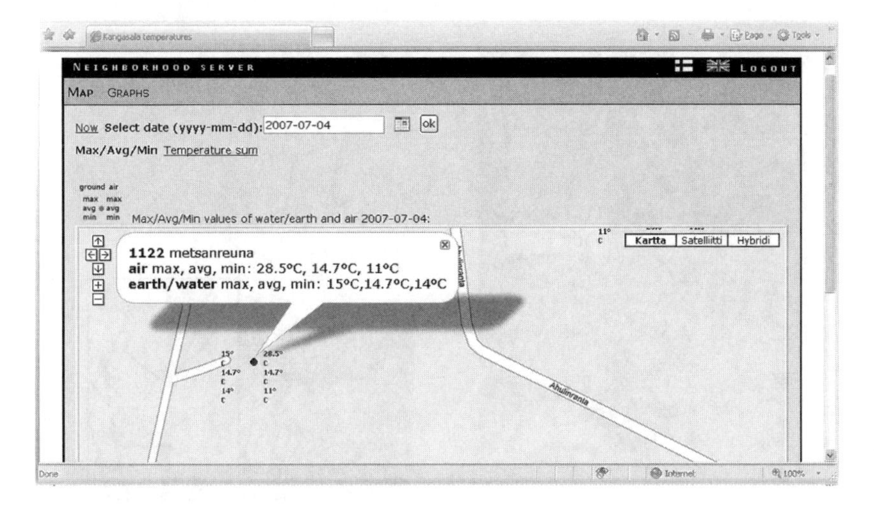

Figure 21.12 Web interface for local residents showing a map view and latest temperature values. Multipoint temperature nodes show air and earth/water temperatures separately.

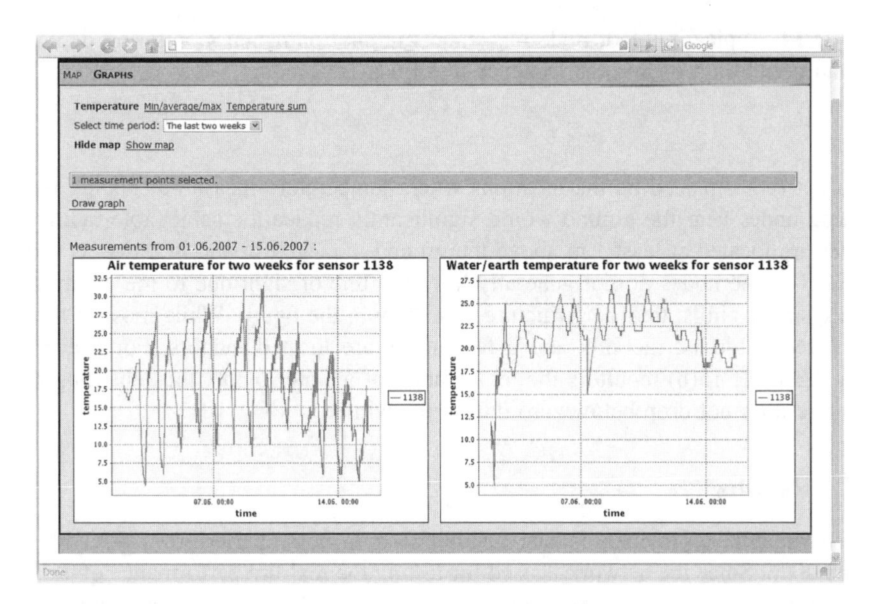

Figure 21.13 Temperature history of selected nodes in Web interface.

measurement history of an individual sensor or a group of sensors. This functionality is depicted in Figure 21.13.

Deployment

The nodes are deployed over a large geographic area of a few square kilometers. As the purpose has been to test network configurability, the node placements are not specifically

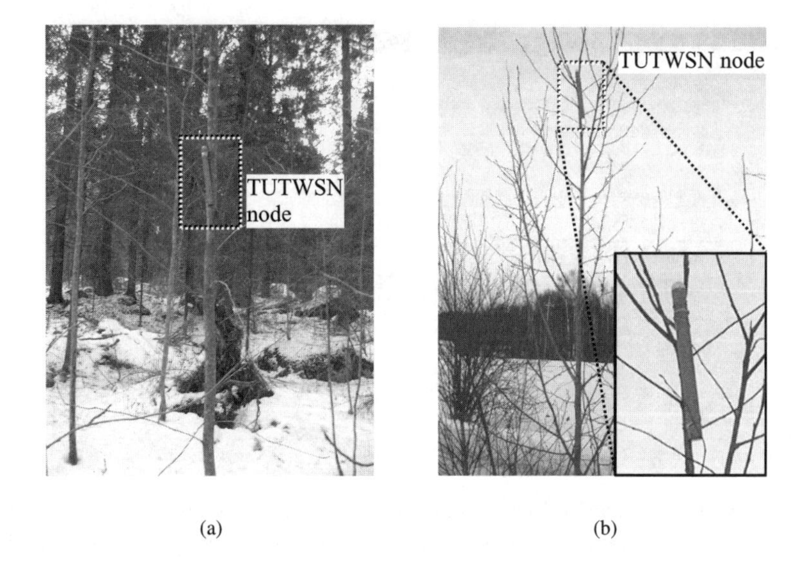

Figure 21.14 Typical outdoor node placements in environmental monitoring deployment. With kind permission of Springer Science and Business Media.

planned but rather placed on the locations where temperature measurements are required. As putting nodes near the ground would significantly reduce the achievable radio ranges, the nodes are located at least 1 m above the ground, typically bound in a tree as shown in Figure 21.14. The nodes do not generally have the line of sight due to the environmental obstacles, such as hills, trees, or structures, as show in the figure. While most of the nodes are deployed outdoors, the sink and a few nodes are located indoors. For example, the node in Figure 21.15(b) monitors the temperature of a water pump, thus ensuring that the temperature does not drop below zero during the winter.

21.3.3 Results*

The environmental deployment is analyzed over a four month period from November 2005 to March 2006. Since the network is used for testing the real-life performance of TUTWSN with different configurations, the time range was selected so that the protocol stack, the node placements, and the used interest settings were unchanged.

During the analyzed time period, the outdoor deployment consisted of 19 nodes covering 2 km^2 area. Two of the nodes were subnodes, while the rest of the nodes acted as cluster heads. Each node was configured to send both its temperature and diagnostics information twice per minute.

*With kind permission of Springer Science and Business Media.

(a) (b)

Figure 21.15 Varying environmental conditions during the deployment: (a) a node surrounded by vegetation and trees during the summer; and (b) an indoor placement.

Distribution of traffic

Geographic locations of deployed sensor nodes (identified with numbers 1–18) and the distribution of transmitted traffic on selected nodes are shown in Figure 21.16. The average successfully transmitted traffic per node is presented in Figure 21.17. The bandwidth usage between temperature and self-diagnostics data was equal. Control traffic (route advertisements and interests) used less than 1% of bandwidth. Since nodes originate the same amount of traffic, the difference in traffic volumes is caused by forwarded data. The nodes located in the edge of the network transmit less data, since routing algorithm tries to minimize required energy and hops; thus, preferring routes through centrally located nodes. Nodes 17 and 18 do not forward data because they were configured as subnodes. Node 6 experienced high link-error rates due to bad location, which resulted into low traffic. A significant portion of the traffic to the sink is forwarded via node 2. Node 2 sent 91% of its traffic (8.7 bit/s) to the sink, which corresponds to over onethird of the traffic received by the sink (17.8 bit/s). Although node 2 is located relatively close to the sink, other nodes have access to the sink through it because the sink is inside a building while node 2 is outside. Thus, nodes have better connection to node 2.

A node had only one active next-hop route at a time. Route changes are caused by a broken next-hop link due to communication errors, or changes in network conditions that caused routing to change the next-hop node. An average time between route changes was 30 minutes, caused typically by the routing algorithm balancing the network load. Typical hop count from a node to the sink was four, while the maximum count was eight.

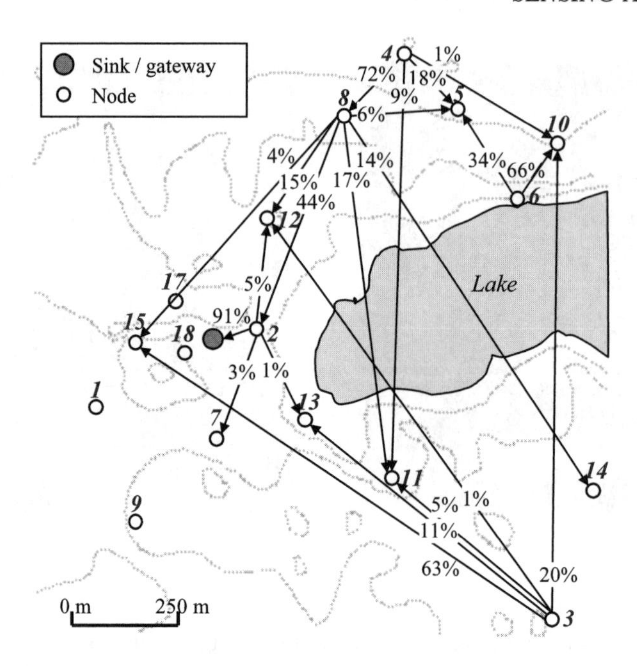

Figure 21.16 Geographic locations of deployed sensor nodes and distribution of transmitted traffic in selected nodes. With kind permission of Springer Science and Business Media.

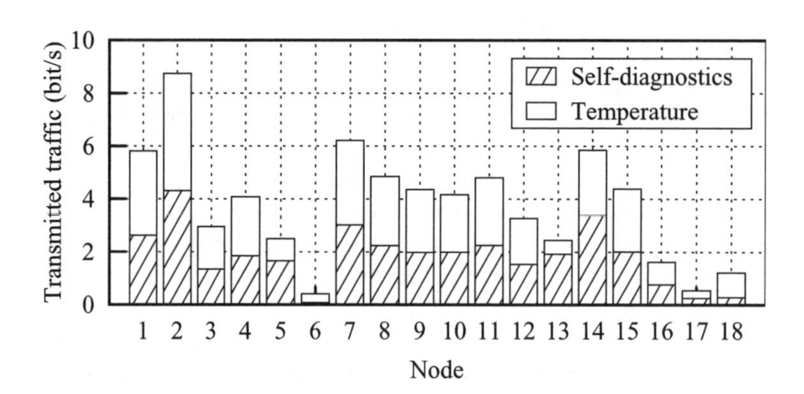

Figure 21.17 Average transmitted traffic per node. With kind permission of Springer Science and Business Media.

Reflections from the ground and buildings have a significant affect on the radio wave propagation. The deployment environment contains cliffs, icy surface of the lake, and other elements of terrain that can even enhance the radio wave propagation. The longest link is 1.1 km from node 3 to node 15 (Suhonen et al. 2006a).

Temperatures

Day temperatures during the measurement period are shown in Figure 21.18. The temperatures are averaged over readings from all sensor nodes. The temperature changes significantly and often rapidly. For example, on January 23, 2006, the lowest temperature was −21.8°C, while the highest temperature on the next day was −5.2°C. The rapid changes can be seen in Figure 21.19, which shows temperature per hour on a selected sensor (Suhonen et al. 2006a).

Temperature changes introduce challenges for the equipment and protocols. As the temperature alternates between below zero and above zero, the casing must be compact to prevent water damage. The MAC protocol must compensate clock drift, since the oscillating frequency of crystals depends slightly on temperature. On the deployment region, temperature does not change evenly and some nodes might be inside buildings.

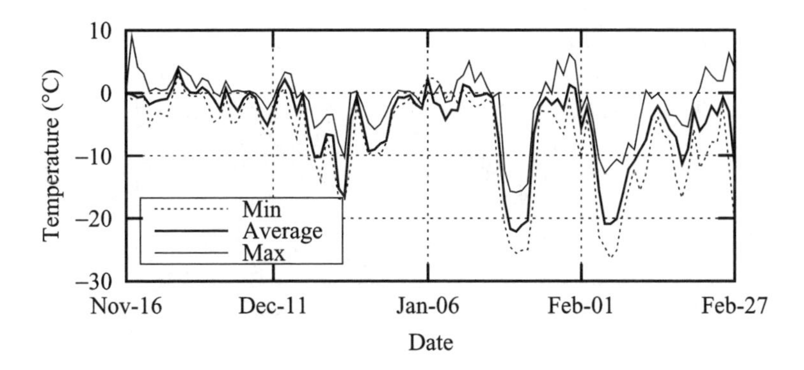

Figure 21.18 Measured minimum, maximum, and average day temperatures. With kind permission of Springer Science and Business Media.

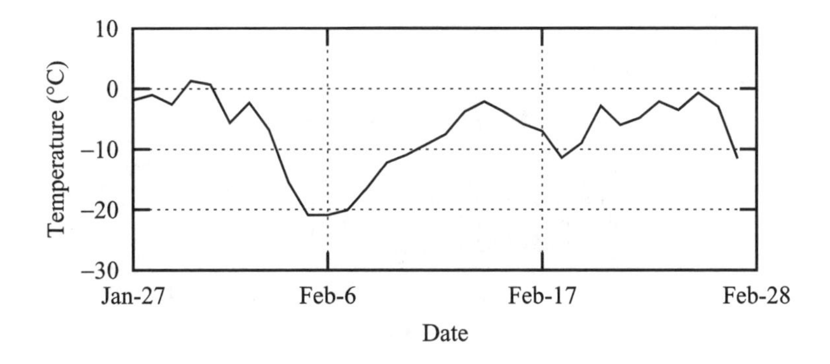

Figure 21.19 Rapid environmental temperature changes on a selected sensor (node 8). With kind permission of Springer Science and Business Media.

Energy consumption

The energy consumption is calculated based on the drop in the battery voltage. The voltage is measured on each node with an ADC and sent to the sink in the self-diagnostics packets. For reliable results, the voltage development is examined over a long time interval, as traffic changes due to the dynamic routing affect the energy consumption. In addition, the temperature has a significant impact on the voltage levels, as show in Figure 21.20 (Suhonen et al. 2006a).

Figure 21.21 presents voltages of two sensor nodes (2 and 8) and their average voltage drops. Although the voltage levels vary according to the temperature, the long-term drop can easily be observed. The steeper voltage drop on node 2 is attributed to the heavier traffic load shown in Figure 21.17. During the measurement period, load balancing by finding alternative routes did not work simply because the node 2 was the only feasible next-hop choice for many nodes. The results emphasize the importance of the automatic load balancing, but also show that the balancing is not always possible due to absence of

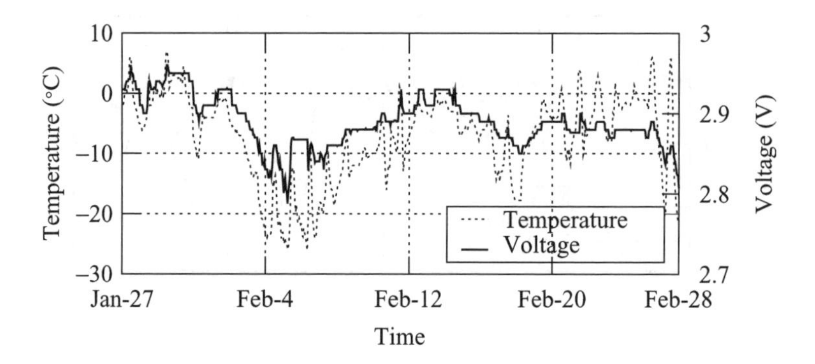

Figure 21.20 The effect of temperature to voltage (node 8). With kind permission of Springer Science and Business Media.

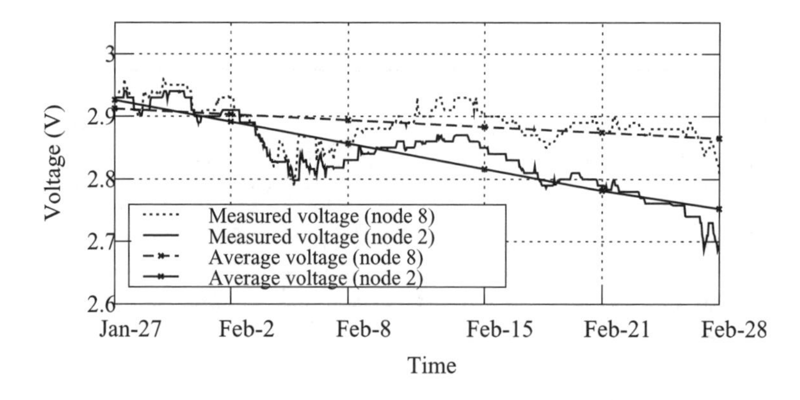

Figure 21.21 Decrease in the battery voltage of two selected nodes. With kind permission of Springer Science and Business Media.

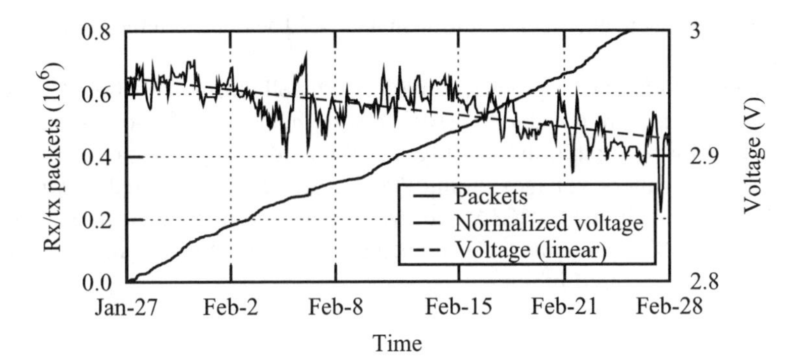

Figure 21.22 Voltage and transmitted/received packets in node 8. The effects of temperature are removed from the voltage curve. With kind permission of Springer Science and Business Media.

alternative nodes. The network analyzation software is used to detect these situations, thus allowing the correction of the situation, e.g. by adding new nodes (Suhonen et al. 2006a).

Figure 21.22 presents the voltage and the incremental sum of transmitted and received packets of node 8. For removing the effects of temperature changes, the voltage is normalized to 0°C based on the calculated regression. By assuming similar load and linear voltage drop until the 2.4 V terminal voltage, a rough lifetime evaluation yields 180 days. The resulting value is consistent with the real-world experiments.

Discussion

The experiments provide a vast amount of information about the real operation of WSN nodes and radio links in a forested, low-temperature outdoor environment. According to the experiments, a forest attenuates radio wave propagation significantly. Achieved radio range in a forest has been below 100 meters, while the longest measured range has been nearly 1.5 km. The edge of the forest seems to operate as a reflector causing a notable gain in antenna radiation pattern. Also, radio wave propagation has been notably affected by snowfall, rain, humidity, temperature, and the frost and snow in trees, on the ground, and around the nodes. Hence, the quality of radio links and the network topology changes dynamically although nodes are stationary (Suhonen et al. 2006a).

Dynamic network topology significantly affects the routing protocol operation. The experiments depict that the entire route to a sink must be considered in the route selection. As the difference between link qualities is very high, only examining next-hop quality when determining a route leads to unsatisfactory performance. Since the environment significantly affects radio wave propagation, cost-effective routing paths do not typically follow geometrically reasonable routes. The utilized cost-gradient based routing seems to work well in outdoor multihop networks without line-of-sight.

The outdoor temperatures until March 2006 ranged from −31.5 °C to 12.0 °C. The high temperature variation significantly reduced the accuracy of the crystals and thus, the accuracy of time synchronization. In the worst-case scenario, some nodes were inside buildings and others were outdoors, resulting in nearly 50 degrees difference in the operation temperature.

The implemented hardware prototypes performed well during the whole test period. In some locations, nodes were not able to associate with the network for long periods of time. This was caused by poor radio-link quality, not the hardware prototype itself. According to the reduction of battery voltages during the test period, the expected network lifetime is around 6 months with one CR123A battery. It should be noted that the network traffic consisted not only of temperature measurements but also diagnostics information, which increased load and decreased lifetime.

22

Transfer Applications

Instead of maximising a network's lifetime, the transfer application targets delay-critical operations with high throughput. There are several benefits of using a sensor network in such an application scenario. Firstly, the network self-configuration allows the rapid deployment of a large-scale sensor network, which is useful in temporary networks, e.g. during emergencies when the normal communication infrastructure is unavailable. Secondly, sensor networks are very energy efficient during their sleep mode. Thus, when a high network capacity is required only periodically, the network is able to achieve a long lifetime.

22.1 TCP/IP for TUTWSN*

This use case shows the configuration of the TUTWSN protocol stack to allow TCP/IP communication through the network. TUTWSN does not support native TCP/IP communication with the WSN, thus the communication must be adapted at proxies providing access to the network. Hence, this use case does not implement access to TUTWSN data using TCP/IP but configures WSN to forward TCP data. Such a feature allows the host PCs to utilize WSN as an alternative communication media when an infrastructure network, e.g. Ethernet or WLAN, is not present.

22.1.1 Problem Statement

The TCP/IP stack is a *de facto* standard on the Internet but as such it is generally not considered as a well-suited technology for WSN communications. Considerable header overheads, connection management, as well as end-to-end flow and congestion control lead to poor performance in low-power WSNs (Chonggang et al. 2005; Kuorilehto et al. 2006; Tian et al. 2005).

There are several conceptual differences in the communication paradigms, flow control, and predictability of WSNs and TCP/IP networks. In contrast to the data-centric WSNs,

TCP communications is connection-oriented between specific endpoints. The end-to-end connection maintenance and flow control of TCP generates a considerable amount of control traffic between the endpoints. In contrast, WSN flow control is performed over a single hop in order to avoid end-to-end retransmissions as communication over several WSN links consumes considerable amount of energy. TCP flow control assumes that all communication errors are caused by congestion whereas the main reasons for errors in WSNs are random bit errors, topology changes, and temporarily unavailable nodes. Furthermore, TCP assumes a symmetric uplink and downlink for each connection but in WSNs their delay and throughput may differ considerably.

Due to the limited resources in WSN nodes, the additional features required for supporting TCP/IP communication should exploit the existing WSN infrastructure and protocol stack conventions as much as possible. Thus, additional code and data memory usage should be minimized. A more frequent duty cycle or higher transmit power results in an increased energy consumption. This cannot be avoided but it can be adjusted as a tradeoff between throughput and energy. Due to the energy tradeoff, the number of nodes that are affected by the TCP/IP communication should be minimized. Thus, if possible the configurations should be restricted only to the nodes that are part of active TCP connections. This minimizes the effect of the adaptation on the overall network performance.

22.1.2 Implementation

TCP/IP communication is integrated into TUTWSN using the WSN as a transparent medium that passes TCP/IP traffic on top of its infrastructure. The developed architecture for TCP/IP networking over a TUTWSN is depicted in Figure 22.1. Two or more TCP/IP endpoints (user/server) communicate over a WSN through gateways and intermediate nodes. The base WSN infrastructure, sensing operations, and data relaying are supported simultaneously with the TCP/IP communication. The path for the TCP/IP communication is referred to as a *tcp route*. WSN data can be sent along that route as well (Kuorilehto et al. 2006).

The implementation consists of a software component for Linux PCs operating as WSN gateways and TUTWSN configuration to a throughput-optimized mode. A Linux PC is connected to a gateway WSN node through a serial port.

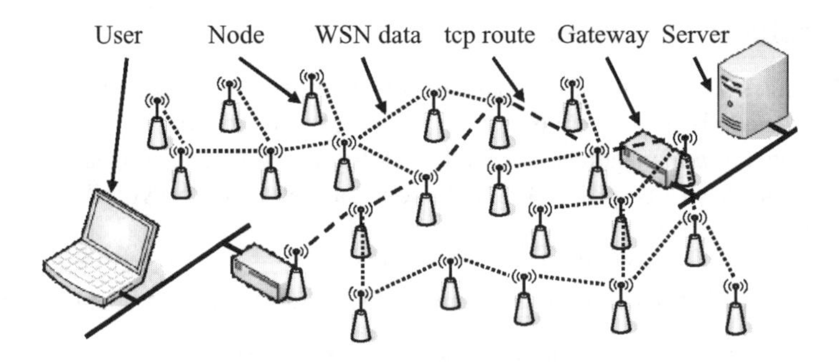

Figure 22.1 Networking architecture for TCP/IP communication over WSN. © 2006 IEEE. Reprinted, with permission, from *Proceedings of 2006 IEEE 17th International Symposium on Personal, Indoor and Mobile Radio Communications.*

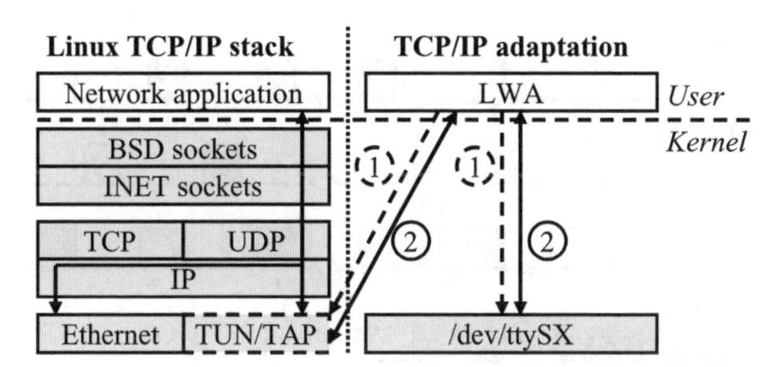

Figure 22.2 Software architecture of the TCP/IP WSN adaptation implementation on Linux. © 2006 IEEE. Reprinted, with permission, from *Proceedings of 2006 IEEE 17th International Symposium on Personal, Indoor and Mobile Radio Communications*.

TCP/IP adaptation layer implementation in Linux

Our TCP/IP adaptation layer is implemented as the Linux Wireless sensor network Adaptation (LWA) application on Ubuntu Linux with kernel 2.6.8.1. The software architecture for Linux is depicted in Figure 22.2. A network application (or IP router) communicates through a legacy Linux networking stack consisting of sockets and protocol layers. A Universal TUN/TAP (TUN as in network TUNnel; TAP as in network TAP) driver passes IP packets to LWA, which fragments and relays data to the WSN through a serial port interface. LWA performs addressing adaptation and IP packet fragmentation on WSN frames whereas legacy Linux flow control is used for TCP. An IP address is mapped to a WSN node address based on the information gathered during the connection initiation.

In the startup phase (1), LWA creates an instance of the TUN/TAP driver and a file handle to the serial port. The TUN/TAP driver network interface is automatically configured to a separate subnet with the `ifconfig` program. The data communication between the TUN/TAP driver and LWA during the active phase (2) is performed through another file handle retrieved during the initialization. When the LWA writes to the file handle, data are passed to the IP layer. Correspondingly, data sent by the IP layer can be read from the file handle.

TUTWSN protocol configuration

In TUTWSN, a cluster headnode maintains its own access cycle and communicates with the next-hop cluster during the Superframe (SF) of that cluster. The rest of the time is spent in the sleep mode and periodically sending network beacons. However, this idle time can be utilized for configuring the network for higher throughput and shorter delay.

The idle time can be used efficiently if the start times of SFs are adjusted so that the access cycle of the next-hop cluster follows immediately after the cluster's own access cycle. The TUTWSN access cycle is configured to a high-throughput mode by including an additional SFs into the idle period, as depicted in Figure 22.3. The original SFs are used for the WSN communication whereas the additional SFs are for the TCP data. Because the same SF structure is used, additional program code is needed only for the control. The additional SF is started by a beacon, which contains timing and slot reservation information. The reservation slots of the SF are allocated for the associated nodes that are part of the active tcp routes.

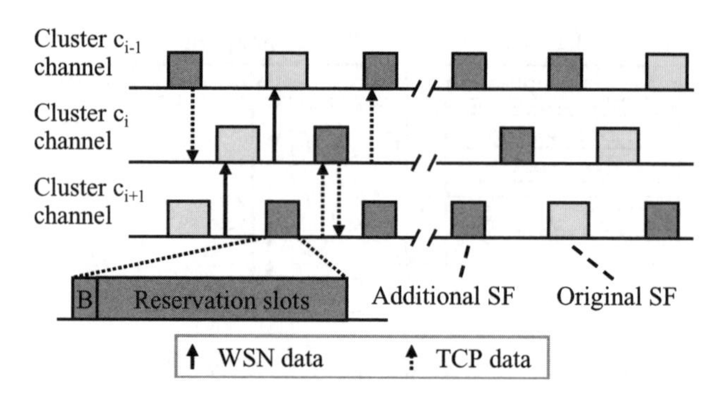

Figure 22.3 Principle of access cycle adaptation. © 2006 IEEE. Reprinted, with permission, from *Proceedings of 2006 IEEE 17th International Symposium on Personal, Indoor and Mobile Radio Communications.*

22.1.3 Results

The evaluation of the TCP/IP adaptation is performed by measuring TCP and WSN performance with the TUTWSN temperature-sensing platform nodes (Section 4.3.1) and two Linux laptops. The WSN performance was evaluated with Hypertext Transfer Protocol (HTTP) and File Transfer Protocol (FTP) applications. For evaluation purposes, TUTWSN nodes on a tcp route utilize modified SFs. A SF consists of cluster beacons, one ALOHA slot, and eight reservation slots. The interval between SFs is set to 500 ms. Per-hop acknowledgments are used for guaranteeing frame delivery. The TCP version used in the experiments was the default Linux kernel TCP Reno.

Round Trip Time

From a human-user point of view, the performance of an HTTP application depends mainly on the Round Trip Time (RTT) of the network. In order to estimate HTTP application performance on top of adapted TUTWSN, RTTs for a tcp route in TUTWSN are measured. The measurements are done with `ping` program. Figure 22.4 depicts RTTs with IP packet sizes of 50, 100, 200, 500, and 1000 bytes for different number of hops. The packet size includes 20-byte IP and 8-byte Internet Control Message Protocol (ICMP) headers. RTTs contain the processing time of the Linux PC but it is negligible. Due to the TDMA scheme of the WSN, there may be at most a 250 ms variation in the delay at the beginning of each transmission (Kuorilehto et al. 2006).

As shown, RTTs increase steadily as the number of hops grows. For the same number of hops, the RTT depends on the number of SFs that are required for transmitting the fragments of a packet. Because 50 B and 100 B packets can be sent within a single SF, their results are similar. In general, the results are expected and can be considered acceptable for transferring, e.g. textual Web pages.

Throughput

In FTP applications, data flow is mainly unidirectional and the objective is to transfer data as quickly as possible. Therefore, for an FTP applications, the throughput of the network is the

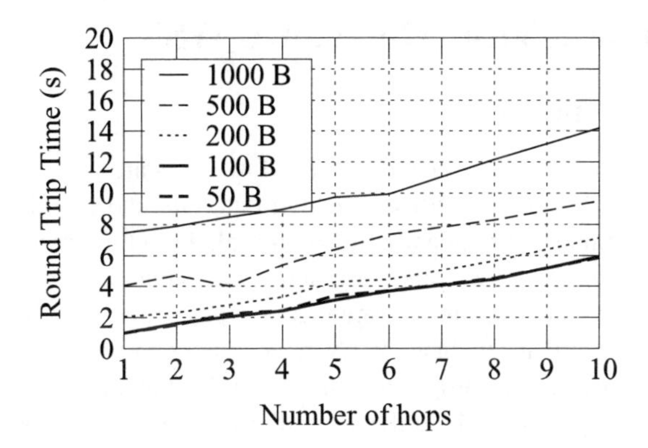

Figure 22.4 RTTs for varying packet sizes and different number of hops. © 2006 IEEE. Reprinted, with permission, from *Proceedings of 2006 IEEE 17th International Symposium on Personal, Indoor and Mobile Radio Communications.*

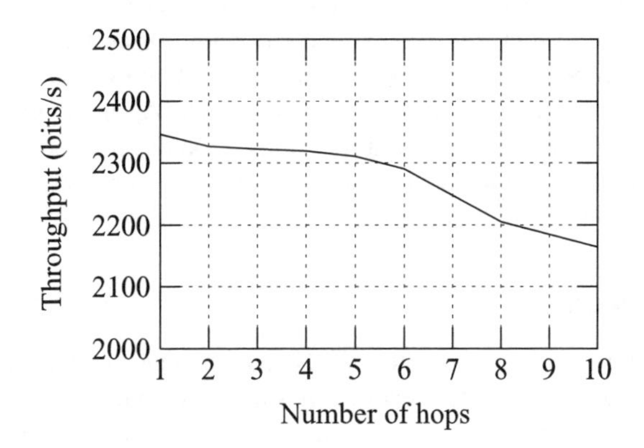

Figure 22.5 Measured throughput for a TCP route for different number of hops. © 2006 IEEE. Reprinted, with permission, from *Proceedings of 2006 IEEE 17th International Symposium on Personal, Indoor and Mobile Radio Communications.*

key property. With the chosen TUTWSN configuration, the theoretical maximum throughput for a tcp route is 2432 bps. The measured throughput of an FTP application through a tcp route for different number of hops is depicted in Figure 22.5 (Kuorilehto et al. 2006).

The FTP application is modeled with the `netcat` program and the throughput is measured according to successfully received IP packets at the receiving Linux PC; thus, discarded packets are not included. This combined with the usage of acknowledgments and retransmissions prevents reaching the theoretical maximum throughput.

As depicted by the results, the effect of the number of hops to the throughput is not considerable. The throughput decreases quite steadily as the number of hops grows due to the higher probability of retransmissions. The maximum throughput is quite low, but also it can be considered acceptable for infrequent downloads of relatively small amounts of data.

22.2 Realtime High-performance WSN*

22.2.1 Problem Statement

Although, WSNs are usually considered as very low data-rate networks, there is a great potential to utilize the benefits of WSNs for high data rate and low delay demanding applications, such as media streaming and critical control. The combination of extremely low power consumption and complex computing and networking set very high challenges for the design of hardware architecture, communication protocols, and data processing algorithms. Hence, it is useful to evaluate WSN node functionality on flexible development platforms. Besides processor re-programmability, a fully reconfigurable processing architecture is important. This allows the utilization of several parallel processors together with custom hardware accelerators providing the widest design space and most probably the highest performance for a given application (Kohvakka et al. 2006a).

The platform is targeted for demanding WSN applications for surveillance and control in an indoor environment. The platform operates in large mesh networks, where around 2 Mbps network throughput, high security, very low routing delay, and very high interference tolerance are required. High re-configurability is achieved by the selection of one to four processors, one to four transceivers and an option for custom hardware accelerator implementations. Unused radios are either switched to low-power sleep mode or removed from the platform. The implementations of TDMA, FDMA, and contention-based MAC protocols are required. High re-configurability allows the development and verification of the highest performance communication architectures for given WSN applications, which is not possible with current WSN platforms.

22.2.2 Implementation

Reliable data transfer in a large multihop WSN requires that the data is protected against routing errors caused by node malfunction, message corruption, and the possible denial of service attacks. The protection against these threats is achieved by transmitting securely encrypted data frames using multiple parallel routes, while the use of multiple parallel transceivers provides several benefits. WSN nodes may perform simultaneous data exchanges with multiple neighbors, or multiply network performance on critical routing paths. Also, hop delays can be minimized by receiving and transmitting a forwarded data frame simultaneously. For achieving the highest performance of large and dense networks, narrow band transceivers with a large number of selectable frequency channels are crucial. For achieving the highest throughput, each transceiver utilizes a unique frequency channel. The control of multiple transceivers requires time-accurate parallel data processing capability (Kohvakka et al. 2006a).

The use of Field Programmable Gate-Array (FPGA) enables high re-configurability, timely accurate parallel processing, and high processing performance. The optimization of the hardware architecture for a given application or networking task is automated by the use of modern software/hardware co-design tools. Accurate performance comparisons between various multiprocessor and hardware accelerator implementations are possible by monitoring internal signals on FPGA. In addition, the processing architecture can be changed in-field after deployment. As the design is performed at a very high level of abstraction, most parts of the design are technology independent, and thus, reusable in ASIC implementations.

Data gathering in nodes is performed using sensor devices connected to an ADC and digital I2C (Inter-Integrated Circuit), SPI, and PWM (Pulse-Width Modulation) interfaces. At least temperature, humidity, acceleration, pressure, acoustic, light, and gas sensors are required. In addition, connectivity with various actuators, such as relays, is required.

Connectivity to the host is implemented by an RS-232 serial port and general purpose I/O connections. Various network interfaces can be implemented on FPGA and custom expansion boards connected to I/O connections.

Due to the required full re-configurability, the use of lowest power technology is very difficult, and energy efficiency can be considered as a secondary priority. Although the development of very energy efficient (MIPS/mW) processing architectures is required, a high static power consumption and limited support for low-power sleep modes is acceptable.

For platform expandability, a modular design approach is needed. A compact-sized mother board with separate radio and I/O boards can be used for various end-applications as such. Expandability for applications is achieved by application-specific radio and I/O boards. The a desired platform size is around $10 \times 10 \times 5$ cm. The platform is typically mains powered obtaining its supply power from a low-voltage DC connector. In addition, a purely wireless operation with small AA-type batteries is required.

Hardware implementation

The hardware architecture of the high-performance, multi-radio platform is presented in Figure 22.6. The architecture is divided into four subsystems: computing, radio, power, and I/O subsystems.

The computing subsystem executes communication protocols and sensor application, and manages node operations. As a high degree of re-programmability and reconfigurability are required, an FPGA is selected. According to experiments, the use of an FPGA can provide even two orders of magnitude higher energy efficiency in the implementations of the protocol stack compared to embedded microprocessor or microcontroller solutions (Rabaey et al. 2000). The Altera Cyclone EP1C20 FPGA is large enough for implementing several embedded processor cores, system memory, and custom logic for controlling the radios and accelerating data processing (Kohvakka et al. 2006a). At reset, the FPGA configuration data is loaded from a FLASH based serial configuration device allowing stand-alone operation. Also, a flexible development environment for various processor and hardware architectures is provided. The FPGA is 19×19 mm in size and has 301 user I/O pins, which are used for the connections to radio and I/O boards.

The radio subsystem provides wireless communication links to neighboring nodes. To allow for a high degree of freedom for the development of link layer protocols, four radio transceivers are used. The selected Nordic Semiconductor nRF2401A radios comprise the physical layers of the communication protocol, which allows the free development for upper protocol layers. The radios operate in the 2.4000–2.4835 GHz license-free frequency band and have 1 Mbps data rate. Due to a narrow 1 MHz radio bandwidth, a total of 83 frequency channels can be used. Thus, a locally unique frequency channel may be assigned for each radio link, which may be utilized for developing highly scalable link layer protocols. In

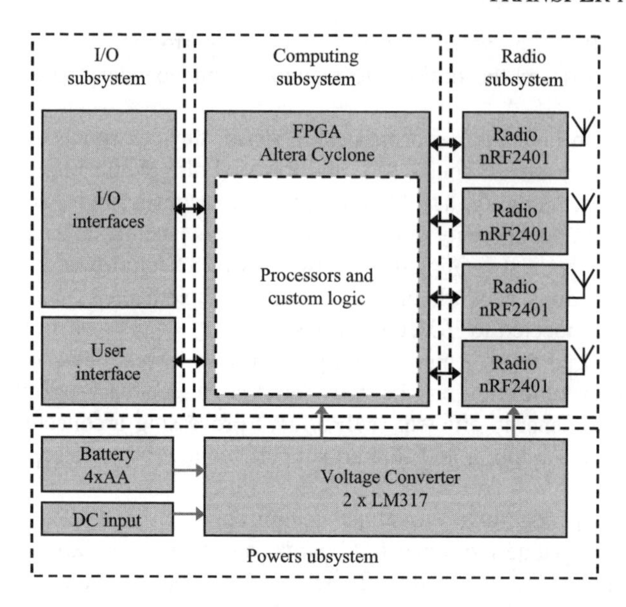

Figure 22.6 Platform hardware architecture. © 2006 ACM. Adapted by permission.

addition, the capability for parallel processing on the computing subsystem allows for all radios to be accessed simultaneously and independently of each other.

The utilization of multiple radios close to each other sets high demands for antenna design, since inter-radio interferences may be severe. This point is critical when transmitting and receiving simultaneously on adjacent channels. To minimize the interference, radios are implemented on separate radio boards and arranged orthogonally to each other. The utilization of full-wave loop antennae further reduces interference, since the antennae have a dipole-type radiation pattern with around +3 dB gain orthogonally to the radio board. The use at a loop antenna fits well for the utilized radio, since they both have high impedance. Hence, the power loss in a simple impedance matching network is low.

The measured range of the radio board at the maximum of 0 dBm transmission power and good conditions is around 100 m. With four simultaneous transmissions and receptions, an applicable radio range reduces to half that distance. Measured radio current consumption in reception mode is 18.2 mA. Transmission mode current consumptions at 0 dBm and −20 dBm power levels are 12.2 mA and 7.8 mA, respectively.

The I/O subsystem consists of a user interface and analog and digital I/O interfaces for sensors and actuators. Two push-buttons, LEDs, and a 7-segment display can be used for implementing a basic user interface. Also, a UART is available for debugging purposes.

The power subsystem regulates system supply voltage. The subsystem gets input voltage from either four AA-type batteries or a mains-power adapter, which is converted to 1.5 V and 3.3 V supply voltages by two LM317 linear voltage regulators (Kohvakka et al. 2006a). Linear regulators are selected due to low electromagnetic interference, small size, and a simple and robust structure.

The multi-radio WSN platform is implemented with three different types of printed circuit boards, namely as a mother board, a radio board and an I/O board. The mother board

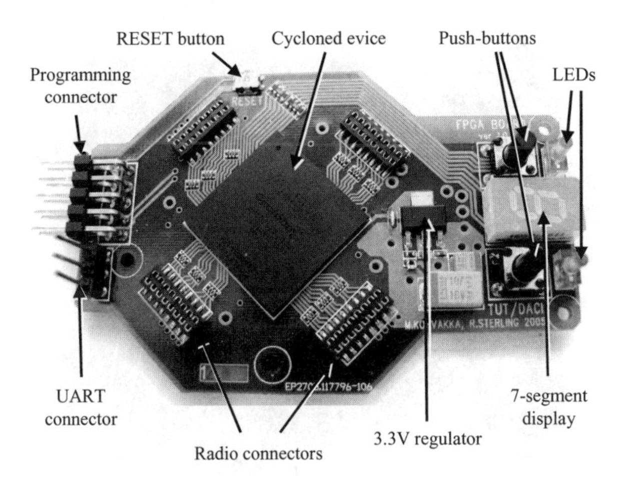

Figure 22.7 Upperside of a mother board.

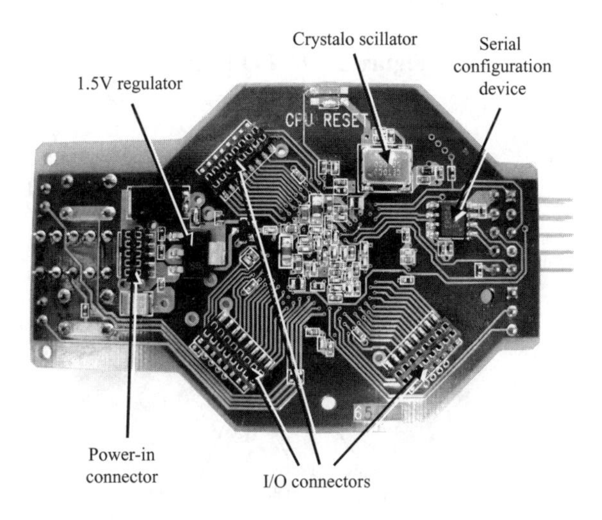

Figure 22.8 Underside of a mother board.

consists of the FPGA chip, the user interface, and the voltage converters, as presented in Figure 22.7 and Figure 22.8. The connectors for radio and I/O boards are on the upper and underside of the board, respectively.

The I/O board is presented in Figure 22.9. The upperside of the board contains signal pre-amplification stages, Analog-to-Digital and Digital-to-Analog Converters and the required voltage regulators. Connectors are provided for a DC supply and a 15-pin general purpose analog and digital I/O. The underside of the board contains terminals for four AA-type batteries.

The complete WSN platform is presented in Figure 22.10 (Kohvakka et al. 2006a). The platform size is 125 × 112 × 30 mm.

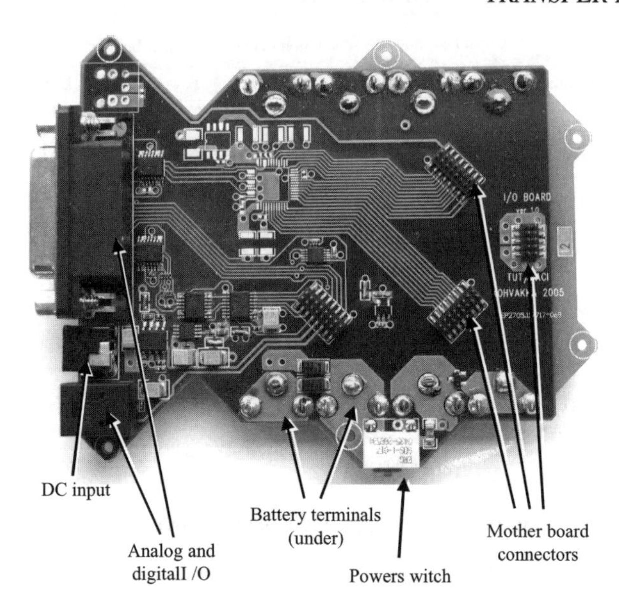

Figure 22.9 I/O board.

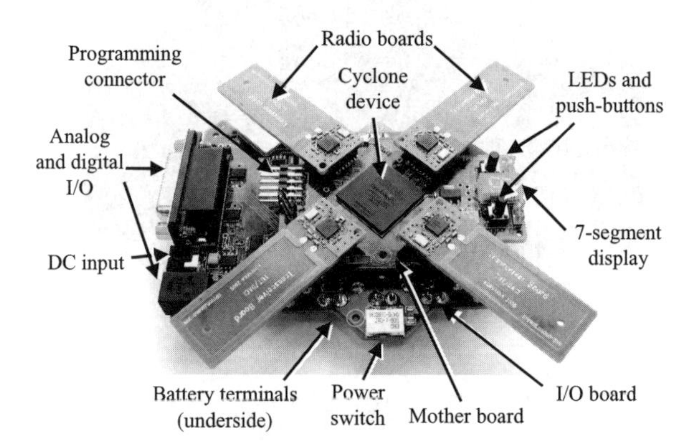

Figure 22.10 Implemented platform. © 2006 ACM. Adapted by permission.

Implementation of a multiprocessor architecture on the FPGA

The Cyclone FPGA contains 20 060 Logic Elements (LEs) and 294 912 bits of embedded dual-port RAM. These resources are suited for implementing a versatile multiprocessor architecture comprised of processors to concurrently execute computationally intensive tasks required by a general WSN application. Furthermore custom hardware accelerators can be synthesized into the FPGA for accelerating functions that are performed faster using dedicated logic rather than executing them on a processor. In addition, power dissipation per operation ratio can be reduced using dedicated logic, which is an important aspect when developing WSN systems (Kohvakka et al. 2006a). For instance, the AES is one of

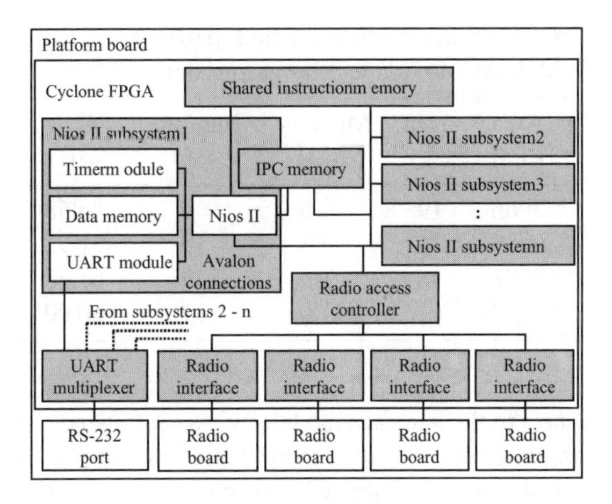

Figure 22.11 Implemented multiprocessor architecture. © 2006 ACM. Adapted by permission.

the functions suitable for acceleration. According to Hämäläinen et al. (2005), up to 48 times speedup in AES calculation is achievable through custom logic implementation on an FPGA compared to an ARM9 processor.

The implemented multiprocessor architecture composed of multiple Nios II processor subsystems is presented in Figure 22.11 (Kohvakka et al. 2006a). Nios II is a 32-bit RISC softcore processor, which is specifically targeted at Altera FPGAs. There are three core variants differing on the pipelines, caches, and arithmetic logic units. A Nios II processor subsystem includes a Nios II core connected to a timer module, UART module, and local data memory. The system peripherals are connected together by Avalon switch fabric. In Avalon, each connected master-slave pair has dedicated wires within each other, leading to a point-to-point connected network.

The IPC is performed via a shared memory. The required memory footprint is reduced by using shared instruction memory, and thus, the same program code for all the processors. Hence, it is required that each processor has a unique PID in logic or non-volatile memory, which is fetched by the processor at the system boot-up. In this architecture, the PIDs are implemented in logic. Clearly, data memory spaces are local for each processor.

The UART module of each subsystem is connected to the UART multiplexer in order to share the single physical RS-232 port on the platform board. This solution is beneficial for debugging and software downloading purposes.

Further, the architecture includes four custom logic radio interface modules for accessing radio front-ends on the platform board. The interface modules are designed to use the radios in *ShockBurst* mode. In this mode of operation, the framing of transmitted data, frame integrity checking using CRC, and terminal addressing are covered by the radio circuitry. However, when operating in this mode, the payload lengths of individual transfers are limited to 256 bits. This shortcoming could be addressed by using the other mode of operation provided by the radio, called *DirectMode*. In this mode, the data framing as well as the frame integrity checking of transmitted bit streams are left for the user's concern. As a consequence, using the DirectMode would increase the flexibility in radio usage, but would also require additional functionality to be implemented by utilizing FPGA resources.

Table 22.1 Resource utilization and performance for 1 to 4 CPUs.

Nios II CPUs	Area (LEs)	Area (%)	Memory (kB)	Memory (%)	F_{max} (MHz)	I (mA)
1	3966	19	2.5	7	85.58	149.9
2	6988	34	5.5	15	71.92	233.3
3	9686	48	8.5	24	67.78	280.7
4	12 463	62	11.0	31	60.00	338.3

The processors connect to the interface modules via a centralized radio access controller, which forwards data and interrupts between radio interfaces and processors. The controller is seen by the processors as a set of addressed registers. There are separate 8-bit wide lines for register address, readable data, and writeable data. Further, these lines are accessed by the processors in a shared manner via Avalon Parallel IO (PIO) interfaces. By using a centralized accessing mechanism of this kind, the number of parallel lines as well as PIO interfaces required is reduced in comparison to the mechanism in which radio interfaces are directly connected to the processors. Currently, the CPUs negotiate the access times to radios using IPC. However, this functionality could be implemented on the radio access controller using the mutex-based reservation mechanism. Furthermore, support for direct data forwarding between radio interfaces can be implemented in the radio access controller, which would be beneficial in very low latency WSN applications.

The software API of a processor is comprised of radio and IPC functions combined with off-the-shelf library routines for accessing UART and timer peripherals. The IPC procedure is implemented as follows. Each processor has a dedicated memory space in the IPC memory for sending messages to other processors. A processor that is sending a message to a remote processor first writes the message and information about the message length to this memory space and issues an IPC request interrupt to the recipient processor. The recipient processor then reads the data from the buffer and issues an IPC confirmation interrupt to the sending processor, which can thereafter perform a new transfer.

The architecture described was implemented on the FPGA using Nios II fast cores with 512 B of instruction cache. Table 22.1 (Kohvakka et al. 2006a) lists the total number of LEs and on-chip memory bits used (excluding memories allocated for program codes), maximum operating frequency (F_{max}) and measured current consumption of the FPGA with instances of the architecture comprising 1–4 processors. The LE consumption measurements show that more than four processors could be synthesized on the FPGA. However, adding a new Nios II core to the system requires 2 kB of on-chip memory at minimum (excluding the data memory demand), and as a consequence, the available program memory for the processors gets too low. Hence, architecture instances of up to four processors are preferred and the unutilized LEs are allocated for implementing parts of communication protocols.

22.2.3 Results

The critical path analysis shows that the system can be driven at 60 MHz clock frequency. However, the radio interface modules are designed to be driven at 25 MHz, and thus,

Table 22.2 Memory footprints for software API functions.

Software part	Code (shared) (B)	RW-data/CPU (B)
Radio API	2344	39
IPC API	628	3
Other SW	1539	1854
Total	4511	1896

separate clock domains are used for Nios II subsystems and radio interface modules by utilizing an on-chip Phase Locked Loop (PLL).

The current consumption of the FPGA board was measured with different numbers of processors synthesized on the FPGA. During the measurements, the processors executed a simple ROM-monitor program at 50 MHz. The radio interfaces were present as well, operating at 25 MHz. The radio boards were not connected to the FPGA board during the measurements. The operating voltage fed to the FPGA was 1.5 V. According to the measurements, the current consumed by the FPGA is practically linearly dependent on the number of processors used. However, the margin in current consumption between the single-processor and dual-processor configurations is distinguishably large. This is due to the fact that the single-processor configuration does not include a UART multiplexer or IPC memory used solely for multiprocessing purposes.

The memory footprints of the object codes required by the software API functions on a single processor are listed in Table 22.2. According to the results, four Nios II CPUs together with the presented software API functions allocate 22.8 kB of the total 32 kB on-chip memory leaving 9.2 kB for application code. In comparison, with two CPUs the free memory for application code increases to 18.4 kB. This should be sufficient assuming that large parts of communication protocols are implemented in hardware logic.

23

Tracking Applications

In a target-tracking application, a specific object is tracked on realtime as it moves around a sensor field. This chapter presents two approaches to target tracking, namely a surveillance system and an indoor-positioning system. In the surveillance system, a node senses its environment and detects movement based on the environmental changes. The movement information is used to detect an unwanted trespasser and alert a security service. The indoor-positioning approach differs from the surveillance system by attaching a positioning node to the tracked object. Obviously, this kind of tracking cannot be used to track hostile targets, but it has the benefit of providing more accurate positioning information. Furthermore, as an individual object is identified based on the tracking node identifier, the movement of a specific object can be easily observed.

In addition to the two tracking approaches, a use case is presented that concentrates on the utilization of the gathered positioning data rather than the actual WSN operation. In the example scenario, WSN is used for positioning players in a team game. A high level API parses and adapts the information so that it can automatically update the state of the game and be used for the game visualization through external UIs.

23.1 Surveillance System*

An indoor-surveillance WSN detects motion and monitors conditions for HVAC control. The example scenario is realized by the implementation of a TUTWSN that measures temperature and detects motion in the public premises of a building. This deployment started in early 2007. The motion-detection alerts are forwarded to a security service.

In addition to implementation and deployment of node platforms, this case illustrates the use of WISENES for the design of application-specific WSN. The high-abstraction level models are first designed in a graphical environment and then implemented on the top of SensorOS on the node platforms.

*With kind permission of Springer Science and Business Media.

Ultra-Low Energy Wireless Sensor Networks in Practice: Theory, Realization and Deployment

23.1.1 Problem Statement

Current burglar alarm systems are integrated in to buildings. The systems use wired connections that cause considerable installation and maintenance costs. While this infrastructure could also be utilized for other building automation, this is rarely the case due to the different suppliers and implementation approaches.

WSNs have several benefits compared to traditional wired surveillance systems. First of all, wireless communication medium allows more flexibility during the installation phase. Further, the maintenance is easier since each unit can basically be freely located without a need to consider other infrastructure. This is beneficial, for example, in cases where the monitoring requirements change. Moreover, WSN nodes typically have integrated sensors that can be utilized by building automation.

23.1.2 Surveillance WSN Design

Several different kinds of sensors can be utilized for detecting motion. These could measure, for example, infrared, vibration, humidity, or temperature and conclude motion detection or the presence of a heat source from a measurement result or their combination. In the presented implementation, infrared sensors are used.

The surveillance WSN is first designed with WISENES. In the WISENES design, the initial selection of the communication protocols and their configuration parameters is based on the designer experience. The suitability of the configuration is then evaluated by simulations, after which the chosen selections can be altered and new simulations run. The final implementation is not started until an acceptable configuration is found in simulations (Kuorilehto et al. 2007c).

The implementation of the surveillance WSN under discussion has three active tasks: motion detection, temperature sensing, and a sink task for data gathering. The relations between tasks together with the basic network architecture is illustrated in Figure 23.1 (Kuorilehto et al. 2007c).

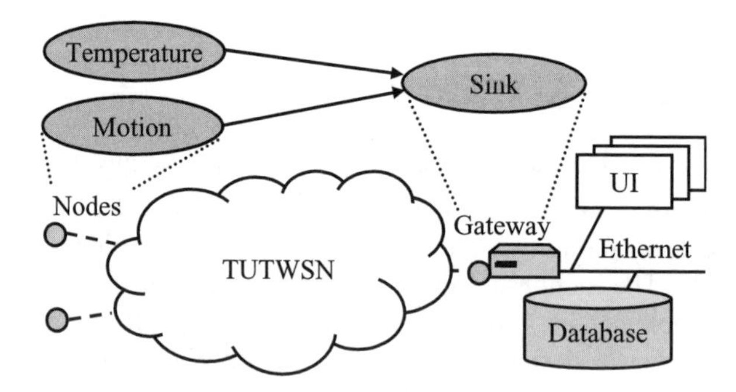

Figure 23.1 The architecture of the surveillance system WSN deployment. With kind permission of Springer Science and Business Media.

WSN requirements

The motion-detection task interfaces with a PIR sensor for generating movement alerts, which are forwarded to the sink task. The temperature-sensing task measures the surrounding temperature periodically and sends it to the sink task. The temperature-sensing task is activated once per minute in all nodes. The motion-detection task is present only on nodes located in public premises of the building, such as aisles. The sink task is executed on a gateway. The sink task stores data to a database and forwards alerts to monitoring UIs.

The requirements for the two sensing tasks differ significantly. The motion detection task is event based and activated by movement. Generated alerts have high priority, are delay critical, and need reliable transmission. Instead, periodic temperature measurements are low-priority packets, and occasional data losses are acceptable. In order to avoid constant maintenance, the WSN should operate approximately one year without battery replacements.

WISENES model design

WSN design in WISENES starts with the definition of the design models. The environment model is defined in XML. For the indoor surveillance use case, it specifies a slightly error-prone communication environment, stationary node locations, typical average values for phenomena, and few target objects with random mobility patterns. A node model is implemented for TUTWSN PIC node by describing its physical characteristics in WISENES XML configuration files. The main functionality for the designed WSN is defined in application and communication models (Kuorilehto et al. 2007c).

The application model consists of the three tasks. Their parameters are given in XML configuration files. Figure 23.2 shows configuration parameters for the motion-detection task. The functionality of tasks is implemented as SDL statecharts. WISENES implementation of the motion-detection task is depicted in Figure 23.3. The task is activated by two events; a motion-detection event from a PIR sensor and a timer event for PIR sensor reactivation. A motion-detection event triggers a data transmission and the initialization of a timeout, while a timer event reactivates the PIR sensor; timeout is needed to avoid continuous alerts. The periodical sensing task initiates a temperature measurement on a timer event, and sends data and initializes the timer after a sensor event. The sink task stores received data to a database.

The TUTWSN communication model configuration is based on the surveillance system requirements. TUTWSN MAC protocol uses a 2 s access cycle to balance energy efficiency, scalability, and delay-critical operation. The number of ALOHA slots is set to four and reservation slots to eight. Bandwidth allocation parameters are explored to obtain the most suitable configuration. Two different cost functions are defined for routing: a delay optimized for motion alerts and a network lifetime optimized for temperature measurements.

WISENES simulation results

The performance of the surveillance WSN with the presented model implementations is evaluated with WISENES simulations. A network of 150 nodes is semi-randomly deployed

```
<application_model>
  <task id="MOTION">
    <interval ms="0"/>
    <priority level="1"/>
    <sensors>
      <sensor id="PIR"/>
    </sensors>
    <data>
      <target task="SINK"/>
      <priority level="1"/>
      <reliable set="yes"/>
      <urgent set="yes"/>
      <optimization target="DELAY"/>
    </data>
  </task>
  <task id="TEMPERATURE">
    ...
  </task>
</application_model>
```

Figure 23.2 WISENES application model XML configuration parameters for motion-detection task. With kind permission of Springer Science and Business Media.

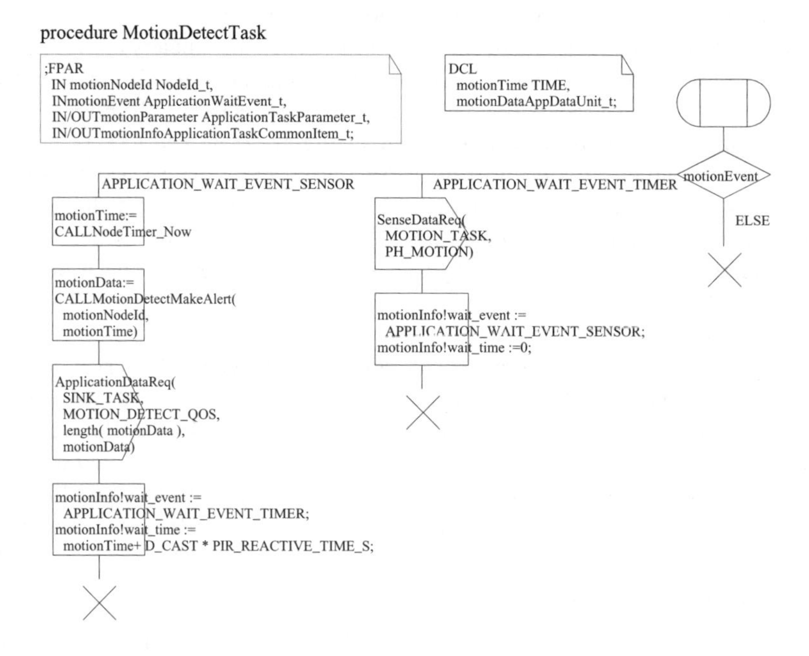

Figure 23.3 WISENES SDL implementation of the motion-detection task. With kind permission of Springer Science and Business Media.

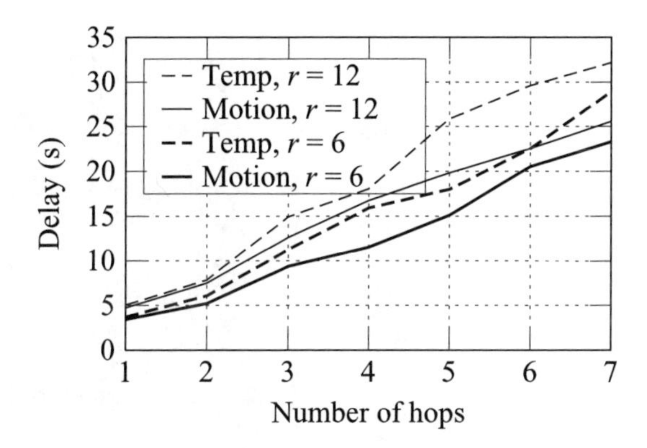

Figure 23.4 Simulated delays for motion alerts and temperature measurements with varying number of hops from the sink node. With kind permission of Springer Science and Business Media.

to the monitored area. Fifty nodes include a PIR sensor and three operate as a sink, while the rest measure temperature and perform data routing. Different configurations are evaluated by parameterizing bandwidth allocation algorithm. A default reservation slot interval (r) sets the maximum time between granted reservation slots for a member node. In simulations r is set to 6 and 12 seconds (Kuorilehto et al. 2007c).

Application requirements are verified by monitoring the delay of motion alerts, and node power consumption. The delay of temperature measurements is considered for comparison. Figure 23.4 shows the average delays of motion alerts and temperature measurements for different r values as the function of the number of hops from a sink node (hop count). As shown, with the same hop count, the delay of alerts is slightly less than that of the temperature measurements. Furthermore, alert packets are typically routed through less hops. The average power consumptions of ten randomly selected headnodes are 650 μW and 635 μW, and of ten subnodes with PIR sensor 434 μW and 443 μW for $r = 6$, and $r = 12$, respectively.

The simulation results of the initial communication model configuration are acceptable. The configuration with $r = 6$ obtains a slightly better performance and balances networking reactiveness and lifetime. Assuming that 90% of the battery capacity can be exploited, the simulated power consumptions obtained using a 1600 mAh CR 123A battery indicate that a lifetime of 276 and 414 days is achievable for the TUTWSN headnode and subnode, respectively. By rotating the headnode and subnode roles a network lifetime of one year is achievable (Kuorilehto et al. 2007c).

23.1.3 WSN Prototype Implementation

After the design has been validated by the simulations, the prototype implementation is made. The surveillance WSN application tasks and protocols are implemented according to the WISENES application and communication models on top of SensorOS. SensorOS offers a congruent programming interface with the WISENES models, which makes a fluent transition between phases possible (Kuorilehto et al. 2007c).

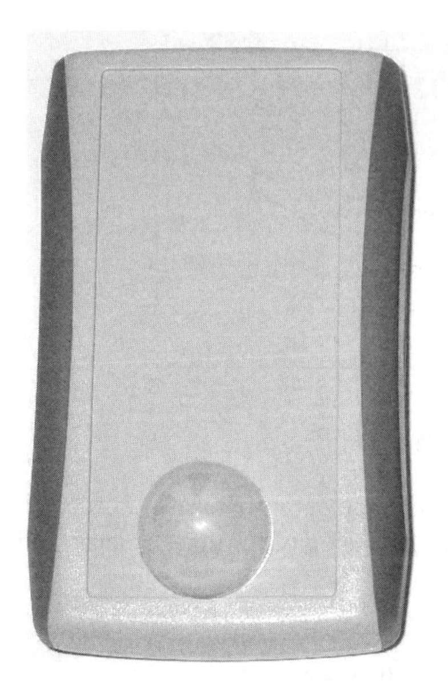

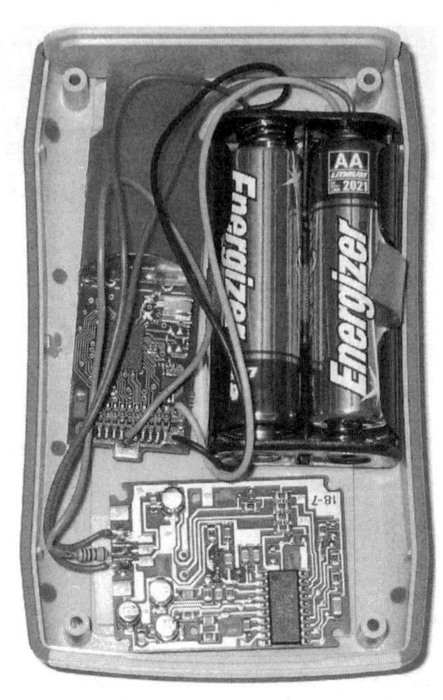

Figure 23.5 TUTWSN temperature-sensing platform equipped with larger batteries and a PIR sensor.

23.1.4 Surveillance WSN Implementation on TUTWSN Prototypes

The prototype implementation of the surveillance WSN is realized in a limited scale with 28 nodes (10 with PIR sensors) on a realistic deployment environment. The TUTWSN temperature-sensing platform is equipped with a PIR sensor and two AA-type batteries, which are encapsulated in plastic enclosures, as depicted in Figure 23.5. The full version of SensorOS is used in nodes equipped with PIR sensors to guarantee reactiveness, while the rest of the nodes have a lightweight kernel to allow longer packet queues. The topology and environment for the prototyped surveillance WSN is depicted on a TUTWSN UI screen capture in Figure 23.6 (Kuorilehto et al. 2007c).

Prototype implementation

Application tasks are implemented as SensorOS threads. Figure 23.7 lists the code of the thread implementing the motion-detection task. For readability, the details of PIR-sensor interfacing and data-message construction are left out. Implementations of other application tasks are similar. The TUTWSN protocol stack is implemented in four threads. API, data routing, and MAC channel access are implemented as separate threads similar to WISENES. MAC and routing layer management operations are integrated to a same thread in order to diminish IPC messaging. TUTWSN protocols are parameterized with the values obtained in WISENES design.

The subnode implementation on a TUTWSN PIC node with full-feature SensorOS consumes 38.1 kB of code and 2253 B of data memory. These take up 60% and 57% of the available resources, respectively. The data memory consumption does not include a heap reserved for dynamic memory. The implementation of temperature measurement

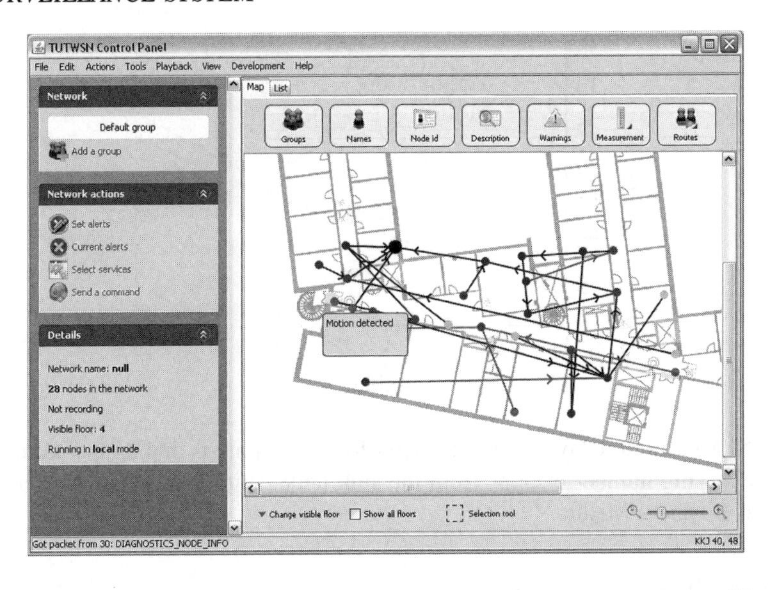

Figure 23.6 A screen capture from TUTWSN UI illustrating prototyped surveillance WSN operation. With kind permission of Springer Science and Business Media.

```
void motion_detect void) {
  os_eventmask_t event;
  os_ipc_message_t *msg;

  activate_pir ();
  while (1) {
    event = os_wait_event(
              EVENT_ALARM | EVENT_PIR_INTERRUPT);
    if (event & EVENT_ALARM) {
      activate_pir();
    } else if (event & EVENT_PIR_INTERRUPT) {
      msg= make_motion_alert_msg (SINK_TASK);
      os_msg_send (API_PID, msg);
      os_set_alarm (PIR_REACTIVATE_TIMEOUT_MS);
    }
  }
}
```

Figure 23.7 The implementation of the motion-detection application task as a SensorOS thread. With kind permission of Springer Science and Business Media.

application and TUTWSN protocols on top of a lightweight SensorOS kernel takes 58.2 kB of code and 2658 B of data memory, which take up 91% and 67% of the available memory, respectively.

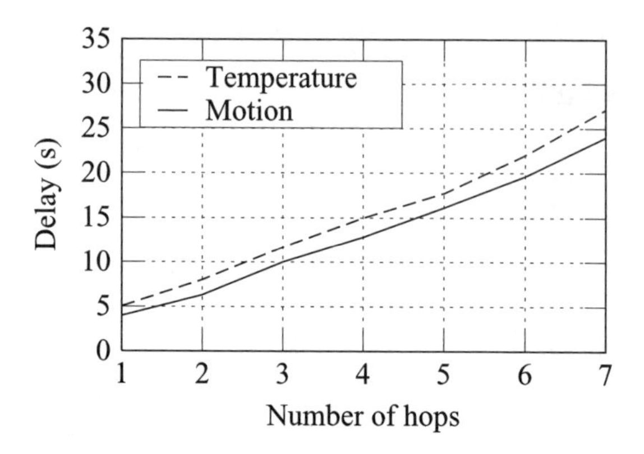

Figure 23.8 Measured delays with prototypes for motion alerts and temperature measurements with varying number of hops from the sink node. With kind permission of Springer Science and Business Media.

Prototype results

The same performance metrics gathered from WISENES simulations are also evaluated for the prototype implementation in order to verify the accuracy of WISENES models and to validate the implementation for final deployment of the surveillance system. The delays of motion alerts and temperature measurement data as the function of hop count are depicted in Figure 23.8. In the prototype implementation r is 6.

The results correspond closely to those obtained from WISENES, with the average difference being 8.9% for motion alerts and 10.2% for temperature data. The slightly better performance obtained in WISENES is due to less retransmissions, which results from the more optimistic environment model used in the simulations. The larger number of nodes in WISENES simulations is balanced by three sinks. The averages of measured power consumptions are 693 μW for a headnode and 467 μW for a subnode equipped with a PIR sensor. For comparison, the measured power consumption of a subnode running a lightweight kernel without a PIR sensor in the same WSN averages 257 μW. These findings are also analogous with WISENES results.

Both simulated and measured results from the prototyped surveillance WSN show that TUTWSN implementation meets the requirements set for the application. The increase of delay with a larger number of hops can be balanced by adding intermediate gateways near nodes located on long communication paths. To conclude, TUTWSN can also be configured for applications with burst-type data profiles.

23.2 Indoor Positioning

Basically, the knowledge of WSN node positions provides added value to every application scenario. Typically in static deployments, the position information can be gathered once and stored to a database for future queries. However, in dynamic WSNs including moving objects, such as humans, additional solutions or algorithms for node localization and positioning need to implemented. GPS offers a standard technology for positioning but is not suitable for all applications. A widely adopted alternative solution is *relative positioning* based on the RSSI values or ultrasound traverse times.

This use case presents a positioning system for multi-object tracking. Example application scenarios for the system include the monitoring of children in a kindergarten, positioning of valuable assets, and several military and training systems. The positioning system is implemented with TUTWSN and uses RSSI values from the radio transceiver for determining the node positions in respect to stationary reference nodes.

23.2.1 Problem Statement

Since a GPS receiver requires a satellite connection it cannot be used, for example, inside buildings. Therefore, for the purposes of indoor positioning another approach is needed. Dense WSN deployments with a rich set of sensors provide the means for the accurate positioning of one or more moving objects.

Buildings are a challenging environment for WSN communication and positioning implementation. The environments are typically heterogeneous with several interference sources, such as walls, windows, metallic pipeworks, and electrical equipment. Walls can vary from concrete walls to lightweight walls or glass partitions. Furthermore, there may be several sources causing radio interference, such as WLANs or other communication networks operating on the same frequency band.

Since each WSN node has an integrated radio transceiver, an RSSI based positioning system is a straightforward solution. Due to the utilization of radio transmissions, the heterogeneity of an indoor environment and the unideal characteristics of the nodes need to be considered. For example, the antenna design and node alignment have a significant effect on the receiver gain.

Security issues are critical in positioning of humans or assets. The information must be protected against interception and the operation of the network should be guaranteed in spite of interruption attempts. This set tight requirements for authentication and encryption.

23.2.2 Implementation

The configuration of TUTWSN for indoor positioning consists of MAC protocol adaptation for realtime positioning, a dedicated node platform implementation for human positioning, and a TUTWSN Application Panel modification for position visualization. Due to the scalability and relative simplicity of the positioning algorithm, it can be either distributed to nodes or executed on a central monitoring station.

WSN requirements

The positioning does not require any extra components on nodes, but a node needs to be able to measure or estimate RSSIs of received transmissions. Since RSSI evaluation requires additional circuitry on the transceiver, it is not often available in commercial low-cost radios. Therefore, some other mechanisms need to be implemented.

The TUTWSN positioning system estimates the location of a *mobile node* relative to other nodes. Thus, the system requires a *backbone WSN* with stationary *reference nodes* at known locations. The reference nodes need to be accurately synchronized, since the relative positioning combines information originating from different sources.

Since the backbone WSN consists of nodes that are stationary located, they can be mains powered. Thus, the MAC protocol design and implementation does not need to optimize energy efficiency. Instead, it is a key design objective for mobile nodes, since their constant maintenance is difficult considering that, for example, nodes are integrated to valuable assets, such as multimedia projectors.

Positioning algorithm

As the reception of data is typically more energy consuming than data sending in WSNs, the positioning algorithm attempts to minimize needed data reception at the mobile nodes. Therefore, these nodes send periodical *location beacons* received by the reference nodes. On reception of a beacon, a reference node estimates the distance to the source of the location beacon. The distance estimates are then combined to obtain a location estimate of a mobile node.

The main principle of the positioning algorithm is presented in Figure 23.9. In the figure, nodes 1, 2, and 3 are reference nodes, while the dark grey node is a mobile node. Each reference node generates an imaginary circle around itself from the RSSI value of the received location beacon. The radius r_X of the circle is based on the theoretical models for radio wave propagation. Once the circles of several reference nodes are combined, the location of a mobile node is in the intersection of the circles.

As the computation capacity of WSN nodes is very limited, the arithmetic operations required for calculating the intersection of the circles are relatively complex. Therefore, the positioning algorithm is simplified by utilizing squares instead of circles. The length of a square edge is $2r_X$, as illustrated in Figure 23.10. As depicted, the accuracy of the positioning does not suffer considerably.

In order to avoid the errors caused by the challenging indoor communication environment, the positioning algorithm needs to perform error correction, and data filtering and

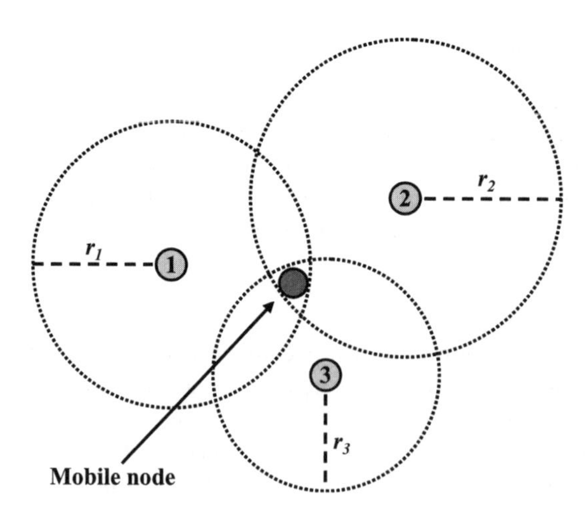

Figure 23.9 The principle of positioning algorithm based on the distance estimates according to RSSIs.

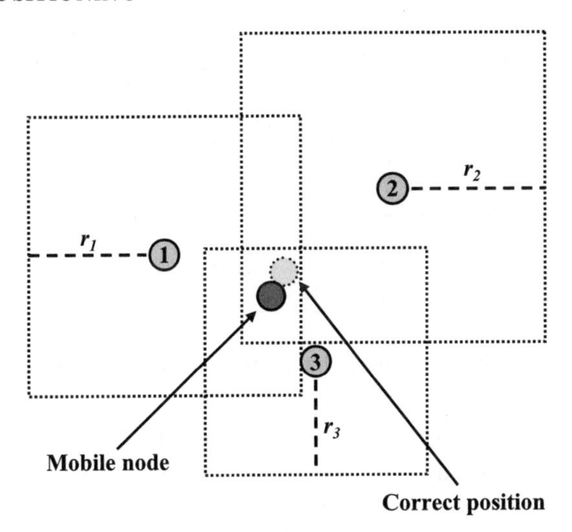

Figure 23.10 The simplification of the positioning algorithm by using squares instead of circles in order to diminish calculation requirements.

averaging. Still, walls and reflections may generate errors that can be avoided only by using a WSN with a dense enough backbone.

To conclude, the main phases of the positioning algorithm are

1. A mobile node sends periodic location beacons.
2. Reference nodes that receive a location beacon calculate a distance estimate and timestamp it.
3. Distance estimates are used for generating squares at a specific time window.
4. The estimates are filtered and averaged through collaborative processing.
5. The intersection of the squares is calculated to obtain a location estimate.

TUTWSN node platform for human positioning

This section presents the hardware implementation of a TUTWSN badge node designed for access control and localization applications. The platform is designed according to the following two main requirements. The platform should be small-sized and easy to carry. On the other hand, the platform should be maintenance-free, such that battery lifetime extends to several years. In addition, the platform should have a simple user interface consisting of Light Emitting Diodes (LEDs) and push-buttons for indicating application status information and for allowing interactivity with users.

The hardware architecture of the prototype is presented in Figure 23.11. The platform utilizes the 2.4 GHz Nordic Semiconductor nRF24L01 transceiver for providing compatibility with other TUTWSN nodes. The antenna is implemented as a dipole along two orthogonal edges of the PCB, which results in a nearly omnidirectional radiation pattern.

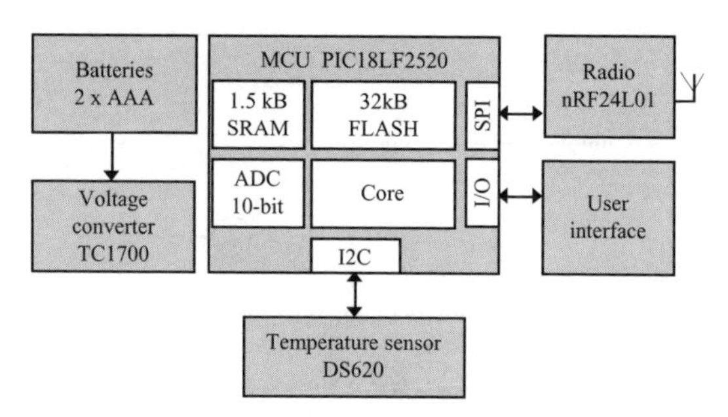

Figure 23.11 TUTWSN badge node hardware architecture.

The operation of the TUTWSN badge node is controlled by a Microchip PIC18LF2520 MCU. The MCU is selected due to its small physical size, low cost, adequate memory resources, and compatible processor architecture with other utilized PIC MCUs providing good reusability of software. The MCU has 32 kB program memory and 1.5 kB data memory, which is sufficient for the access control and localization applications. Internal 256 B EEPROM is used for nonvolatile configuration data. The controller has high energy efficiency and versatile power-saving modes allowing accurate and low-energy wake-up timing with external 32.768 kHz clock crystal. The active mode operation is clocked by an internal adjustable clock source. The utilized clock frequency is 4 MHz resulting in 1 MIPS performance. An internal 10-bit ADC is utilized for monitoring the battery energy status. A user interface is implemented with two push-buttons and three LEDs.

The sensing subsystem consists of a Dallas Semiconductor DS620 temperature sensor and a digital 3-axis VTI SCA3000 accelerometer. The accelerometer can be used for determining whether a user is walking or sitting, which assists with the positioning application to determine user locations more accurately.

The accelerometer is equipped with an SPI and it has internal memory buffers of 64 samples per axis for output acceleration data. The active power consumption of the sensor is only 120 μA. Due to the low power consumption, it is possible to maintain the accelerometer over long periods of active time, for example during an entire working day. The sensor is configured to set an interrupt signal to wake up the MCU as the acceleration exceeds a predetermined threshold value. Thus, all movements of the node are detected even if the MCU is in sleep mode most of the time.

The supply energy is obtained from two AAA batteries resulting in 2.4 V supply voltage. The battery voltage is regulated to 2.0 V by TC1700 linear regulators for minimizing node current consumption.

The implemented TUTWSN badge node prototype encased in a plastic container is presented in Figure 23.12.

The antenna structure has to fit a narrow, L-shaped area, which is limited by other components. Omnidirectional gain was also an objective. The final structure is a modified dipole, the properties of which are presented in Figure 23.13.

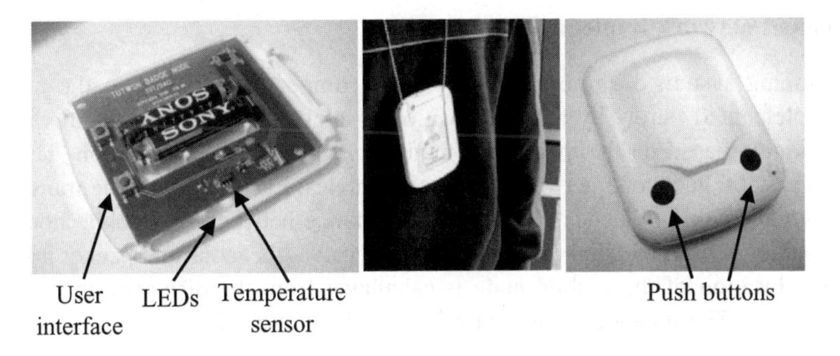

Figure 23.12 TUTWSN badge node for human positioning.

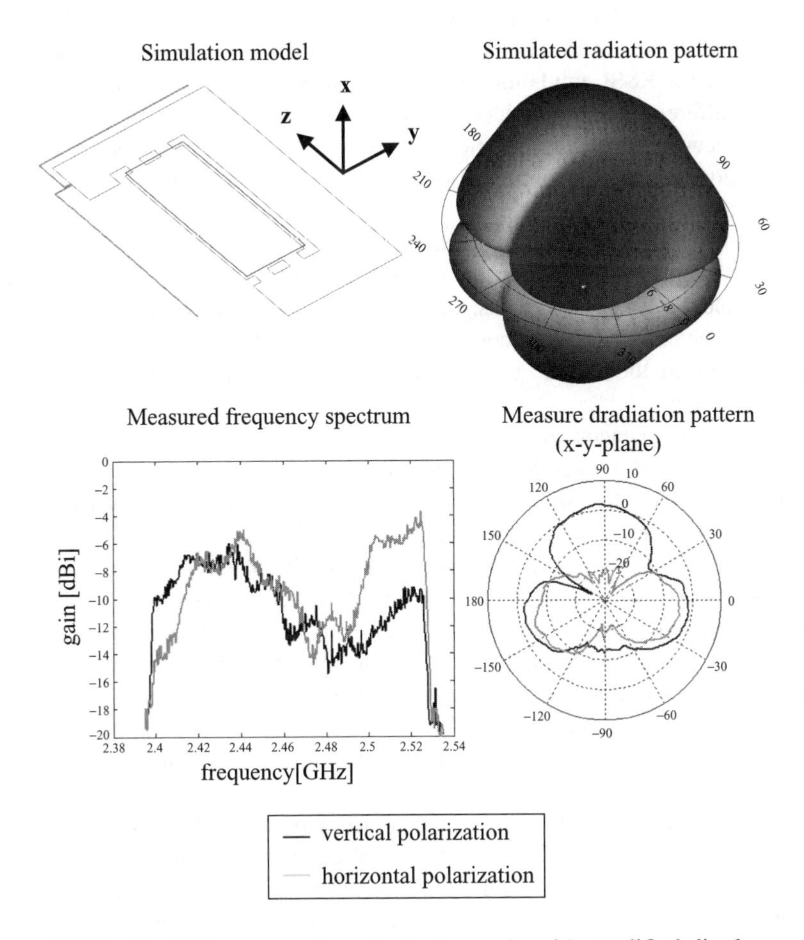

Figure 23.13 Radiation properties of a badge node with modified dipole antenna.

Measurement network implementation

The positioning system does not need any extra components other than the nodes. It is instead implemented purely by relative positioning algorithms. The algorithms implement node positioning based on the theoretical models of radio wave propagation. The network consists of a static backbone network and mobile nodes. Each mobile node transmits short beacons, which backbone nodes listen to. The distance between the mobile node and the backbone nodes is derived from the power of the received beacons and packet information. Finally, the location of the mobile node is calculated from the distances and locations of backbone nodes. The location information is visualized in a UI.

Positioning implementation with TUTWSN

The radio transceivers used in TUTWSN node platforms do not support the RSSI evaluation of a received transmission. Therefore, an approach similar to that of normal TUTWSN operation is used for RSSI emulation of the location beacons. Thus, a mobile node sends beacons with different TX power levels and indicates the used power level in the beacon payload. Security is included in the implementation as an option that enables a MIC for authentication of location beacons. Data encryption is not supported in the current implementation.

Mobile nodes are implemented with the TUTWSN badge node platform. Due to the limited resources and the requirement of a very low power operation, mobile nodes do not support full-scale TUTWSN protocols. Instead, they send only periodic location beacons to a predefined channel. Each beacon is sent with four different TX power levels. An optional downlink to the mobile nodes is enabled by a short ALOHA period after the beacon transmissions.

If a single location beacon channel is overloaded causing frequent collisions, the mobile nodes can be configured to operate on several different channels that are either randomized dynamically or set statically. Similarly, the reference nodes can either negotiate the listened channels during runtime or the channel can be set statically.

The reference nodes communicate with each other by using TUTWSN protocols in a normal configuration. A reference node may operate either as a headnode or as a subnode. Headnodes maintain their own access cycle, while every node communicates with its parents during the active period of the parent cluster. The rest of the time, instead of entering in to a sleep mode, the nodes listen to the location beacon channel for detecting beacons. Reference nodes are synchronized to a global time source that is a gateway to the network. The synchronization is updated with the network maintenance communication (cluster and network beacons).

A communication overview with two reference and two mobile nodes is depicted in Figure 23.14. Mobile nodes randomize the interval of their location beacons in predefined limits in order to avoid constant collisions. As shown in the figure, the reference nodes do not continuously receive at the location beacon channel. Thus, some beacons might be missed, but this can be balanced by additional reference nodes.

Even though the positioning algorithm is simple enough to be executed on reference WSN nodes, the current implementation gathers the received location beacons to a gateway for further processing. The gateway first makes the distance estimates, combines the

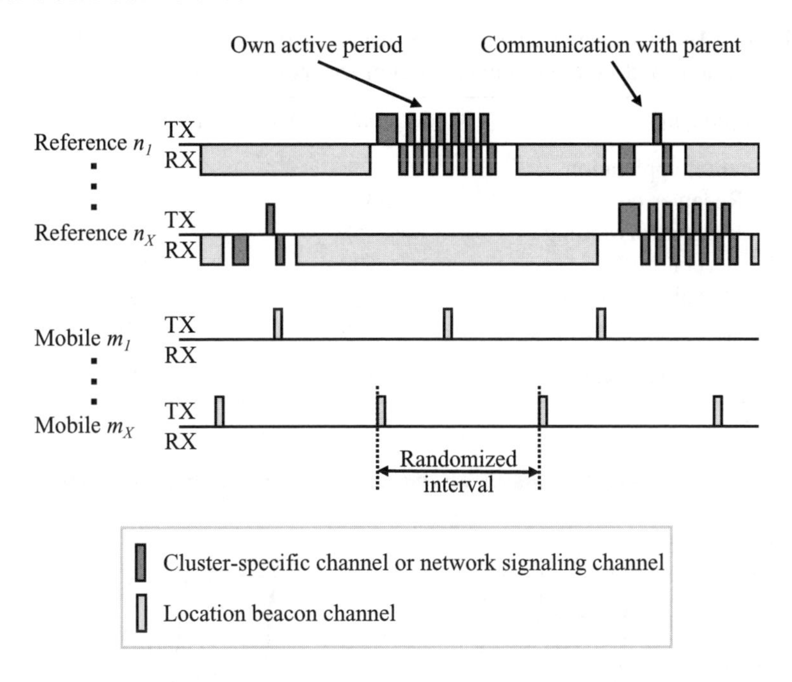

Figure 23.14 The integration of location beacons to the backbone reference WSN communication.

simultaneous estimates from different reference nodes, and calculates a location estimate separately for each mobile node.

Prototype results

The listening of location beacons in reference nodes is implemented as a separate location application that enters a receive mode and handles location beacons. Due to the more restricted resources of the mobile nodes, a completely tailored implementation is needed. The memory consumptions of location application and tailored implementation are summarized in Table 23.1

The code memory usage of the location application totals 3486 B. The implementation uses 281 B of data memory. These figures include beacon and downlink data buffers for five packets. A tailored implementation of the mobile node consumes 10.5 kB of code and 558 B of data memory.

The estimated locations of mobile nodes are visualized in a TUTWSN Application Panel. An example screen capture from the UI is depicted in Figure 23.15. In order to visualize the principles of the positioning, the screen capture also shows the squares defining the distance estimates that are used for the location estimation.

Currently, the positioning system achieves a room-level accuracy. Thus, a mobile node can be positioned to a correct room with a relatively high probability. Yet, the experiments show that the antenna of the mobile node platform has a significant effect on the accuracy. This implies that the accuracy can be tailored application specifically by modifying the antenna and adjusting the density of the backbone WSN.

Table 23.1 Code and data memory usage of different indoor-positioning applications and tailored implementation for mobile node.

Component	Code memory (B)	Data memory (B)
Location application		
Receiver core	2302	18
Beacon forwarding	58	0
Beacon buffer	702	176
Downlink data buffer	424	87
TOTAL	3486	281
Mobile node (total)	10 754	558

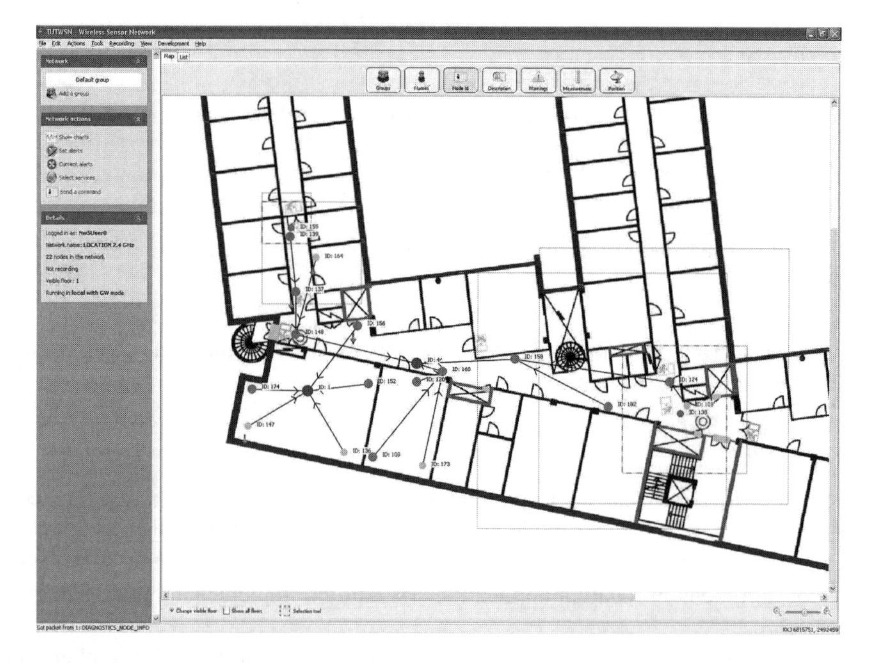

Figure 23.15 The visualization of the locations of positioned nodes in TUTWSN UI with the imaginary squares used in the calculation.

23.3 Team Game Management

A deployed WSN measures and gathers a wide variety of information. Typically, a small subset of the information is utilized by client applications, while the rest of the data are discarded. However, by refining a larger set of gathered information, several new conclusions may be drawn. This is especially true for monitoring WSNs. Interpretation of a single value only determines a temporal condition at a single location, but by combining several different aspects more high level results can be obtained.

This use case describes an example application scenario that utilizes WSN data in the field of sports and physical activities. The main objective is to implement an interface to the WSN data in a way, which allows a scalable and adaptive data gathering for different purposes. Thus, the use case does not concentrate on the WSN configuration but to an application domain-specific interface.

23.3.1 Problem Statement

When considering a typical team game, the interesting aspects that can instrumented with WSNs are the locations of players and, for example, the ball, physical condition of active players, and the occurrences of game-specific events, such as goals. In most cases, the instrumentation can be done with common, low-cost sensors that are already available in TUTWSN nodes. Additional sensors are required, if for example, the heart rate of a player should be monitored.

From a WSN point of view, a team game does not cause any additional requirements to the network. Thus, if the main objective is the positioning of players, the operation is similar to any other positioning application. Similarly, the network itself does not regard whether other sensor data relates to a team game or some other target environment.

Due to the similarities to other application scenarios, existing WSN configurations are suitable for the team game management use case. Thus, the main challenge is the refining and utilization of WSN data. As an example, gathered node locations and sensory data could be automatically combined to generate several interesting statistics that describe the progress of a game. Such statistics include, for example, ball possession, attack and defence proportions, average heart rate of individual players, etc.

The combining of sensory data to higher layer results is application specific. The usability of an API suffers if it needs to be implemented separately for each team game and for each WSN data usage model. Therefore, the API need to be scalable for both different kind of games and different types of visualization and utilization applications.

23.3.2 Implementation

The use-case implementation describes and visualizes the utilization of location information in a team game. The positioning of objects, including players, a ball, and other relevant game equipment, is implemented by GPS for outdoor games and by the algorithms and methods presented in Section 23.2 for indoor games. Each positioned object is integrated with a node designed for athletes.

The implemented API, referred to as TUTWSN *Sports API*, gathers, refines, and publishes the results relating to the game. The Sports API connects to Ethernet GW nodes through a socket connection. This approach also makes it possible to input WSN data from WISENES.

TUTWSN node platform for players

The design objectives of the platform utilized in the team game application are a very small size, around a one year battery lifetime, and at least 50 m radio range. The node should be carried easily on the wrist or in the pocket of a player. Most preferably, the shape of

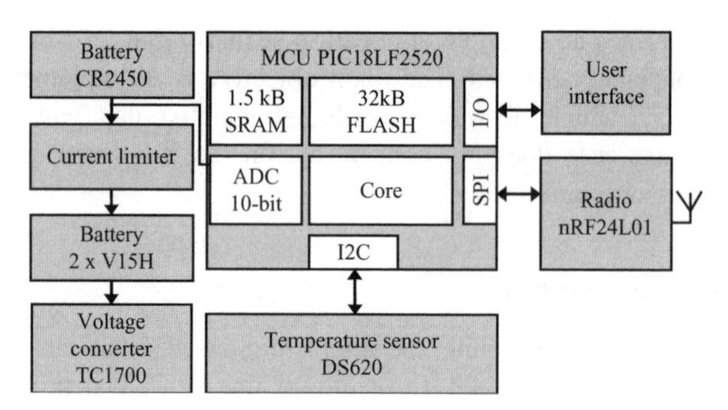

Figure 23.16 The hardware architecture of a TUTWSN watch node.

the node is round like a wrist watch. Accordingly, a new platform called TUTWSN watch node is designed.

The hardware architecture of the platform is presented in Figure 23.16. The hardware architecture is quite similar to the TUTWSN badge node. For compatibility with other TUTWSN nodes, the platform utilizes the Nordic Semiconductor nRF24L01 transceiver. The computing subsystem consists of PIC18LF2520 MCU having an adequate memory resources and small size, as required. The platform is equipped with a Dallas DS620 digital temperature sensor. The temperature sensor can be used, for example, to determine whether the sensor is being carried in the hand or pocket, or is not in active use. A 500 mAh CR2450 primary lithium battery is selected as the primary power source due to its high capacity, small size, and suitable shape. Since the maximum output current of the lithium battery is limited to around 200 μA, a small battery pack consisting of two Varta V15H NiMH rechargeable batteries is used as an energy buffer. A simple user interface is implemented by an LED and push-button.

The implemented TUTWSN watch node is presented in Figure 23.17. The PCB is circular shaped having a diameter of 30 mm. The PCB is enclosed in a plastic casing presented in the figure. The upperside of the board contains the antenna, radio, temperature sensor, push-button, and LED. The underside of the board contains the MCU and the power subsystem.

The small circular shape of the platform sets high requirements of the antenna design. The whole antenna structure can take only a very little segment of the circular PCB. Antenna should also be as isotropic as possible because a orientation of antenna can not be known in a operation situation. A double sided antenna structure was selected due to good simulation results. Properties of the antenna are presented in Figure 23.18.

Sports API implementation

The basic architecture of the Sports API is depicted in Figure 23.19. The API connects through a sockets interface directly to the Ethernet GW nodes and utilizes TUTWSN API messages for communication. The Sports API constantly listens for incoming data and sends required configuration messages to the WSN.

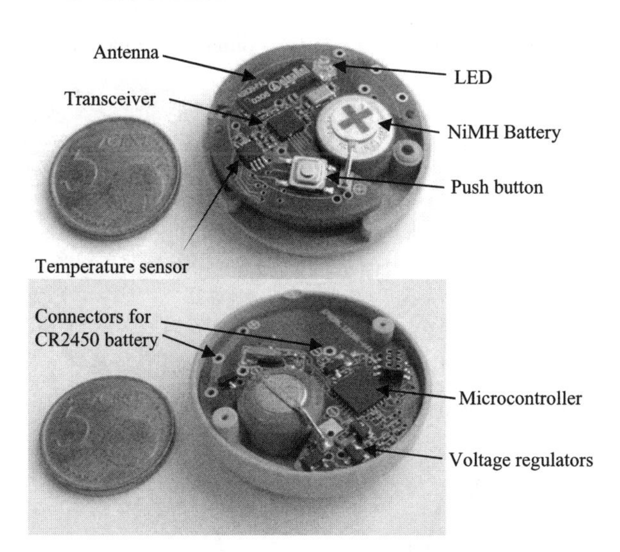

Figure 23.17 Implemented TUTWSN watch node.

The Sports API is initialized by reading the game and application parameters from two XML files that are referred to as `init` XML and `rules` XML. The `init` XML file defines the relations of players and other objects to the WSN nodes. In addition, the file describes the composition of teams and other relevant combinations of objects. The `rules` XML file defines the relations of the WSN data to the internal state and higher level results of the game.

The Sports API maintains the internal state of the game according to the received data and `rules` XML. The API offers an interface to UIs for gathering the game state and results. Currently, the Sports API supports Web-based UIs and implements the visualization with Asynchronous JavaScript and XML (AJAX) architecture (Garrett 2005). The information is passed to UIs in a format that allows a straightforward integration of, for example, Google Maps®.

23.3.3 Example Application Scenario

The operation of the Sports API is illustrated by implementing an example application on top of it. The application, called *Grand Theft Sensor*, is a simplified version of a traditional *capture the flag* game, in which two teams attempt to protect their own flag and capture the flag of the other team.

In the Grand Theft Sensor, teams score points if they succeed in capturing a flag. A flag either belongs to a team or is neutral, in which case it can be captured by either of the teams. Team members and all flags are tagged with WSN nodes that use GPS for positioning. The game state is maintained by the Sports API.

The players, teams, and flags are initialized by the `init` XML file depicted in Figure 23.20. The XML links the objects to WSN node IDs. The game-state management

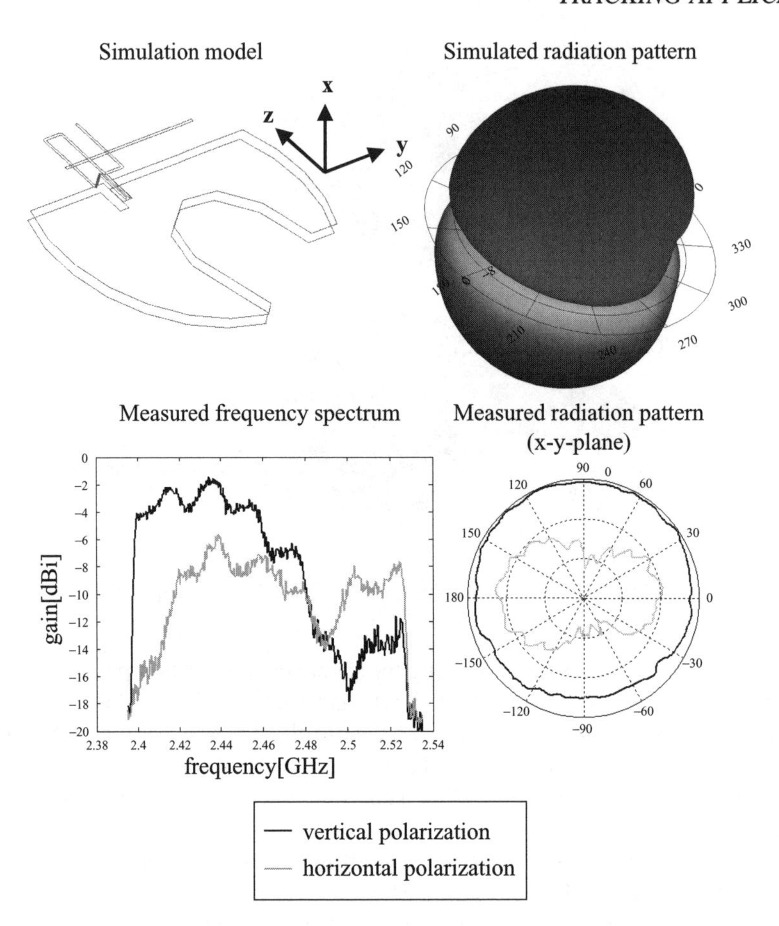

Figure 23.18 Radiation properties of a watch node with double-sided antenna.

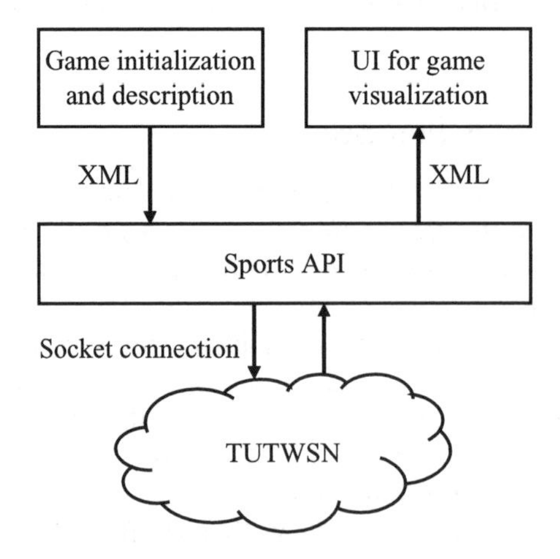

Figure 23.19 Overview of the Sports API architecture and communication.

```
<init>
  <teams>
    <team id="1"name="Red">
      <object type="player" node="1"
              name="Jorma"role="captain"/>
      <object type="player" node="2"
              name="Kaaleppi" role="offensive"/>
    </team>
    <team id="2" name="Blue">
      ...
    </team>
  </teams>

  <field>
    <object type="flag" node="6"owner="none"
            name="forest" role="normal"/>
    <object type="flag" node="7"owner="Red"
            name="Red Base" role="base"/>
  </field>
</init>
```

Figure 23.20 The init XML for the example Grand Theft Sensor game.

```
<rules>
  <game rounds="3"round_time="10"/>
  <contact diameter="10"/>

  <event name="encounter" teampoints="0"
         personalpoints="0"
         message="encountered"/>
  <event name="protect" teampoints="0"
         personalpoints="0"
         message="is now protecting own flag"/>
  <event name="capture" teampoints="5"
         personalpoints="shared"
         message="captured the flag"/>
  <event name="discover" teampoints="3"
         personalpoints="3"
         message="discovered the flag"/>
</rules>
```

Figure 23.21 The `rules` XML defining the events in Grand Theft Sensor game.

is based on the `rules` XML presented in Figure 23.21. For Grand Theft Sensor, the XML defines the events that can occur during the game and how the events affect the game state, i.e. current score.

In Grand Theft Sensor, each event originates from the GPS positions of nodes. An event is generated if a player *encounters* another player, *protects* a flag of their own team, *captures* an opponent's flag, or *discovers* a neutral flag. An event occurs if GPS coordinates

Figure 23.22 A screen capture of a Web browser UI illustrating the example game.

indicate that two objects are less than 10 m apart (the diameter can be adjusted from `rules` XML).

The initialization of Sports API and the visualization of the game state according to the information provided by the API is implemented in a Java Servlet. After initialization, the Servlet polls Sports API at an adjustable rate for game state updates. As a response it receives an XML file that is utilized for visualizing the game state through Google Maps® and another XML file that lists occurred events. A screen capture of the Web-browser UI generated by the Servlet is depicted in Figure 23.22.

In general, Sports API implements an architecture that allows a straightforward game visualization and state management based on the gathered WSN data. The implementation of Grand Theft Sensor indicates that TUTWSN is suitable for instrumenting players in games with a moderate tempo, but for fast indoor games, such as floorball (similar to a sort of indoor hockey), more bandwidth and computing capacity are required. The accuracy and response times required for positioning in such cases set significant requirements for network performance.

Part VI

CONCLUSIONS

24

Conclusions

For a while, wireless sensor networks have been an open battlefield of opinions and narrow-minded truths. On the other hand, the visions and dreams about wireless sensor networks realizing ubiquitous, intelligent environments have been very high compared to the art of current technology. In brief, the enabling technology base has been missing to turn the dreams into a reality.

We focused on embedded wireless sensor network nodes, approximately the scale of 50–200 cm^3 and up to 200g (grams) for a targeted node. With our physical nodes we demonstrated that current off-the-self components are already feasible for very good sensor networks, but naturally more integrated single-chip solutions would improve the platforms further.

The size of the nodes is an interesting and controversial issue in general. Many applications require the smallest possible size, but it is simply not feasible to implement, for example, antennas and energy source in a minuscule casing. There is also the fallacy of size and simplicity: fully capable WSN nodes could contain more complex networking functionality than mobile phones.

Energy efficiency has been a major objective since the number of WSN nodes is expected to be very high. Replacing batteries for a 1000-node network every three months in a building would mean at least 300 hours of maintenance work per year and a lot of wastage. For this reason, the nodes should harvest their energy from the environment or the battery lifetime should be the same as the node itself. Much more research is required into power sources and developing more energy efficient radio and processor technologies.

The WSN MAC protocol plays a key role in energy consumption minimization. As data frames are typically very short and a radio consumes practically the same amount of power whether transmitting or receiving a data frame or listening for possible incoming data, it is essential to keep the radio active only when actually transmitting or receiving data. The idle listening function is one of the major deficiencies of current WSN MAC proposals. As presented in this book, TUTWSN minimizes the idle listening by synchronization of each transmitted and received frame. As local synchronization only with direct neighbors is applied, adequate synchronization accuracy is easily achieved with low-power MCUs.

Ultra-Low Energy Wireless Sensor Networks in Practice: Theory, Realization and Deployment

An important issue is also the behavior of the radio environment in real WSN applications. Experimental studies indicated that the network topology is dynamic even with static nodes due to very low transmission power levels and altering radio link qualities. WSN protocols should update network topology energy efficiently for maximizing link qualities and routing performance. In the current proposals, the network is typically assumed to be static and neighbor-discovery energy is omitted. The minimization of energy consumption in static network leads to unsatisfactory results in practical implementations. Subsequently, it means that the focus in WSN research should be rapidly shifted from static to dynamic networks.

TUTWSN utilizes several techniques for maximizing network energy efficiency and performance in dynamic networks. As examples, TUTWSN utilizes contention-based slots in each superframe for network associations, a neighbor-discovery protocol for signaling neighborhood information in network, and a network signaling channel for minimizing neighbor-discovery energy. In addition, network topology is formed as a multicluster tree improving network robustness against link failures and allowing cost-based multipath routing.

Once the base technology has been matured, the question is what comes next. When networks really can work in thousands of nodes deployed everywhere, the focus is again on identifying applications and how to develop them. The most important issue is that users should not interact with the network but that it should work in background. The same applies to the development of applications. Instead of programming a single node, there must be methods to abstract the applications and let automated tools take care of the distribution of tasks to the nodes. The challenge is to efficiently implement such methods to optimize energy and memory capacity.

This book has discussed our work covering the most demanding features of wireless sensor networks – very low energy consumption, multihopping mesh networking, free mobility of nodes, and truly autonomous operation, all at the same time. We can claim that we have succeeded in solving some of the major obstacles that have stood in our way but look forward to the day when our dream of ubiquitous computing finally comes true. However, the research never stops and we must admit to having witnessed one of the great milestones in WSN development with the creation of TUTWSN.

References

Aberer K, Hauswirth M and Salehi A 2006 The global sensor network middleware for efficient and flexible deployment and interconnection of sensor networks. Technical Report 2006-006, LSIR. Available at: `http://infoscience.epfl.ch/`.

Akyildiz IF and Kasimoglu IH 2004 Wireless sensor and actor networks: Research challenges. *Ad Hoc Networks (Elsevier)* **2**(4), 351–367.

Akyildiz IF, Su W, Sankarasubramaniam Y and Cayirci E 2002a A survey on sensor networks. *IEEE Communications Magazine* **40**(8), 102–114.

Akyildiz IF, Su W, Sankarasubramaniam Y and Cayirci E 2002b Wireless sensor networks: A survey. *Computer Networks (Elsevier)* **38**(4), 393–422.

Andel TR and Yasinsac A 2006 On the credibility of manet simulations. *IEEE Computer* **39**(7), 48–54.

Archer W, Levis P and Regehr J 2007 Interface contracts for TinyOS. *Proc. 6th Int'l Conf. on Information Processing in Sensor Networks (IPSN'07)*, pp. 158–165, Cambridge, MA, USA.

Arvind DK and Wong KJ 2004 Speckled computing: Disruptive technology for networked information appliances. *Proc. IEEE Int'l Symposium on Consumer Electronics (ISCE'04)*, pp. 334–338, Reading, UK.

Avancha S, Undercoffer J, Joshi A and Pinkston J 2004 Security for wireless sensor networks. *Wireless Sensor Networks*, 1st edn, Springer pp. 253–275.

Bachir A and Barthel D 2005 Localized max-min remaining energy routing for WSN using delay control. *Proc. Int'l Conf. on Communications (ICC)*, vol. 5, pp. 3302–3306.

Baker D and Ephremides A 1981 The architectural organization of a mobile radio network via a distributed algorithm. *IEEE Transactions on Communications* **29**(11), 1694–1701.

Bakshi A and Prasanna VK 2004 Algorithm design and synthesis for wireless sensor networks. *Proc. 2004 Int'l Conf. on Parallel Processing*, pp. 423–430, Montreal, Quebec, Canada.

Baldwin P, Kohli S, Lee EA, Liu X and Zhao Y 2004 Modeling of sensor nets in ptolemy II. *Proc. 3rd Int'l Symposium on Information Processing in Sensor Networks (IPSN'04)*, pp. 359–368, Berkeley, CA, USA.

Bao L and Garcia-Luna-Aceves J 2001 A new approach to channel access scheduling for ad hoc networks. *Proc. 7th Annual Int'l Conf. on Mobile Computing and Networking (Mobicom'01)*, pp. 210–221, Rome, Italy.

Barr R, Bicket JC, Dantas DS, Du B, Kim TD, Zhou B and Sirer EG 2002 On the need for system-level support for ad hoc and sensor networks. *ACM SIGOPS Newsletter on Operating Systems Review* **36**(2), 1–5.

Batalin MA, Rahimi M, Yu Y, Liu D, Kansal A, Sukhatme GS, Kaiser WJ, Hansen M, Pottie GJ, Srivastava M and Estrin D 2004 Call and response: Experiments in sampling the environment

Proc. 2nd Int'l Conf. on Embedded Networked Sensor Systems (SenSys'04), pp. 25–38. ACM Press, Baltimore, MD, USA.

Beckwith R, Teibel D and Bowen P 2004 Report from the field: Results from an agricultural wireless sensor network. *Proc. 29th Annual IEEE Int'l Conf. on Local Computer Networks (LCN'04)*, pp. 471–478, Tampa, FL, USA.

Bein D and Datta AK 2004 A self-stabilizing directed diffusion protocol for sensor networks. *Proc. Int'l Conf. on Paraller Processing Workshops (ICPP)*, pp. 69–76.

Benini L, Bogliolo A and Micheli GD 2004 A survey of design techniques for system-level dynamic power management. *IEEE Transactions on Very Large Scale Integration Systems* **8**(3), 299–316.

Beutel J 2006 Fast-prototyping using the BTnode platform. *Proc. Design, Automation and Test in Europe (DATE'06)*, pp. 977–982, Munich, Germany.

Beutel J, Dyer M, Hinz M, Meier L and Ringwald M 2004a Next-generation prototyping of sensor networks. *Proc. 2nd Int'l Conf. on Embedded Networked Sensor Systems (SenSys'04)*, pp. 291–292. ACM Press, Baltimore, MD, USA.

Beutel J, Kasten O, Mattern F, Römer K, Siegemund F and Thiele L 2004b Prototyping wireless sensor network applications with BTnodes. *Proc. 1st European Workshop on Wireless Sensor Networks (EWSN'04)*, pp. 323–338 number 2920 in *LNCS*. Springer-Verlag, Berlin, Germany.

Beutel J, Kasten O, Ringwald M, Siegemund F and Thiele L 2003 Bluetooth smart nodes for ad-hoc networks. Technical Report 167, Computer Engineering and Networks Laboratory, ETH Zurich.

Bharghavan V, Demers A, Shenkar S and Zhang L 1994 MACAW: A media access protocol for wireless LANs. *Proc. ACM SIGCOMM'94*, pp. 212–225, London, UK.

Bhatti S, Carlson J, Dai H, Deng J, Rose J, Sheth A, Shucker B, Gruenwald C, Torgerson A and Han R 2005 MANTIS OS: An embedded multithreaded operating system for wireless micro sensor platforms. *Mobile Networks and Applications* **10**(4), 563–579.

Blu 2004 *Specification of the Bluetooth System.*

Bokareva T, Bulusu N and Jha S 2004 A performance comparison of data disseminating protocols for wireless sensor networks. *Proc. Global Telecommunications Conf. Workshops*, pp. 85–89.

Bonivento A, Carloni LP and Sangiovanni-Vincentelli A 2006 Platform based design for wireless sensor networks. *Mobile Networks and Applications* **11**(4), 469–485.

Boulis A, Han CC and Srivastava MB 2003 Design and implementation of a framework for efficient and programmable sensor networks. *Proc. 1st Int'l Conf. on Mobile Systems, Applications, and Service (MobiSys'03)*, pp. 187–200, San Francisco, CA, USA.

Bush LA, Carothers CD and Szymanski BK 2005 Algorithm for optimizing energy use and path resilience in sensor networks. *Proc. Second European Workshop on Wireless Sensor Networks (EWSN)*, pp. 391–396.

Cha H, Choi S, Jung I, Kim H, Shin H, Yoo J and Yoon C 2007 RETOS: Resilient, expandable, and threaded operating system for wireless sensor networks. *Proc. 6th Int'l Conf. on information Processing in Sensor Networks (IPSN'07)*, pp. 148–157, Cambridge, MA, USA.

Chan H, Perrig A and Song D 2004 Key distribution techniques for sensor networks. *Wireless Sensor Networks*, 1st edn. Springer, pp. 277–303.

Chen G, Branch J, Pflug MJ, Zhu L and Szymanski BK 2004 SENSE: A sensor network simulator. In *Advances in Pervasive Computing and Networking* (ed. Szymanksi BK and Yener B), Springer, pp. 249–267.

Cheong E, Lee EA and Zhao Y 2006 Viptos: A graphical development and simulation environment for tinyos-based wireless sensor. Technical Report UCB/EECS-2006-15, UCB.

Chlamtac I and Farag'o A 1994 Making transmission schedules immune to topology changes in multihop packet radio networks. *IEEE/ACM, Trans. Networking* **2**(1), 23–29.

Chong CY and Kumar SP 2003 Sensor networks: Evolution, opportunities, and challenges. *Proceedings of the IEEE* **91**(8), 1247–1256.

Chonggang W, Sohraby K, Yueming H, Bo L and Weiwen T 2005 Issues of transport control protocols for wireless sensor networks. *Proc. Int'l Conf. on Communications, Circuits and Systems (ICCCAS'05)*, vol. 1, pp. 422–426, Hong Kong, China.

Chu D, Lin K, Linares A, Nguyen G and Hellerstein JM 2006 sdlib: a sensor network data and communications library for rapid and robust application development. *Proc. 5th Int'l Conf. on Information Processing in Sensor Networks (IPSN'06)*, pp. 432–440, Nashville, TN, USA.

Crossbow Technology 2004a Micaz wireless measurement system. Available at: `xbow.com/Products/Product_pdf_files/Wireless_pdf/6020-0060-01_A_MICAz.pdf`.

Crossbow Technology 2004b Stargate x-scale processor platform. Available at: `http://www.xbow.com/Products/Product_pdf_files/Wireless_pdf/6020-0049-01_B_STARGATE.pdf`.

Crossbow Technology 2006 Crossbow technology–wireless sensor networks, inertial & gyro systems, smart dust, advanced sensors. Available at: `http://www.xbow.com`.

Crossbow Technology 2007a Mica2 wireless measurement system. Available at: `http://www.xbow.com/Products/Product_pdf_files/Wireless_pdf/MICA2_Datasheet.pdf`.

Crossbow Technology 2007b Mica2dot wireless microsensor mote. Available at: `www.xbow.com/Products/Product_pdf_files/Wireless_pdf/MICA2DOT_Datasheet.pdf`.

Crowley K, Frisby J, Murphy S, Roantree M and Diamond D 2005 Web-based real-time temperature monitoring of shellfish catches using a wireless sensor network. *Sensors and Actuators A: Physical* **122**(2), 222–230.

Culler DE, Hill J, Buonadonna P, Szewczyk R and Woo A 2001 A network-centric approach to embedded software for tiny devices. *Lecture Notes in Computer Science* vol. 2211. Springer, pp. 114–130.

Curino C, Giani M, Giorgetta M, Giusti A, Murphy AL and Picco GP 2005 TinyLIME: Bridging mobile and sensor networks through middleware. *Proc. 3rd IEEE Int'l Conf. on Pervasive Computing and Communications (PerCom'05)*, pp. 61–72, Kauai Island, HI, USA.

Diaz M, Rubio B and Troya JM 2005 A coordination middleware for wireless sensor networks. *Proc. 2005 Systems Communications (ICW'05, ICHSN'05, ICMCS'05, SENET'05)*, pp. 377–382, Montreal, Quebec, Canada.

Doherty L, Warneke B, Boser B and Pister K 2001 Energy and performance considerations for smart dust. *Int'l Journal of Parallel and Distributed Systems and Networks* **4**(3), 121–133.

Du W, Deng J, Han YS and Varshney PK 2006 A key predistribution scheme for sensor networks using deployment knowledge. *IEEE Transactions on Dependable and Secure Computing* **3**(1), 62–77.

Dubois-Ferriere H, Meier R, Fabre L and Metrailler P 2006 Tinynode: A comprehensive platform for wireless sensor network applications. *Proc. 5th Int'l Conf. on Information Processing in Sensor Networks (IPSN'06)*, pp. 358–365, Nashville, TN, USA.

Dulman S and Havinga PJM 2003 A simulation template for wireless sensor networks. *Proc. IEEE Int'l Symposium on Autonomous Decentralized Systems (ISADS'03)*, Pisa, Italy. Fast abstract.

Dunkels A, Grönvall B and Voigt T 2004 Contiki–a lightweight and flexible operating system for tiny networked sensors. *Proc. 29th Annual IEEE Int'l Conf. on Local Computer Networks (1st IEEE Workshop on Embedded Networked Sensors) (EmNetS'04)*, pp. 455–462, Tampa, FL, USA.

Dunkels A, Schmidt O, Voigt T and Ali M 2006 Protothreads: Simplifying event-driven programming of memory-constrained embedded systems. *Proc. 4th Int'l Conf. on Embedded Networked Sensor Systems (SenSys'06)*, pp. 29–42, Boulder, CO, USA.

El-Hoiydi A 2002a Aloha with preamble sampling for sporadic traffic in ad hoc wireless sensor networks. *Proc. IEEE Int'l Conf. on Communications (ICC'02)*, vol. 5, pp. 3418–3423, New York City, NY, USA.

El-Hoiydi A 2002b Spatial TDMA and CSMA with preamble sampling for low power ad hoc wireless sensor networks. *Proc. IEEE Int'l Symposium on Computers and Communications (ISCC'02)*, pp. 685–692, Taromina, Italy.

El-Hoiydi A, Decotignie JD and Hernandez J 2004 Low power MAC protocols for infrastructure wireless sensor networks. *Proc. Fifth European Wireless Conf.*, Barcelona, Spain.

Elson J and Estrin D 2004 *Sensor Networks: A Bridge to the Physical World.* Springer, pp. 3–20.

Elson J, Girod L and Estrin D 2004 EmStar: Development with high system visibility. *IEEE Wireless Communications* **11**(6), 70–77.

Elson J and Parker A 2006 Tinker: A tool for designing data-centric sensor networks. *Proc. 5th Int'l Conf. on Information Processing in Sensor Networks (IPSN'06)*, pp. 350–357, Nashville, TN, USA.

Enea 2006 OSE – Enea. Available at: `http://www.enea.com`.

Enz CC, El-Hoiydi A, Decotignie JD and Peiris V 2004 WiseNET: An ultralow-power wireless sensor network solution. *Computer* **37**(8), 62–70.

Eschenauer L and Gligor VD 2002 A key-management scheme for distributed sensor networks. *Proc. 9th ACM Conf. on Computer and Communications Security (CCS'02)*, pp. 41–47, Washington D.C., USA.

Eswaran A, Rowe A and Rajkumar R 2005 Nano-RK: An energy-aware resource-centric RTOS for sensor networks. *Proc. 26th IEEE Int'l Real-Time Systems Symposium (RTSS'05)*, pp. 256–265, Miami, FL, USA.

Finn G 1987 Routing and addressing problems in large metropolitan-scale internetworks. Technical Report, University of Southern California.

Flowers D and Yang Y 2007 AN1066–MiWi wireless networking protocol stack. Available at: `http://www.microchip.com/downloads/en/AppNotes/01066a.pdf`.

Fok CL, Roman GC and Lu C 2005 Rapid development and flexible deployment of adaptive wireless sensor network applications. *Proc. 25th IEEE Int'l Conf. on Distributed Computing Systems (ICDCS'05)*, pp. 653–662, Columbus, OH, USA.

Fraboulet A, Chelius G and Fleury E 2007 Worldsens: Development and prototyping tools for application specific wireless sensors networks. *Proc. 6th Int'l Conf. on Information Processing in Sensor Networks (IPSN'07)*, pp. 176–185, Cambridge, MA, USA.

FreeRTOS 2006 FreeRTOS–a free RTOS for ARM7, ARM9, cortex-M3, MSP430, MicroBlaze, AVR, x86, PIC18, H8S, HCS12 and 8051. Available at: `http://www.freertos.org`.

Fuggetta A, Picco GP and Vigna G 1998 Understanding code mobility. *IEEE Transactions on Software Engineering* **24**(5), 342–361.

Fullmer C and Garcia-Luna-Aceves J 1997 Solutions to hidden terminal problems in wireless networks. *Proc. ACM SIGCOMM'97*, pp. 39–49, Cannes, France.

Gang Lu, Krishnamachari B and Raghavendra CS 2004 An adaptive energy-efficient and low-latency MAC for data gathering in wireless sensor networks. *Proc. Parallel and Distributed Processing Symposium (IPDPS'04)*, pp. 224–231.

Garcia RR 2006 Understanding the zigbee stack. Available at: `http://www.eetasia.com/`.

Garrett JJ 2005 Ajax: A new approach to web applications. Available at: `http://www.adaptivepath.com/publications/essays/archives/`.

Gay D, Levis P, Behren Rv, Welsh M, Brewer E and Culler D 2003 The nesC language: A holistic approach to networked embedded systems. *Proc. ACM SIGPLAN 2003 Conf. on Programming Language Design and Implementation (PLDI'03)*, pp. 1–11, San Diego, CA, USA.

Gehrmann C 2002 Bluetooth security white paper. White Paper 1.01, Bluetooth SIG Security Expert Group.

Gelernter D 1985 Generative communication in linda. *ACM Programming Languages and Systems* **7**(1), 80–112.

Gerla M, Kwon T and Pei G 2000 On demand routing in large ad hoc wireless networks with passive clustering. *Proc. IEEE WCNC'00*, pp. 100–105, Chicago, IL, USA.

Goussevskaia O, do Val Machado M, Mini RAF, Loureiro AAF, Mateus GR and Nogueira JM 2005 Data dissemination based on the energy map. *IEEE Communications Magazine* **43**(7), 134–143.

Gozdecki J, Jajszczyk A and Stankiewicz R 2003 Quality of service terminology in IP networks. *IEEE Communications Magazine* **41**(3), 153–159.

Gracanin D, Eltoweissy M, Wadaa A and DaSilva LA 2005 A service-centric model for wireless sensor networks. *IEEE Journal on Selected Areas in Communications* **23**(6), 1159–1166.

Gu L and Stankovic JA 2006 t-kernel: Providing reliable OS support to wireless sensor networks. *Proc. 4th Int'l Conf. on Embedded Networked Sensor Systems (SenSys'06)*, pp. 1–14, Boulder, CO, USA.

Gummadi R, Gnawali O and Govindan R 2005 Macro-programming wireless sensor networks using kairos. *Lecture Notes in Computer Science*, vol. 3560. Springer, pp. 126–140.

Guo C, Zhong L and Rabaey J 2001 Low power distributed mac for ad hoc sensor radio networks. *Global Telecommunications Conf. (GLOBECOM'01)*, vol. 5, pp. 2944–2948, San Antonio, TX, USA.

Gupta V, Wurm M, Zhu Y, Millard M, Fung S, Gura N, Eberle H and Shantz SC 2005 Sizzle: A standards-based end-to-end security architecture for the embedded internet. *Pervasive and Mobile Computing (Elsevier)* **1**(4), 425–445.

Gura N, Patel A, Wander A, Eberle H and Shantz SC 2004 Comparing elliptic curve cryptography and RSA on 8-bit CPUs. In *Lecture Notes in Computer Science* (ed. Joye M and Quisquater JJ) vol. 3156, Springer, pp. 119–132.

Gutiérrez JA, Callaway EH and Barrett RL 2004 *Low-Rate Wireless Personal Area Networks: Enabling Wireless Sensors with IEEE 802.15.4*, vol. 1, IEEE Press.

Hadim S and Mohamed N 2006 Middleware challenges and approaches for wireless sensor networks. *IEEE Distributed Systems Online* **7**(3), 1–15.

Hämäläinen P 2006 Cryptographic security designs and hardware architectures for wireless local area networks. PhD Thesis, Tampere University of Technology.

Hämäläinen P, Heikkinen J, Hännikäinen M and Hämäläinen TD 2005 Design of transport triggered architecture processors for wireless encryption. *Proc. 8th Euromicro Conf. on Digital System Design (DSD '05)*, pp. 144–152, Porto, Portugal.

Hämäläinen P, Kuorilehto M, Alho T, Hännikäinen M and Hämäläinen TD 2006 Security in wireless sensor networks: Considerations and experiments. In *Lecture Notes in Computer Science* (ed. Vassiliadis S, Wong S and Hämäläinen TD) vol. 4017, Springer, pp. 167–177.

Han CC, Kumar R, Shea R, Kohler E and Srivastava M 2005 A dynamic operating system for sensor nodes. *Proc. 3rd Int'l Conf. on Mobile Systems, Applications, and Services (MobiSys'05)*, pp. 163–176, Seattle, WA, USA.

Hart JK and Martinez K 2006 Environmental sensor networks: A revolution in the earth system science? *Earth-Science Reviews* **78**(3-4), 177–191.

Havinga P and Smit G 2000 Design techniques for low power systems. *Journal of Systems Architecture* **46**(1), 1–21.

He T, Krishnamurthy S, Luo L, Yan T, Gu L, Stoleru R, Zhou G, Cao Q, Vicaire P, Stankovic JA, Abdelzaher TF, Hui J and Krogh B 2006 VigilNet: An integrated sensor network system for energy-efficient surveillance. *ACM Transactions on Sensor Networks* **2**(1), 1–38.

Heile B 2006 Wireless sensors and control networks: Enabling new opportunities with ZigBee. Available at: `http://www.zigbee.org`.

Heinzelman WB, Chandrakasan AP and Balakrishnan H 2002 An application-specific protocol architecture for wireless microsensor networks. *IEEE Wireless Communications* **1**(4), 660–670.

Heinzelman WB, Murphy AL, Carvalho HS and Perillo MA 2004 Middleware to support sensor network applications. *IEEE Network* **1**(18), 6–14.

Heinzelman WR, Chandrakasan A and Balakrishnan H 2000 Energy-efficient communication protocols for wireless microsensor networks. *Proc. Hawaii Int'l Conf. on Systems Sciences*, Hawaii.

Hill J and Culler D 2002 Mica: A wireless platform for deeply embedded networks. *IEEE Micro* **22**(6), 12–24.

Hill J, Horton M, Kling R and Krishnamurthy L 2004 Wireless sensor networks: The platforms enabling wireless sensor networks. *Communications of the ACM* **6**(47), 41–46.

Hill J, Szewczyk R, Woo A, Hollar S, Culler D and Pister K 2000 System architecture directions for networked sensors. *Proc. 9th ACM Int'l Conf. on Architectural Support for Programming Languages and Operating Systems (ASPLOS'00)*, pp. 94–103, Cambridge, MA, USA.

Hong S and Kim TH 2003 SenOS: State-driven operating system architecture for dynamic sensor node reconfigurability. *Proc. Int'l Conf. on Ubiquitous Computing (ICUC)*, pp. 201–204, Seoul, Korea.

Huang Q, Bhattacharya S, Lu C and Roman GC 2005 FAR: Face-aware routing for mobicast in large-scale sensor networks. *ACM Transactions on Sensor Networks (TONS)* **1**(2), 240–271.

Huang Q, Lu C and Roman GC 2003 Spatiotemporal multicast in sensor networks. *Proc. 1st Int'l Conf. on Embedded networked sensor systems (SenSys'03)*, pp. 205–217, Los Angeles, CA, USA.

Hwang LJ, Sheu ST, Shih YY and Cheng YC 2005 Grouping strategy for solving hidden node problem in IEEE 802.15.4 LR-WPAN. *Proc. First Int'l Conf. on Wireless Internet (WICON)*, pp. 26–32, Visegrad-Budapest, Hungary.

IEE 1997 *IEEE Std 802.11-1997 Information Technology – Telecommunications and Information Exchange Between Systems – Local And Metropolitan Area Networks – specific Requirements – Part 11: Wireless Lan Medium Access Control (MAC) and Physical Layer (PHY) Specifications.*

IEE 2002 *IEEE Standard for Information Technology – Telecommunications and Information Exchange Between Systems – Local and Metropolitan Area Networks – Specific Requirements – Part 15.1: Wireless Medium Access Control (MAC) and Physical Layer (PHY) Specifications for Wireless Personal Area Networks (WPANs).*

IEE 2003a *IEEE Standard for Information Technology – Telecommunications and Information Exchange Between Systems – Local and Metropolitan Area Networks – Specific Requirements – Part 15.3: Wireless Medium Access Control (MAC) and Physical Layer (PHY) Specifications for High Rate Wireless Personal Area Networks (WPANs).*

IEE 2003b *IEEE Standard for Information Technology – Telecommunications and Information Exchange Between Systems – Local and Metropolitan Area Networks – Specific Requirements – Part 15.4: Wireless Medium Access Control (MAC) and Physical Layer (PHY) Specifications for Low-Rate Wireless Personal Area Networks (LR-WPAN).*

IEEE 2006 IEEE p1451.5 project. Available at: `http://grouper.ieee.org/groups/1451/5/`.

Intanagonwiwat C, Govindan R, Estrin D, Heidemann J and Silva F 2003 Directed diffusion for wireless sensor networking. *IEEE/ACM Transactions on Networking* **11**(1), 2–16.

ITU 2002 *Specification and Description Language (SDL).*

Jaikaeo C, Srisathapornphat C and Shen CC 2000 Querying and tasking in sensor networks. *Proc. SPIE's 14th Annual Int'l Symposium on Aerospace/Defense Sensing, Simulation, and Control (Digitization of the Battlespace V)*, vol. 4037, pp. 184–197, Orlando, FL, USA.

Jain N and Das S 2001 A multichannel CSMA MAC protocol with receiver-based channel selection for multihop wireless networks. *Proc. 9th Int'l Conf. on Computer Communications and Networks (ICCCN'01)*, pp. 432–439, Scottsdale, AZ, USA.

Jakobsson M and Wetzel S 2001 Security weaknesses in Bluetooth. *Proc. Cryptographer's Track at RSA Conf. 2001 (CT-RSA 2001)*, pp. 176–191, San Francisco, CA, USA.

James J and Hall P 1989 *Handbook of Microstrip Antennas* vol. 1, Peter Peregrinus Ltd.

JCP 2006 *JSR 256 Mobile Sensor API Specification Version 0.27.*

J-Sim 2006 J-Sim home page. Available at: `http://www.j-sim.org`.

Juntunen JK, Kuorilehto M, Kohvakka M, Kaseva VA, Hännikäinen M and Hämäläinen TD 2006 WSN API: Application programming interface for wireless sensor networks. *Proc. The 17th Annual IEEE Int'l Symposium on Personal, Indoor and Mobile Radio Communications (PIMRC'06)*, Helsinki, Finland.

Kahn R, Gronemeyer S, Burchfiel J and Kunzelman R 1978 Advances in packet radio technology. *Proceedings of the IEEE* **66**(11), 1468–1496.

Kangas T 2006 Methods and implementations for automated system on chip architecture exploration. PhD Thesis, Tampere University of Technology.

Karir M, Polley J, Blazakis D, McGee J, Rusk D and Baras JS 2004 ATEMU: A fine-grained sensor network simulator. *Proc. 1st IEEE Int'l Conf. on Sensor and Ad Hoc Communication Networks (SECON'04)*, pp. 145–152, Santa Clara, CA, USA.

Karl H and Willig A 2005 *Protocols and Architectures for Wireless Sensor Networks*. John Wiley & Sons Ltd, Chichester.

Karlof C, Sastry N and Wagner D 2004 TinySec: A link layer security architecture for wireless sensor networks. *Proc. 2nd Int'l Conf. on Embedded Networked Sensor Systems (SenSys'04)*, pp. 162–175, Baltimore, MD, USA.

Karlof C and Wagner D 2003 Secure routing in wireless sensor networks: Attacks and countermeasures. *Ad Hoc Networks (Elsevier)* **1**(2–3), 293–315.

Karn P 1990 MACA–A new channel access method for packet radio. *Proc. ARRL/CRRL Amateur Radio 9th Computer Networking Conf.*, pp. 134–140.

Karp B and Kung HT 2000 GPSR: Greedy perimeter stateless routing for wireless networks. *Proc. 6th Annual Int'l Conf. on Mobile Computing and Networking (MobiCom'00)*, pp. 243–254, Boston, MA, USA.

Kelly IV C, Ekanayake VN and Manohar R 2003 SNAP: A sensor network asynchronous processor. *Proc. 9th Int'l Symposium on Asynchronous Circuits and Systems (ASYNC'03)*, pp. 24–35, Vancouver, BC, Canada.

Kleinrock L and Tobagi F 1975 Packet switching in radio channels: Part I – The carrier sense multiple access modes and their throughput-delay characteristics. *IEEE Transactions on Communications* **COM-23**(12), 1400 –1416.

Kohl J and Neuman C 1993 The Kerberos network authentication service (V5) RFC 1510.

Kohvakka M, Arpinen T, Hännikäinen M and Hämäläinen TD 2006a High-performance multi-radio wsn platform. *Proc. 2nd Int'l Workshop on Multi-hop Ad Hoc Networks: From Theory to Reality (REALMAN'06)*, pp. 95–97, Italy.

Kohvakka M, Hännikäinen M and Hämäläinen T 2003 Wireless sensor prototype platform. *Proc. IEEE Int'l Conf. on Industrial Electronics, Control and Instrumentation (IECON'03)*, pp. 860–865, Virginia, USA.

Kohvakka M, Hännikäinen M and Hämäläinen T 2005a Wireless sensor network implementation for industrial linear position metering. *Proc. 8th Euromicro Conf. on Digital System Design (DSD'05)*, pp. 267–273, Portugal.

Kohvakka M, Hännikäinen M and Hämäläinen TD 2005b Energy optimized beacon transmission rate in a wireless sensor network. *Proc. 16th Int'l Symposium on Personal Indoor and Mobile Radio Communications (PIMRC'05)*, pp. 1269–1273, Germany.

Kohvakka M, Hännikäinen M and Hämäläinen TD 2005c Ultra low energy wireless temperature sensor network implementation. *Proc. 16th Int'l Symposium on Personal Indoor and Mobile Radio Communications (PIMRC'05)*, pp. 801–805, Germany.

Kohvakka M, Kuorilehto M, Hännikäinen M and Hämäläinen T 2006b Performance analysis of IEEE 802.15.4 and Zigbee for large-scale wireless sensor network applications. *Proc. 3rd ACM Int'l Workshop on Performance Evaluation of Wireless Ad Hoc, Sensor, and Ubiquitous Networks (PE-WASUN 2006)*, pp. 48–57, Spain.

Kohvakka M, Suhonen J, Hännikäinen M and Hämäläinen T 2006c Transmission power based path loss metering for wireless sensor networks. *Proc. 17th Annual IEEE Int'l Symposium on Personal, Indoor and Mobile Radio Communications (PIMRC'06)*, Finland.

Krco S, Cleary D and Parker D 2005 P2P mobile sensor networks. *Proc. 38th IEEE Annual Hawaii Int'l Conf. on System Sciences (HICSS 2005)*, p. 324c, Big Island, HI, USA.

Kulik J, Heinzelman W and Balakrishnan H 2002 Negotiation-based protocols for disseminating information in wireless sensor networks. *Wireless Networks (Kluwer)* **8**(2), 169–185.

Kumar R, Wolenetz M, Agarwalla B, Shin J, Hutto P, Paul A and Ramachandran U 2003 DFuse: A framework for distributed data fusion. *Proc. 1st Int'l Conf. on Embedded Networked Sensor Systems (SenSys'03)*, pp. 114–125, Los Angeles, CA, USA.

Kuorilehto M, Alho TA, Hännikäinen M and Hämäläinen TD 2007a SensorOS: A new operating system for time critical WSN applications. In *Lecture Notes in Computer Science* (ed. Vassiliadis S, Wong S and Hämäläinen TD) vol. 4599, Springer, pp. 431–442.

Kuorilehto M, Hännikäinen M and Hämäläinen TD 2005a A middleware for task allocation in wireless sensor networks. *Proc. 16th Annual IEEE Int'l Symposium on Personal Indoor and Mobile Radio Communications (PIMRC'05)*, pp. 821–826, Berlin, Germany.

Kuorilehto M, Hännikäinen M and Hämäläinen TD 2005b A survey of application distribution in wireless sensor networks. *EURASIP Journal on Wireless Communications and Networking, Special Issue on Ad Hoc Networks: Cross-Layer Issues* **2005**(5), 774–788.

Kuorilehto M, Hännikäinen M and Hämäläinen TD 2006a Rapid design and evaluation framework for wireless sensor networks. /Ad Hoc Networks (Elsevier)/2007, doi:10.1016/j.adhoc.2007.08.003.

M. Kuorilehto, M. Hännikäinen, T. D. Hämäläinen, "Rapid Design and Evaluation Framework for Wireless Sensor Networks," *Elsevier Ad Hoc Networks*, 2007b, accepted, DOI: 10.1016/j.adhoc.2007.08.003.

Kuorilehto M, Suhonen J, Hännikäinen M and Hämäläinen TD 2007c Tool-aided design and implementation of indoor surveillance wireless sensor network. In *Lecture Notes in Computer Science* (ed. Vassiliadis S, Wong S and Hämäläinen TD) vol. 4599, Springer, pp. 396–407.

Kuorilehto M, Suhonen J, Kohvakka M, Hännikäinen M and Hämäläinen TD 2006 Experimenting TCP/IP for low-power wireless sensor networks. *Proc. 17th Annual IEEE Int'l Symposium on Personal Indoor and Mobile Radio Communications (PIMRC'05)*, Helsinki, Finland.

Levis P and Culler D 2002 Maté: A tiny virtual machine for sensor networks. *Proc. 10th ACM Int'l Conf. on Architectural Support for Programming Languages and Operating Systems (ASPLOS'02)*, pp. 85–95, San Jose, CA, USA.

Levis P, Lee N, Welsh M and Culler D 2003 TOSSIM: Accurate and scalable simulation of entire TinyOS applications. *Proc. 1st Int'l Conf. on Embedded Networked Sensor Systems (SenSys'03)*, pp. 126–137, Los Angeles, CA, USA.

Li S, Lin Y, Son SH, Stankovic JA and Wei Y 2004 Event detection services using data service middleware in distributed sensor networks. *Telecommunication Systems* **26**(2), 351–368.

Lifton J, Seetharam D, Broxton M and Paradiso J 2002 Pushpin computing system overview: A platform for distributed, embedded, ubiquitous sensor networks. *Proc. 1st Int'l Conf. on Pervasive Computing (Pervasive'02)*, pp. 139–151, Zurich, Switzerland.

Lin C and Gerla M 1997 Adaptive clustering for mobile wireless networks. *IEEE Journal of Selected Areas in Communications* **15**(7), 1265–1275.

Liu J, Chu M, Liu J, Reich J and Zhao F 2003 State-centric programming for sensor-actuator network systems. *IEEE Pervasive Computing* **2**(4), 50–62.

Liu J, Perrone LF, Nicol DM, Liljenstam M, Elliott C and Pearson D 2001 Simulation modeling of large-scale ad-hoc sensor networks. *Proc. 2001 Simulation Interoperability Workshop*, Harrow, Middlesex, UK.

Liu J, Zhao F and Petrovic D 2005 Information-directed routing in ad hoc sensor networks. *IEEE Journal on Selected Areas in Communications* **23**(4), 851–861.

Liu T and Martonosi M 2003 Impala: A middleware system for managing autonomic, parallel sensor systems. *Proc. 9th ACM SIGPLAN Symposium on Principles and Practice of Parallel Programming (PPoPP'03)*, pp. 107–118, San Diego, CA, USA.

Luo H, Ye F, Cheng J, Lu S and Zhang L 2005 TTDD: Two-tier data dissemination in large-scale wireless sensor networks. *Wireless Networks (Kluwer)* **11**(1-2), 161–175.

Madden S, Franklin MJ, Hellerstein JM and Hong W 2003 The design of an acquisitional query processor for sensor networks. *Proc. ACM Int'l Conf. on Management of Data (SIGMOD'03)*, pp. 491–502, San Diego, CA, USA.

Mallanda C, Suri A, Kunchakarra V, Iyengar S, Kannan R, Durresi A and Sastry S 2006 Simulating wireless sensor networks with OMNeT++. Available at: `http://csc.lsu.edu/sensor_web/simulator.html`.

Maltz DA, Broch J, Jetcheva J and Johnson DB 1999 The effects of on-demand behavior in routing protocols for multihop wireless ad hoc networks. *IEEE Journal on Selected Areas in Communications* **17**(8), 1439–1453.

Mangione-Smith B 1995 Low power communications protocols: paging and beyond. *Proc. IEEE Symposium on Low Power Electronics*, pp. 8–11, San Jose, CA, USA.

Martinez K, Padhy P, Riddoch A, Ong R and Hart J 2005 Glacial environment monitoring using sensor networks. *Proc. REALWSN 2005*, Stockholm, Sweden.

Menezes AJ, Oorschot PC and Vanstone SA 1996 *Handbook of Applied Cryptography*. CRC Press.

Mic 2005 *Application Note 953: Data Encryption Routines for the PIC18.*

Microchip Technology Inc. 2004 PIC18F8722 family data sheet. Available at: `http://ww1.microchip.com/downloads/en/DeviceDoc/39646b.pdf`.

Min R, Bhardwaj M, Cho SH, Ickes N, Shih E, Sinha A, Wang A and Chandrakasan A 2002 Energy-centric enabling technologies for wireless sensor networks. *IEEE Wireless Communications* **9**(4), 28–39.

Mochocki BC and Madey GR 2003 H-MAS: A heterogeneous, mobile, ad-hoc sensor network simulation environment. *Proc. 7th Annual Swarm Users/Researchers Conf.*, Notre Dame, IN, USA.

Morris S 2002 Recommendations to early implementers: Encrypting broadcast transmissions in Bluetooth piconets. White Paper 1.0, Bluetooth SIG Security Expert Group.

Muller T 1999 Bluetooth security architecture. White Paper 1.C.116/1.0, Bluetooth SIG.

Murphy AL and Picco GP 2001 LIME: A middleware for physical and logical mobility. *Proc. 21st Int'l Conf. on Distributed Computing Systems (ICDCS-21)*, pp. 524–533, Phoenix, AZ, USA.

Nasipuri A and Das SR 2000 Multichannel CSMA with signal power-based channel selection for multihop wireless networks. *Proc. IEEE Vehicular Technology Conf. (VTC'00)*, pp. 211–218.

Nasipuri A, Zhuang J and Das SR 1999 A multichannel CSMA MAC protocol for multihop wireless networks. *Proc. IEEE Wireless Communications and Networking Conf. (WCNC '99)*, vol. 3, pp. 1402–1406, New Orleans, LA, USA.

Nelson R and Kleinrock L 1985 Spatial TDMA: A collision-free multihop channel access protocol. *IEEE Transactions on Communications* **COM-33**(9), 934–944.

Networked & Embedded Systems Laboratory 2006 sQualnet–home. Available at: http://nesl.ee.ucla.edu/projects/squalnet.

Newton R, Arvind and Welsh M 2005 Building up to macroprogramming: An intermediate language for sensor networks. *Proc. 4th Int'l Conf. on Information Processing in Sensor Networks (IPSN'05)*, Los Angeles, CA, USA.

Niculescu D and Nath B 2003 Trajectory based forwarding and its applications. *Proc. 9th Annual Int'l Conf. on Mobile Computing and Networking (MobiCom'03)*, pp. 260–272, San Diego, CA, USA.

NIS 2001 *Advanced Encryption Standard (AES)*.

NIST Advanced Network Technologies Division 2006 Wireless ad hoc networks: Smart sensor networks. Available at: http://www.antd.nist.gov/wahn_ssn.shtml.

Nordic Semiconductor ASA 2006 Single chip 2.4 GHz transceiver nRF2401A. Available at: http://www.nordicsemi.no/files/Product/data_sheet/Product_Specification_nRF2401A_1_1.pdf.

Ns-2 2006 The network simulator–ns-2. Available at: http://www.isi.edu/nsnam/ns.

Oh S, Chen P, Manzo M and Sastry S 2006 Instrumenting wireless sensor networks for real-time surveillance. *Proc. 2006 IEEE Int'l Conf. on Robotics and Automation (ICRA'06)*, pp. 3128–3133, Orlando, FL, USA.

OMNeT++ 2006 OMNeT++ community site. Available at: http://www.omnetpp.org.

OPNET 2006 Modeler wireless suite for defense. Available at: http://www.opnet.com/products/modeler/home.html.

Park H and Srivastava MB 2003 Energy-efficient task assignment framework for wireless sensor networks. Technical Report 0026, CENS. Available at: http://research.cens.ucla.edu.

Park S, Savvides A and Srivastava MB 2001 Simulating networks of wireless sensors. *Proc. Winter Simulation Conf. 2001 (WSC'01)*, pp. 1330–1338, Arlington, VA, USA.

Pei G and Chien C 2001 Low power TDMA in large wireless sensor networks. *Proc. IEEE Military Communications Conf. (MILCOM'01)*, pp. 347–351.

Perrig A, Stankovic J and Wagner D 2004 Security in wireless sensor networks. *Communications of the ACM* **47**(6), 53–57.

Perrig A, Szewczyk R, Tygar JD, Wen V and Culler DE 2002 SPINS: Security protocols for sensor networks. *Wireless Networks (Kluwer)* **8**(5), 521–534.

Perrone LF and Nicol DM 2002 A scalable simulator for TinyOS applications. *Proc. Winter Simulation Conf. 2002 (WSC'02)*, pp. 679–687, San Diego, CA, USA.

Piguet C, Masgonty JM, Arm C, Durand S, Schneider T, Rampogna F, Scarnera C, Iseli C, Bardyn JP, Pache R and Dijkstra E 1997 Low-power design of 8-b embedded coolrisc microcontroller cores. *IEEE Journal of Solid-State Circuits* **32**(7), 1067–1078.

Polastre J, Hill J and Culler D 2004a Versatile low power media access for wireless sensor networks. *Proc. 2nd Int'l Conf. on Embedded Networked Sensor Systems (Sensys'04)*, pp. 95–107, Baltimore, MD, USA.

Polastre J, Szewczyk R and Culler D 2005 Telos: Enabling ultra-low power wireless research *Proc. Symposium in Information Processing in Sensor Networks*, pp. 364–369, Los Angeles, California, USA.

Polastre J, Szewczyk R, Mainwaring A, Culler D and Anderson J 2004b Analysis of wireless sensor networks for habitat monitoring. *Wireless Sensor Networks* 1st edn, Springer, pp. 399–425.

Ptolemy II 2006 Ptolemy II. Available at: http://ptolemy.eecs.berkeley.edu/ptolemyII.

Qi H, Xu Y and Wang X 2003 Mobile-agent-based collaborative signal and information processing in sensor networks. *Proceedings of the IEEE* **91**(8), 1172–1183.

Rabaey J, Ammer J, da Silva Jr. J, Patel D and Roundy S 2000 Picoradio supports ad hoc ultra-low power wireless networking. *IEEE Computer Magazine* **33**(7), 42–48.

Raghunathan V, Schurgers C, Park S and Srivastava M 2002 Energy-aware wireless microsensor networks. *IEEE Signal Processing Magazine* **19**(2), 40–50.

Rajendran V, Obraczka K and Garcia-Luna-Aceves J 2003 Energy-efficient, collision-free medium access control for wireless sensor networks *Proc. 1st Int'l Conf. on Embedded Networked Sensor Systems (SenSys'03)*, pp. 181–192, Los Angeles, CA, USA.

Rappaport T 1989 Indoor radio communications for factories of the future. *IEEE Communications Magazine* **27**(5), 15–24.

Rappaport T 1996 *Wireless Communications–Principles and Practice* 2 edn, Prentice Hall, Chapter 1.

Reason JM and Rabaey JM 2004 A study of energy consumption and reliability in a multi-hop sensor network. *ACM SIGMOBILE Mobile Computing and Communications Review* **8**(1), 84–97.

Reijers N, Halkes G and Langendoen K 2004 Link layer measurements in sensor networks. *Proc. 1st IEEE Int'l Conf. on Mobile Ad-hoc and Sensor Systems (MASS'04)*, pp. 224–233, Florida, USA.

Rhee I, Warrier A, Aia M and Min J 2005 Z-MAC: A hybrid MAC for wireless sensor networks. *Proc. 3rd ACM Conf. on Embedded Networked Sensor Systems (Sensys '05)*, pp. 90–101, New York, NY, USA.

Römer K 2004 Tracking real-world phenomena with smart dust. *Proc. 1st European Workshop on Wireless Sensor Networks (EWSN'04)*, pp. 28–43 number 2920 in *LNCS*. Springer-Verlag, Berlin, Germany.

Römer K, Kasten O and Mattern F 2002 Middleware challenges for wireless sensor networks. *ACM SIGMOBILE Mobile Computing and Communications Review* **6**(4), 59–61.

Römer K and Mattern F 2004 The design space of wireless sensor networks. *IEEE Wireless Communications* **11**(6), 54–61.

Roberts L 1975 ALOHA packet system with and without slots and capture. *ACM SIGCOMM Computer Communication Review* **5**(2), 28–42.

Roundy S, Steingart D, Frechette L, Wright P and Rabaey J 2004 Power sources for wireless sensor networks. *Proc. 1st Eur. Workshop on Wireless Sensor Network*, pp. 1–17, Berlin, Germany.

Roundy S, Wright PK and Rabaey J 2003 A study of low level vibrations as a power source for wireless sensor nodes. *Computer Communications* **26**(11), 1131–1144.

Sachs R 2006 Z-Wave, the standard in wireless home control. Available at: `http://www.zen-sys.com/`.

Sadagopan N, Krishnamachari B and Helmy A 2003 The acquire mechanism for efficient querying in sensor networks. *Proc. 1st Int'l Workshop on Sensor Network Protocols and Applications*, pp. 149–155.

Sastry N and Wagner D 2004 Security considerations for IEEE 802.15.4 networks. *Proc. 2004 ACM Workshop on Wireless Security (WiSE'04)*, pp. 32–42, Philadelphia, PA, USA.

Savvides A and Srivastava MB 2002 A distributed computation platform for wireless embedded sensing. *Proc. IEEE Int'l Conf. on Computer Design: VLSI in Computers and Processors (ICCD'02)*, pp. 220–225, Freiburg, Germany.

Scalable Network Technologies 2006 Qualnet network simulator by scalable network technologies. Available at: `http://www.qualnet.com`.

Schiller J, Liers A and Ritter H 2005 Scatterweb: A wireless sensornet platform for research and teaching. *Computer Communications (Elsevier)* **28**(13), 1545–1551.

Schmid T, Dubois-Ferriere H and Vetterli M 2005 Sensorscope: Experiences with a wireless building monitoring sensor network. *Proc. REALWSN 2005*, Stockholm, Sweden.

Schmidt DC 2006 Model driven engineering. *IEEE Computer* **39**(2), 25–31.

Schurgers C and Srivasta MB 2001 Energy efficient routing in wireless sensor networks. *Proc. IEEE Military Communications Conf. (MILCOM'01)*, pp. 357–361, McLean, VA, USA.

Schurgers C, Tsiatsis V, Ganeriwal S and Srivastava M 2002 Optimizing sensor networks in the energy-latency-density design space. *IEEE Transactions on Mobile Computing* **1**(1), 70–80.

Seidel S and Rappaport T 1992 914 MHz path loss prediction models for indoor wireless communications in multifloored buildings. *IEEE Transactions on Antennas and Propagation* **40**(2), 207–217.

Semtech Corp. 2006 XE8802 sensing machine. Available at: `http://www.semtech.com/`.

Sgroi M, Wolisz A, Sangiovanni-Vincentelli A and Rabaey JM 2005 A service-based universal application interface for ad hoc wireless sensor and actuator networks, *Ambient Intelligence* vol. 2, Springer, pp. 149–172.

Shah RC and Rabaey JM 2002 Energy aware routing for low energy ad hoc sensor networks. *Proc. Wireless Communications and Networking Conf. (WCNC)*, pp. 350–355, Orlando, FL, USA.

Sharp C, Schaffert S, Woo A, Sastry N, Karlof C, Sastry S and Culler D 2005 Design and implementation of a sensor network system for vehicle tracking and autonomous interception. *Proc. 2nd European Workshop on Wireless Sensor Networks (EWSN'05)*, pp. 93–107, Istanbul, Turkey.

Shen CC, Badr C, Kordari K, Bhattacharyya SS, Blankenship GL and Goldsman N 2006 A rapid prototyping methodology for application-specific sensor networks. *Proc. IEEE Int'l Workshop on Computer Architecture for Machine Perception and Sensing (CAMPS'06)*, Montreal, Quebec, Canada.

Shen CC, Srisathapornphat C and Jaikaeo C 2001 Sensor information networking architecture and applications. *IEEE Personal Communications* **8**(4), 52–59.

Shi E and Perrig A 2004 Designing secure sensor networks. *IEEE Wireless Communications* **11**(6), 38–43.

Shong Wu C and Li VO 1987 Receiver-initiated busy-tone multiple access in packet radio networks. *Proc. ACM SIGCOMM'87 Workshop*, pp. vol. 17(5), pp. 336–342, Stowe, Vermont.

Sichitiu M, Ramadurai V and Peddabachagari P 2003 Simple algorithm for outdoor localization of wireless sensor networks with inaccurate range measurements. *Proc. Int'l Conf. on Wireless Networks*, pp. 300–305, Nevada, USA.

Sikora A 2006 ZigBee competitive technology analysis, rev. 1.0. Available at: `http://www.zigbee.org`.

Sikora M, Laneman J, Haenggi M, D.J. Costello J and Fuja T 2004 On the optimum number of hops in linear wireless networks. *Proc. IEEE Information Theory Workshop*, pp. 165–169, Texas, USA.

Simon G, Maróti M, Ákos Lédeczi, Balogh G, Kusy B, Nádas A, Pap G, Sallai J and Frampton K 2004 Sensor network-based countersniper system. *Proc. 2nd Int'l Conf. on Embedded Networked Sensor Systems (SenSys'04)*, pp. 1–12, Baltimore, MD, USA.

Simon G, Völgyesi P, Maróti M and Ákos Lédeczi 2003 Simulation-based optimization of communication protocols for large-scale wireless sensor networks. *Proc. 2003 IEEE Aerospace Conf. (AeroConf'03)*, vol. 3, pp. 1339–1346, Big Sky, MT, USA.

Singh S and Raghavendra CS 1998 PAMAS: Power-aware multi-access protocol with signaling for ad hoc networks. *ACM Computer Communication Review* **28**(3), 5–26.

Sirer EG, Grimm R, Gregory AJ and Bershad BN 1999 Design and implementation of a distributed virtual machine for networked computers. *Proc. 17th ACM Symposium on Operating Systems Principles (AeroConf'03)*, pp. 202–216, Kiawah Island, SC, USA.

Smart-its 2007 Smart-its, home page. Available at: `http://www.smart-its.org/`.

So J and Vaidya N 2004 Multi-channel MAC for ad hoc networks: Handling multi-channel hidden terminals using a single transceiver. *Proc. MobiHoc'04*, pp. 222–233, Roppongi Hills, Tokyo, Japan.

Sobeih A, Hou JC, Chuan Kung L, Li N, Zhang H, Peng Chen W, Ying Tyan H and Lim H 2006 J-Sim: A simulation and emulation environment for wireless sensor networks. *IEEE Wireless Communications* **13**(4), 104–119.

Sohrabi K, Gao J, Ailawadhi V and Pottie GJ 2000 Protocols for self-organization of a wireless sensor network. *IEEE Personal Communications* **7**(5), 16–27.

Sohrabi K and Pottie G 1999 Performance of a novel self-organization protocol for wireless ad-hoc sensor networks. *Proc. 50th IEEE Vehicle Technology Conf. (VTC)*, pp. 1222–1226, The Netherlands.

Song JH, Poovendran R, Lee J and Iwata T 2006 The AES-CMAC algorithm RFC 4493.

Souto E, Guimarães G, Vasconcelos G, Vieira M, Rosa N, Ferraz C and Kelner J 2005 Mires: A publish/subscribe middleware for sensor networks. *Personal and Ubiquitous Computing* **10**(1), 37–44.

ssfnet.org 2006 Scalable simulator framework. Available at: `http://www.ssfnet.org/homePage.html`.

Stallings W 1995 *Network and Internetwork Security: Principles and Practice*. Prentice-Hall.

Stallings W 2004 *Data and Computer Communications* 7 edn. Prentice-Hall.

Stallings W 2005 *Operating Systems Internals and Design Principles* 5 edn. Prentice-Hall.

Stankovic JA, Abdelzaher TF, Lu C, Sha L and Hou JC 2003 Real-time communication and coordination in embedded sensor networks. *Proceedings of the IEEE* **91**(7), 1002–1022.

Steere DC, Baptista A, McNamee D, Pu C and Walpole J 2000 Research challenges in environmental observation and forecasting systems. *Proc. 6th Annual ACM/IEEE Int'l Conf. on Mobile Computing and Networking (MobiCom'00)*, pp. 292–299, Boston, MA, USA.

STI 2003 *Smart Transducer Interface V1.0*.

Stutzman WL and Thiele GA 1998 *Antenna Theory and Design* 2nd edn. John Wiley & Sons, Inc.

Suhonen J, Kohvakka M, Hännikäinen M and Hämäläinen T 2006a Design, implementation, and outdoor deployment for wireless sensor network for environmental monitoring. *Proc. Embedded Computer Systems: Architectures, Modeling, and Simulation (SAMOS VI)*, pp. 109–121, Samos, Greece.

Suhonen J, Kohvakka M, Kuorilehto M, Hännikäinen M and Hämäläinen TD 2007 Cost-aware capacity optimization in dynamic multi-hop wsns. *Proc. Design, Automation and Test in Europe (DATE'07)*, pp. 666–671, Nice, France.

Suhonen J, Kuorilehto M, Hännikäinen M and Hämäläinen TD 2006b Cost-aware dynamic routing protocol for wireless sensor networks–design and prototype experiments. *Proc. 17th Annual IEEE Int'l Symposium on Personal, Indoor and Mobile Radio Communications (PIMRC'06)*, pp. 1–5, Helsinki, Finland.

Sun Microsystems 2002 Java message service. Available at: `http://java.sun.com/products/jms/`.

Sun Microsystems 2006a Trail: Creating a GUI with JFC/swing. Available at: `http://java.sun.com/docs/books/tutorial/uiswing`.

Sun Microsystems 2006b Trail: JDBC database access. Available at: `http://java.sun.com/docs/books/tutorial/jdbc/`.

Sundresh S, WooYoung K and Gul A 2004 SENS: A sensor, environment and network simulator. *Proc. 37th Annual Simulation Symposium (ANSS'04)*, pp. 221–228, Arlington, VA, USA.

Systems QS 2006 QNX realtime operating system (RTOS) software, development tools, and services for embedded applications. Available at: `http://www.qnx.com`.

Szewczyk R, Mainwaring A, Polastre J, Anderson J and Culler D 2004 An analysis of a large scale habitat monitoring application. *Proc. 2nd Int'l Conf. on Embedded Networked Sensor Systems (SenSys'04)*, pp. 214–226, Baltimore, MD, USA.

Tang Z and Garcia-Luna-Aceves JJ 1999 Hop-reservation multiple access (HRMA) for ad-hoc networks. *Proc. 18th Annual Joint Conf. of the IEEE Computer and Communications Societies (INFOCOM'99)*, pp. 194–201, New York, NY, USA.

Telelogic 2006 Telelogic TAU SDL suite–Communications software specification and software development. Available at: `http://www.telelogic.com/products/tau/sdl/index.cfm`.

Terfloth K, Wittenburg G and Schiller J 2006 FACTS: A rule-based middleware architecture for wireless sensor networks. *Proc. 1st Int'l Conf. on Communication System Software and Middleware (Comsware'06)*, New Delhi, India.

Texas Instruments Inc. 2007 CC2420 2.4 GHz IEEE 802.15.4 / ZigBee-ready RF transceiver. Available at: `http://www.ti.com/lit/gpn/cc2420`.

Tian He, Stankovic JA, Lu C and Abdelzaher T 2003 SPEED: A stateless protocol for real-time communication in sensor networks. *Proc. 23rd Int'l Conf. on Distributed Computing Systems*, pp. 46–55, Providence, RI, USA.

Tian Y, Xu K and Ansari N 2005 TCP in wireless environments: Problems and solutions. *IEEE Communications Magazine* **43**(3), S27–S32.

Titzer BL, Lee DK and Palsberg J 2005 Avrora: Scalable sensor network simulation with precise timing. *Proc. 4th Int'l Conf. on Information Processing in Sensor Networks (IPSN'05)*, Los Angeles, CA, USA.

Tobagi FA and Kleinrock L 1975 Packet switching in radio channels: Part II–the hidden terminal problem in carrier sense multiple-access modes and the busy-tone solution. *IEEE Transaction on Communications* **COM-23**(12), 1417–1433.

Tolle G, Polastre J, Szewczyk R, Culler D, Turner N, Tu K, Burgess S, Dawson T, Buonadonna P, Gay D and Hong W 2005 A macroscope in the redwoods. *Proc. 3rd Int'l Conf. on Embedded Networked Sensor Systems (SenSys'05)*, pp. 51–63, San Diego, CA, USA.

Tseng ZC, Ni S and Shih E 2003 Adaptive approaches to relieving broadcast storms in a wireless multihop mobile ad hoc network. *Transactions on Computers* **52**(5), 545–557.

Turau V, Witt M and Weyer C 2006 Analysis of a real multi-hop sensor network deployment: The heathland experiment. *Proc. 3rd Int'l Conf. on Networked Sensing Systems (INSS'06)*, Chicago, IL, USA.

Tzamaloukas A and Garcia-Luna-Aceves J 2001 A receiver-initiated collision-avoidance protocol for multi-channel networks. *Proc. 20th Annual Joint Conf. of the IEEE Computer and Communications Societies (INFOCOM'01)*, pp. 189–198, Anchorage, AK, USA.

van Dam T and Langendoen K 2003 An adaptive energy-efficient mac protocol for wireless sensor networks. *Proc. 1st Int'l Conf. on Embedded Networked Sensor Systems (Sensys'03)*, pp. 171–180, Los Angeles, CA, USA.

Varshney M, Xu D, Srivastava M and Bagrodia R 2007 SenQ: A scalable simulation and emulation environment for sensor networks. *Proc. 6th Int'l Conf. on Information Processing in Sensor Networks (IPSN'07)*, pp. 196–205, Cambridge, MA, USA.

Vinoski S 1997 CORBA: Integrating diverse applications within distributed heterogeneous environments. *IEEE Communications Magazine* **35**(2), 46–55.

Völgyesi P and Ákos Lédeczi 2002 Component-based development of networked embedded applications. *Proc. 28th Euromicro Conf. (EUROMICRO'02)*, pp. 68–73, Dortmund, Germany.

Wan CY, Campbell AT and Krishnamurthy L 2005 Pump-slowly, fetch-quickly (PSFQ): A reliable transport protocol for sensor networks. *IEEE Journal on Selected Areas in Communication* **23**(4), 862–872.

Warneke B, Last M, Leibowitz B, and Pister KSJ 2001 Smart dust: Communicating with a cubic-millimeter computer. *Computer* **34**(1), 43–51.

Weiser M 1993 Hot topics: Ubiquitous computing. *IEEE Computer* **26**(10), 71–72.

Weiser M 1999 The computer for the 21st century. *ACM SIGMOBILE Mobile Computing and Communications Review* **3**(3), 3–11.

Welsh M and Mainland G 2004 Programming sensor networks using abstract regions. *Proc. 1st Symposium on Networked Systems Design and Implementation (NSDI'04)*, San Francisco, CA, USA.

Wibree 2007 Ultra-low power radio technology for small devices. Available at: `http://www.wibree.com/technology/Wibree_2Pager.pdf`.

Wind Driver 2006 VxWorks. Available at: `http://www.windriver.com`.

Wolf M and Kress D 2003 Short-range wireless infrared transmission: The link budget compared to RF. *IEEE Wireless Communications Magazine* **10**(2), 8–14.

Wolf W 2001 *Computers as Components: Principles of Embedded Computing System Design*. Morgan Kaufmann Publishers.

Wong KJ and Arvind D 2006 SpeckMAC: Low-power decentralized MAC protocols for low data rate transmissions in specknets. *Proc. 2nd Int'l Workshop on Multi-hop Ad Hoc Networks: From Theory to Reality*, pp. 71–78, Florence, Italy.

Woo A, Madden S and Govindan R 2004 Networking support for query processing in sensor networks. *Communications of the ACM* **47**(6), 47–52.

Wood A and Stankovic JA 2002 Denial of service in sensor networks. *IEEE Computer* **35**(10), 54–62.

Wu H, Luo Q, Zheng P and Ni LM 2007 VMNet: Realistic emulation of wireless sensor networks. *IEEE Transactions on Parallel and Distributed Systems* **18**(2), 277–288.

Wu SL, Lin CY, Tseng YC and Sheu JP 2000 A new multi-channel MAC protocol with on-demand channel assignment for multi-hop mobile ad hoc networks. *Proc. Int'l Symposium on Parallel Architectures, Algorithms and Networks (ISPAN'00)*, pp. 232–237, Dallas / Richardson, TX, USA.

Wu T and Biswas S 2005 A self-reorganizing slot allocation protocol for multi-cluster sensor networks. *Proc. Fourth Int'l Conf. on Information Processing in Sensor Networks (IPSN'05)*, pp. 309–316, Los Angeles, CA, USA.

Xu N, Rangwala S, Chintalapudi KK, Ganesan D, Broad A, Govindan R and Estrin D 2004 A wireless sensor network for structural monitoring. *Proc. 2nd Int'l Conf. on Embedded Networked Sensor Systems (SenSys'04)*, pp. 13–24, Baltimore, MD, USA.

Yang Liu, Elhanany I and Hairong Qi 2005 An energy-efficient QoS-aware media access control protocol for wireless sensor networks. *Proc. Mobile Adhoc and Sensor Systems Conf.*, p. 3, Washington D.C., USA.

Yannakopoulos J and Bilas A 2005 CORMOS: A communication-oriented runtime system for sensor networks. *Proc. 2nd European Workshop on Wireless Sensor Networks (EWSN'05)*, pp. 342–353, Istanbul, Turkey.

Yao Y and Gehrke J 2002 The Cougar approach to in-network query processing in sensor networks. *ACM SIGMOD Record* **31**(3), 9–18.

Ye F, Chen A, Lu S and Zhang L 2001 A scalable solution to minimum cost forwarding in large sensor networks. *Proc. Tenth Int'l Conf. on Computer Communications and Networks*, pp. 304–309, Scottsdale, AZ, USA.

Ye F, Zhong G, Lu S and Zhang L 2005 GRAdient broadcast: a robust data delivery protocol for large scale sensor networks. *Wireless Networks (Kluwer)* **11**(3), 285–298.

Ye W, Heidemann J and Estrin D 2002 An energy-efficient MAC protocol for wireless sensor networks. *Proc. 21st Annual Joint Conf. of the IEEE Computer and Communications Societies (INFOCOM'02)*, vol. 3, pp. 1567–1576, New York, NY, USA.

Ye W, Heidemann J and Estrin D 2004 Medium access control with coordinated, adaptive sleeping for wireless sensor networks. *IEEE/ACM Transactions on Networking* **12**(3), 493–506.

Yu Y, Hong B and Prasanna VK 2005 Communication models for algorithm design in networked sensor systems. *Proc. 19th IEEE Int'l Parallel and Distributed Processing Symposium (IPDPS'05)*, Denver, CO, USA.

Yu Y, Krishnamachari B and Prasanna VK 2004 Issues in designing middleware for wireless sensor networks. *IEEE Network* **1**(18), 15–21.

Zeng X, Bagrodia R and Gerla M 1998 GloMoSim: A library for parallel simulation of large-scale wireless networks. *Proc. 12th Workshop on Parallel and Distributed Simulations (PADS 1998)*, pp. 154–161, Banff, Alberta, Canada.

Zhang P, Sadler CM, Lyon SA and Martonosi M 2004 Hardware design experiences in zebranet. *Proc. 2nd Int'l Conf. on Embedded Networked Sensor Systems (SenSys'04)*, pp. 227–238, Baltimore, MD, USA.

Zhu S, Setia S and Jajodia S 2003 LEAP: Efficient security mechanism for large-scale distributed sensor networks. *Proc. 10th ACM Conf. on Computer and Communications Security (CCS'03)*, pp. 62–72, Washington D.C., USA.

Zig 2004 *ZigBee Specification Version 1.0.*

Index

μTESLA, 136

Abstract regions, 109, 112
ACQUIRE, 98
Aggregation, 92, 105
Agilla, 108, 111
ALOHA, 77, 88, 219
Analysis, 60
Antenna, 46
  dipole, 49
  isotropic, 344
  loop, 49
  monopole, 49
  omnidirectional, 338
  planar, 48
API, *see* Application Programming Interface
Application, 11
  domain, 11
  event detection, 12
  monitoring, 12, 327
  object classification, 12
  object tracking, 12, 335
  quality of service, 128, 219
Application Programming Interface, 103, 104, 106, 193, 203, 343
Asymmetric traffic, 91
ATEMU, 62, 65
Avrora, 62, 65

B-MAC, 88
Bertha, 120, 122
Bluetooth, 22
  security, 139
Bootloader, 121, 203, 213

Carrier Sense Multiple Access, 77, 79
Cluster head, 80
Collisions, 76
Component, 56
Component-based design, 56, 58
Concurrency, 116
Configuration, 55, 279, 328
Connectivity, 53
Contiki, 120, 122, 123
CORMOS, 120, 122
Cost, 17, 52
Cougar, 107, 111
Coverage, 53
Cross-layer, 217
CSMA, *see* Carrier Sense Multiple Access

Data fusion, 105
DCA, 82
Deployment, 11, 52, 55, 279
  Infrastructure, 280
Design dimension, 51, 54
Design flow, 54
Design methodology, 56
DFuse, 109, 112
Directed diffusion, 94, 256
DMAC, 84
DSWare, 107, 111
DVS, *see* Dynamic voltage scaling
Dynamic source routing (DSR), 93
Dynamic voltage scaling, 116

Embedded system, 115
Emstar, 58, 59
Energy management, 116

Energy-aware routing, 101
Energy-efficient routing, 101
Environmental monitoring, 222, 300–312
Evaluation, 55, 60, 247
Event-driven programming, 118

Face Aware Routing (FAR), 96
FACTS, 109, 112
FAMA, 79
Fault tolerance, 16
FDMA, *see* Frequency Division Multiple Access
Finite state machine, 120, 251
Flooding, 97
Frequency Division Multiple Access, 79
FSM, *see* Finite state machine

Gateway, 10, 193, 279, 280, 282, 344
Global Positioning System, 10, 11, 200, 288, 334, 335, 343, 345
Global sensor network, 110, 112
GPS, *see* Global Positioning System
GRAdient Broadcast (GRAB), 97
GRATIS, 57, 59
Greedy Perimeter Stateless Routing (GPSR), 95
GSN, *see* Global Sensor Network

H-MAS, 61, 64
Headnode, 146
Heterogeneity, 52
Hidden-node problem, 77, 253

Idle listening, 76
IEEE 1451, 19
IEEE 1451.5, 106
IEEE 802.15.4, 22, 260, 261
  Beacon Order (BO), 24
  security, 136
  Superframe Order (SO), 24
Impala, 109, 112
In-network processing, 105
Information-directed routing, 99
Interest, 196, 198

J-Sim, 61, 64

Kairos, 109, 112

LEACH, 85
Lifetime, 16, 108
Localization, 222, 327, 334
Localized max-min energy routing, 101

MAC, *see* Medium Access Control
MACA, 78
Macroprogramming, 109
MagnetOS, 108, 111
MARE, 108, 111
Maté, 107, 108, 111
Medium Access Control, 75, 145
  Contention-based, 77
  Contention-free, 79
  low duty-cycle, 82
  multi channel, 80
  quality of service, 130, 219
  synchronized, 82
  unsynchronized, 87
Microcontroller Unit (MCU), 33
Middleware, 103
  application-driven, 106, 108, 110
  database, 106, 107, 110
  mobile agent, *see* Mobile agent
  programming abstraction, 106
  virtual machine, *see* Virtual machine
MiLAN, 108, 112
Minimum-cost forwarding, 99
Mires, 109, 112
MiWi, 28
MMAC, 81
Mobicast, 96
Mobile agent, 106, 108–110
Mobility, 52
Model-driven design, 56, 58
MOS, 121, 122

NAMA, 85
nano-RK, 121, 122
nesC, 57, 59
Network scan, 83, 151, 232

Operating system, 103, 115, 203
  event-based, 117, 118, 123

microkernel, 115, 117
monolithic, 115, 117
multithreading, 117, 118, 203
preemptive, 117, 118, 123, 203
OS, *see* Operating System
Overhead, 77
Overhearing, 76

PACT, 86
PAMAS, 90
Physical layer
quality of service, 131, 220
Platform-based design, 57, 58
Positioning, 334, 343
POSIX, 121
Proactive routing, 93
Protocol stack, 103
Protothreads, 110
Prowler, 61, 64
Pump-slowly, fetch-quickly (PSFQ), 129

Q-MAC, 131
QoS, *see* Quality of Service
Quality of Service, 53, 105, 108, 125

Reactive routing, 93
Realtime, 17, 53, 116
Received signal strength indicator, 286, 288, 335, 336, 340
Receiver-decided forwarding, 93
Reprogramming, 121, 203, 213
Resource constraints, 11
RETOS, 121, 122
RF wake-up scheme, 87
Robustness, 53
Routing
cartesian, 95
cost field-based, 99
datacentric, 94
geographic, 95
location-based, 95
multipath Routing, 97
negotiation-based, 97
nodecentric, 93
quality of service, 129, 219
query-based, 98
RSSI, *see* Received Signal Strength Indicator
RTS/CTS procedure, 78

S-MAC, 83, 256
Scalability, 16
SDL, *see* Specification and description language
Security, 17, 133
active attacks, 134
architectures, 135
attacks, 133
centralized key distribution, 141
denial of service, 134
experiments, 290
key distribution, 140
node capturing, 134
passive attacks, 134
pre-distributed keys, 140
public-key cryptography, 140
random key pre-distributed, 141
routing attacks, 135
threats, 133
TinySec, 135
traffic analysis, 134
Self-configuration, 224
Sender-decided forwarding, 93
senQ, 61, 64, 247
SENS, 65
SENSE, 61, 64
SENSIM, 61, 64
Sensor Protocols for Information via Negotiation (SPIN), 98
SensorOS, 196, 203, 327, 331
bootloader, 213
lightweight kernel, 203, 211, 333
system call interface, 207
SensorSim, 61, 64, 247
Sensorware, 108, 111
Sequential assignment routing (SAR), 130
Service discovery, 195
Simulation, 60, 61, 247, 329
SINA, 107, 111
Sink, 10, 91
SMACS, 84
SNAP, 62, 65

SNEP, 136
SOS, 120, 122
Specification and description language, 247, 251, 329
SpeckMAC, 88
SPEED, 129
SPINS, 136
SQL, *see* Structured query language
SRSA, 86
STEM, 90
Structured query language, 107, 111
Subnode, 146
Superframe, 315
SWAN, 61, 64
Synchronization inaccuracy, 219, 233, 311
Synchrozation inaccuracy, 230

t-kernel, 121, 122
T-MAC, 84
Task allocation, 104, 108, 109
TCMote middleware, 107, 111
TDMA, *see* Time Division Multiple Access
Time Division Multiple Access, 79
TinyDB, 107, 111
TinyLIME, 107, 111
TinyOS, 56, 57, 62, 120, 122
TML, 110
Topology, 53
TOSSF, 62, 65
TOSSIM, 57, 62, 65
Tracking, 327
Trajectory and Energy-Based Data Dissemination (TEDD), 96
Trajectory-Based Forwarding (TBF), 95
TRAMA, 85
Transceiver
  antenna, 46
Transport
  quality of service, 128
Tuple space, 107, 108
TUTWSN, 5, 256
  API, 193, 260, 282, 344
  API profile, 196
  contention models, 234
  gateway API, 194, 198
  MAC, 259
  node API, 196, 198
  power consumption models, 240
  radio energy models, 230
  Routing, 183, 260
  SensorOS, *see* SensorOS
  throughput models, 238
  user interface, 280, 282–283, 332, 341
TUTWSN MAC, 145
  channel access, 147
  neighbor discovery, 152
  slot allocation, 159
Two-tier data dissemination (TTDD), 97

Ubiquitous computing, 9

Viptos, 57, 59
Virtual machine, 106–108, 110
VisualSense, 57, 61, 64
VM, *see* Virtual machine

Wake-up radio protocol, 89
Wibree, 28
Wireless Local Area Network, 4, 9, 69, 313
Wireless Sensor Network, 4
  characteristics, 9, 313
  requirements, 16
WiseMAC, 88
WISENES, 247, 327, 329
  design environment, 248, 328
  framework, 249
WLAN, *see* Wireless Local Area Network
WMNet, 62, 65
Worldsens, 58, 59
WSN, *see* Wireless Sensor Network

Z-MAC, 89
Z-Wave, 28
ZigBee, 14, 25, 256, 260, 261, 272
  contention models, 235
  power consumption models, 243
  radio energy models, 232
  security, 137
  throughput models, 239